Self-Organization of Molecular Systems

NATO Science for Peace and Security Series

This Series presents the results of scientific meetings supported under the NATO Programme: Science for Peace and Security (SPS).

The NATO SPS Programme supports meetings in the following Key Priority areas: (1) Defence Against Terrorism; (2) Countering other Threats to Security and (3) NATO, Partner and Mediterranean Dialogue Country Priorities. The types of meeting supported are generally "Advanced Study Institutes" and "Advanced Research Workshops". The NATO SPS Series collects together the results of these meetings. The meetings are co-organised by scientists from NATO countries and scientists from NATO's "Partner" or "Mediterranean Dialogue" countries. The observations and recommendations made at the meetings, as well as the contents of the volumes in the Series, reflect those of parti-cipants and contributors only; they should not necessarily be regarded as reflecting NATO views or policy.

Advanced Study Institutes (ASI) are high-level tutorial courses intended to convey the latest developments in a subject to an advanced-level audience

Advanced Research Workshops (ARW) are expert meetings where an intense but informal exchange of views at the frontiers of a subject aims at identifying directions for future action

Following a transformation of the programme in 2006 the Series has been re-named and re-organised. Recent volumes on topics not related to security, which result from meetings supported under the programme earlier, may be found in the NATO Science Series.

The Series is published by IOS Press, Amsterdam, and Springer, Dordrecht, in conjunction with the NATO Public Diplomacy Division.

Sub-Series

A.	Chemistry and Biology	Springer
B.	Physics and Biophysics	Springer
C.	Environmental Security	Springer
D.	Information and Communication Security	IOS Press
E.	Human and Societal Dynamics	IOS Press

http://www.nato.int/science
http://www.springer.com
http://www.iospress.nl

Series A: Chemistry and Biology

Self-Organization of Molecular Systems

From Molecules and Clusters to Nanotubes and Proteins

Edited by

Nino Russo

Department of Chemistry, University of Calabria
Arcavacata di Rende (CS)
Italy

Victor Ya. Antonchenko

Bogolyubov Institute for Theoretical Physics
National Academy of Sciences of Ukraine
Kiev, Ukraine

and

Eugene S. Kryachko

Bogolyubov Institute for Theoretical Physics
National Academy of Sciences of Ukraine
Kiev, Ukraine

 Springer

Published in cooperation with NATO Public Diplomacy Division

Proceedings of the NATO Advanced Research Workshop on
Molecular Self-Organization: From Molecules to Water, to Nanoparticles,
to DNA and Proteins
Kyiv, Ukraine
8–12 June 2008

Library of Congress Control Number: 2009926510

ISBN 978-90-481-2483-1 (PB)
ISBN 978-90-481-2482-4 (HB)
ISBN 978-90-481-2590-6 (e-book)

Published by Springer,
P.O. Box 17, 3300 AA Dordrecht, The Netherlands.

www.springer.com

Printed on acid-free paper

Preface

The NATO ARW "Molecular Self-Organization in Micro-, Nano-, and Macro-Dimensions: From Molecules to Water, to Nanoparticles, DNA and Proteins" to commemorate Professor Alexander S. Davydov was held in Kiev, Ukraine, on 8–12 June, 2008, at the Bogolyubov Institute for Theoretical Physics of the National Academy of Sciences of Ukraine.

The objective of this NATO ARW is to unveil and formulate the principal features that govern myriads of the molecular self-organization processes in micro-, nano-, and macro-dimensions from the following key representatives such as liquid water and aqueous solutions, and molecular liquids, nanodots, nanoparticles including gold, solitons, biomolecules such as DNA and proteins, biopolymers and biosensors, catalysis, molecular modeling, molecular devices, and thin films, and to offer another, more advanced directions in computational, experimental, and technological areas of nano- and bioscience towards engineering novel and powerful molecular self-organized assemblies with tailored properties.

Nanoscience is indeed one of the most important research and development frontiers in modern science. Simplistically, nanoscience is the science of small particles of materials of a size of nanometre. Molecular nanoscience and nanotechnology have brought to us the unprecedented experimental control of the structure of matter with novel extraordinary properties that open new horizons and new opportunities, and new ways to make things, particularly in our everyday life, to heal our bodies, and to care of the environment. Unfortunately, they have also brought unwelcome advances in weaponry and opened yet more ways to foul up the world on an enormous scale. We therefore highly need to unveil the general principles that govern the molecular self-organization at a nanoscale in order to understand the future capabilities of the human race, to have a vision of where the nanotechnology is leading, how it will affect what we are, and what our societies will become. We hence have a lot of work ahead of us in order to harness new nano-dimensions' developments to good ends.

These global challenges of the nano-dimensional world as being transformed into the major objectives of this NATO Advanced Research Workshop are the following:

- To view it from and to juxtapose it at the different dimension's angles, viz., the microscopic scale, on the one hand, and from the macroscopic one, on the other

- From these perspectives of 'meeting' all dimensions, to unveil and formulate the principal features that govern myriads of the molecular self-organization processes in all these dimensions, largely focusing of their key representatives such as liquid water and aqueous solutions, and molecular liquids, nanodots, nanoparticles including the gold ones, solitons, biomolecules including the DNA and proteins, biopolymers and biosensors, catalysts, molecular models, molecular devices, and thin films
- To offer another, more advanced directions in computational, experimental, and technological areas of nanoscience towards engineering novel and powerful molecular self-organized assemblies with tailored properties.

That is why the chief idea and the main purpose of our multi-disciplinary NATO Advanced Research Workshop is to bring together the scientists who work in the fundamental and applied areas of physics, chemistry, and biology from the NATO and Cooperation Partner Countries to share their research expertise, their knowledge of our many-dimensional world that manifests via the myriads of facets. It suffices to recall in this regard a nonlinearity-induced conformational dynamics of molecular complexes as a key issue in soft-matter physics and biophysics, effective trapping of nonlinear localised excitations and how they contribute to macroscopic phenomena, nonlinear charge and energy transport, proton conductivity through water-filled carbon nanotubes—many of them are rooted to the pioneering researches that was conducted in the 1970s and 1980s by the outstanding Soviet physicists and the former director of the Bogolyubov Institute for Theoretical Physics Professor Alexander S. Davydov who was also well known for his works in nuclear physics and molecular crystals (Davydov's splitting). His idea of the energy transport through proteins based on the concept of soliton (Davydov's soliton) was widely recognized and inspired many studies in France, Germany, Denmark, UK., USA, Greece, Israel, Japan, Italy, and China. Another facets that we meet downshifting from a macroscopic scale to the nano one are the extraordinary catalytic properties of gold micro- and nanoparticles that have been confirmed by countless experiments, applications of nanoparticles in the biophysical and biomedical sciences, and in biotechnology, particularly including biosensors, DNA transfection, enzyme encapsulation, and drug delivery, the surface-functionalized nanoparticles, nanotubes, etc.

To achieve the goals of this NATO Advanced Research Workshop, the Workshop program is composed of the lectures presented by the key speakers who review the current state of art of their fields, the participants' talks on the main frontier achievements, the posters with short screenings that helps the audience for a further discussion, and the extended and inspirational roundtable discussions and informal and friendly debates—this orchestrated way of intensive communications, as we do believe, leads to proposing novel ideas and formulating new concepts of another, more advanced directions in the computationally simulated and nano- and biotechnological realizations of nanostructured molecular self-organized devices and materials with tailored characteristics. The planned workshop also provides the important opportunity for the scientists from the countries of the former Soviet Union to have a broad access to the internationally recognized experts from the

Western Europe, USA, Canada, and Japan, to fully participate by presenting their works and sharing their thoughts. By all accounts, this workshop will be definitively an extraordinary scientific event with the anticipated long-range future contacts and cemented scientific collaborations that will bring very real technological benefits, inspire new ideas, and spark further theoretical and experimental efforts in the study of the nanoworld.

This book represents itself the collection of lectures which review the current state of art of the fields of micro-, nano-, and macro-dimensions, present the main frontier achievements in the molecular self-organization processes in liquid water and aqueous solutions, and molecular liquids, nanodots, nanoparticles including gold, solitons, biomolecules such as DNA and proteins, biopolymers and biosensors, catalysis, molecular modeling, molecular devices, and thin films, and offer another, more advanced directions in computational, experimental, and technological areas of nano- and bioscience towards engineering novel and powerful molecular self-organized assemblies with tailored properties.

Italy	*Nino Russo*
Ukraine	*Victor Ya. Antonchenko*
Belgium/Ukraine	*Eugene S. Kryachko*

Contents

Contributors

F. Kh. Abdullaev Physical-Technical Institute of the Uzbek Academy of Sciences, 100084, Tashkent, Uzbekistan

Al. Alijah Departamento de Química, Universidade de Coimbra, 3004-535 Coimbra, Portugal

A. L. Atkinson Department of Chemistry, Northwestern University, Evanston IL 60208-3113 USA

B. B. Baizakov Physical-Technical Institute of the Uzbek Academy of Sciences, 100084, Tashkent, Uzbekistan

A. Bidon-Chanal Departament de Fisicoquímica and Institut de Biomedicina (IBUB), Facultat de Farmàcia, Universitat de Barcelona, AV. Diagonal 643, 08028, Barcelona, Spain

E. J. Brändas Department of Physical and Analytical Chemistry, Quantum Chemistry, Uppsala University, SE-751 20 Uppsala, Sweden

L. S. Brizhik Bogolyubov Institute for Theoretical Physics, 03680 Kyiv, Ukraine

T. Bryk Institute for Condensed Matter Physics, National Academy of Sciences of Ukraine, 1 Svientsitskii Str., 79011 Lviv, Ukraine

A. P. Chetverikov Dept. of Physics, Saratov State University, Astrakhanskaya 83, Saratov-410012, Russia

V. D. Danchuk National Transport University, 1 Suvorov str., 01010 Kyiv, Ukraine, vdanchuk@ukr.net

M. Druchok Institute for Condensed Matter Physics, National Academy of Sciences of Ukraine, 1 Svientsitskii Str., 79011 Lviv, Ukraine

P. D'yachkov Kurnakov Institute of General and Inorganic Chemistry, Russian Academy of Sciences, Leninskii pr. 31, Moscow 119991 Russian Federation

W. Ebeling Institut für Physik, Humboldt-Universität Berlin, Newtonstrasse 15, Berlin-12489, Germany

L. A. Ferreira Instituto de Física de São Carlos, IFSC/USP, São Carlos, SP, Brazil

A. A. Eremko Bogolyubov Institute for Theoretical Physics, 03680 Kyiv, Ukraine, eremko@bitp.kiev.ua

D. A. Estrín Departamento de Química Inorgánica, Analítica y Química Física / INQUIMAE-CONICET, Facultad de Ciencias Exactas y Naturales, Universidad de Buenos Aires. Ciudad Universitaria, Pabellón II, Buenos Aires (C1428EHA), Argentina

R. Grasso Dipartimento di Metodologie Fisiche e Chimiche per l'Ingegneria, Catania University, viale A. Doria 6, I-95125 Catania, & Laboratori Nazionali del Sud—Istituto Nazionale di Fisica Nucleare, Via S. Sofia 44, 95123 Catania (Italy)

M. Gulino Dipartimento di Metodologie Fisiche e Chimiche per l'Ingegneria, Catania University, viale A. Doria 6, I-95125 Catania, & Laboratori Nazionali del Sud—Istituto Nazionale di Fisica Nucleare, Via S. Sofia 44, 95123 Catania (Italy)

D. Hennig Institut für Physik, Humboldt-Universität Berlin, Newtonstrasse 15, Berlin-12489, Germany

P. E. Hoggan LASMEA, UMR 6602 CNRS., University Blaise Pascal, 24 avenue des Landais, 63177 Aubiere Cedex, France

M. Holovko Institute for Condensed Matter Physics, National Academy of Sciences of Ukraine, 1 Svientsitskii Str., 79011 Lviv, Ukraine

A. P. Kravchuk National Transport University, 1 Suvorov str., 01010 Kyiv, Ukraine

E. S. Kryachko Bogolyubov Insitute for Theoretical Physics, Kiev-143, 03680 Ukraine

V. D. Lakhno Institute of Mathematical Problems of Biology, RAS, 142290, Pushchino, Institutskaya str., 4, Russia, lak@impb.psn.ru

L. Lanzano' Dipartimento di Metodologie Fisiche e Chimiche per l'Ingegneria, Catania University, viale A. Doria 6, I-95125 Catania, & Laboratori Nazionali del Sud—Istituto Nazionale di Fisica Nucleare, Via S. Sofia 44, 95123 Catania (Italy)

M. Leopoldini Dipartimento di Chimica and Centro di Calcolo ad Alte Prestazioni per Elaborazioni Parallele e Distribuite-Centro d'Eccellenza MIUR, Universita' della Calabria, I-87030 Arcavacata di Rende (CS), Italy

F. J. Luque Departament de Fisicoquímica and Institut de Biomedicina (IBUB), Facultat de Farmàcia, Universitat de Barcelona, AV. Diagonal 643, 08028, Barcelona, Spain

D. Makaev Kurnakov Institute of General and Inorganic Chemistry, Russian Academy of Sciences, Leninskii pr. 31, Moscow 119991 Russian Federation

T. Marino Dipartimento di Chimica and Centro di Calcolo ad Alte Prestazioni per Elaborazioni Parallele e Distribuite-Centro d'Eccellenza MIUR, Universita' della Calabria, I-87030 Arcavacata di Rende (CS), Italy

M. A. Martí Departamento de Química Inorgánica, Analítica y Química Física/ INQUIMAE-CONICET, Facultad de Ciencias Exactas y Naturales, Universidad de Buenos Aires. Ciudad Universitaria, Pabellón II, Buenos Aires (C1428EHA), Argentina

J. M. McMahon Department of Chemistry, Northwestern University, Evanston IL 60208-3113 USA

F. Musumeci Dipartimento di Metodologie Fisiche e Chimiche per l'Ingegneria, Catania University, viale A. Doria 6, I-95125 Catania, & Laboratori Nazionali del Sud—Istituto Nazionale di Fisica Nucleare, Via S. Sofia 44, 95123 Catania (Italy)

N. Panagiotides Department of Physics, University of Ioannina, P.O. Box 1186, GR-45110 Ioannina, Greece

N. I. Papanicolaou Department of Physics, University of Ioannina, P.O. Box 1186, GR-45110 Ioannina, Greece

B. M. A. G. Piette Department of Mathematical Sciences, University of Durham, Durham DH1 3LE, UK

V. V. Porsev Saint-Petersburg State University, Saint-Petersburg, Russia

G. Privitera Dipartimento di Metodologie Fisiche e Chimiche per l'Ingegneria, Catania University, viale A. Doria 6, I-95125 Catania, & Laboratori Nazionali del Sud—Istituto Nazionale di Fisica Nucleare, Via S. Sofia 44, 95123 Catania (Italy)

G. O. Puchkovska Institute of Physics of National Academy of Sciences of Ukraine, 46 Nauki Prospect, 03022 Kyiv, Ukraine

N. Russo Dipartimento di Chimica and Centro di Calcolo ad Alte Prestazioni per Elaborazioni Parallele e Distribuite-Centro d'Eccellenza MIUR, Universita' della Calabria, I-87030 Arcavacata di Rende (CS), Italy

G. C. Schatz Department of Chemistry, Northwestern University, Evanston IL 60208-3113 USA

A. Scordino Dipartimento di Metodologie Fisiche e Chimiche per l'Ingegneria, Catania University, viale A. Doria 6, I-95125 Catania, & Laboratori Nazionali del Sud—Istituto Nazionale di Fisica Nucleare, Via S. Sofia 44, 95123 Catania (Italy)

M. Salerno Dipartimento di Fisica "E.R. Caianiello" and Consorzio Nazionale Interuniversitario per le Scienze Fisiche della Materia (CNISM), Universit'a di Salerno, I-84081 Baronissi (SA), Italy

M. Tedesco Dipartimento di Metodologie Fisiche e Chimiche per l'Ingegneria, Catania University, viale A. Doria 6, I-95125 Catania, Italy

M. Toscano Dipartimento di Chimica and Centro di Calcolo ad Alte Prestazioni per Elaborazioni Parallele e Distribuite-Centro d'Eccellenza MIUR, Universita' della Calabria, I-87030 Arcavacata di Rende (CS), Italy

A. Triglia Dipartimento di Metodologie Fisiche e Chimiche per l'Ingegneria, Catania University, viale A. Doria 6, I-95125 Catania, Italy

A. V. Tulub Saint-Petersburg State University, Saint-Petersburg, Russia

M. G. Velarde Instituto Pluridisciplinar, Paseo Juan XIII, n. 1, Madrid-28040, Spain

W. J. Zakrzewski Department of Mathematical Sciences, University of Durham, Durham DH1 3LE, UK

Recent Progress on Small Hydrogen Molecular Ions

Alexander Alijah

Abstract The hydrogen molecular ions H_n^+ have been studied intensely by experimentalists and theoreticians. Apart from H_2^+, most work has been dedicated to H_3^+ and mainly concerns the electronic ground state. In my talk, I presented some recent results on the excited electronic singlet and triplet states before turning to the next higher member of the H_n^+ series, H_4^+. This ion, initially characterized in the 1970s and first observed in 1984, has since been largely overlooked. Indeed, recent studies focus on H_5^+ and H_6^+. H_4^+ is interesting as the smallest of the clusters with an H_3^+ core. Various nuclear configurations have been suggested as stable in the literature. We have systematically explored the potential energy surface and thereby examined those structures. Benchmark calculations have been performed at the stationary points. The results of this study were presented at the conference.

Keywords Astrochemistry · Hydrogen clusters · Rovibrational states · Potential energy surfaces

1 H_3^+

H_3^+ is the simplest polyatomic molecule and as such has served as a test system for the development of theoretical and computational methods. Its importance in physics, chemistry and astronomy has made it subject of two Discussion Meetings of the Royal Society [1, 2]. The minimum energy configuration in the electronic ground state is that of an equilateral triangle. H_3^+ is thus the prototype of a three-centre-two-electron bond. The lowest electronic states of H_3^+ can be derived easily from their electronic configurations. Consider a superposition of $1s$-orbitals localised at the three hydrogen atoms in equilateral triangular configuration, giving rise to molecular orbitals of symmetry a' and e'. The electronic ground state is derived from the configuration a'^2 and designated $X\ ^1A'$. The excited electronic

A. Alijah
Departamento de Química, Universidade de Coimbra, 3004–535 Coimbra, Portugal
e-mail: alijah@ci.uc.pt

N. Russo et al. (eds.), *Self-Organization of Molecular Systems: From Molecules and Clusters to Nanotubes and Proteins*, NATO Science for Peace and Security Series A: Chemistry and Biology, © Springer Science+Business Media B.V. 2009

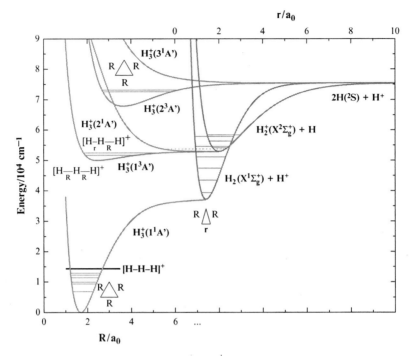

Fig. 1 Cuts through potential energy surfaces of H_3^+

configuration $a'e'$ leads to the electronic states $^1E'$ and $^3E'$, which are however not stable in this configuration according to the Jahn–Teller theorem. The singlet state is decomposed into $B^1\Sigma^+$ and $3\,^1A'$, while the two branches of the triplet state become $a^3\Sigma_u^+$ and $2^3A'$. Conical intersections occur at all configurations of symmetry D_{3h}. Figure 1 shows one-dimensional cuts of the full potential energy surfaces in which the remaining two coordinates are relaxed to mimic paths leading from the minima to dissociation.

1.1 Singlet H_3^+

Quite a number of potential energy surfaces have been reported for the electronic ground state. The most accurate one, the so-called CRJK surface [3], is based on 69 ab initio points obtained to sub-microhartree accuracy by Cencek et al. [4]. It is a local surface valid only to up to 20,000 cm^{-1} which has been used extensively for rovibrational calculations [5–8]. There is also a range of potential energy surfaces which extend into the region of dissociation into $H_2 + H^+$ [9–12].

At even higher energies, some surfaces may not be satisfactory. As they are based on single-surface modelling approaches, they may not describe well the avoided crossing with the next higher electronic singlet state, $B\,^1\Sigma^+$. The two lowest singlet surfaces would be needed, for example, to describe the charge transfer process $H_2 + H^+ \rightleftharpoons H_2^+ + H$ which is non-adiabatic. For the $B\,^1\Sigma^+$ state, no accurate and complete surface has been known in the literature until recently.

We have now presented full potential energy surfaces of the three lowest electronic singlet states [13]. In this work we used the highly accurate ab initio points by Cencek et al. [4] and Polyansky et al. [10], complemented by our own points calculated at the conventional FCI/cc-pV5Z level. The analytical approach is based on the diatomics-in-molecules (DIM) [14] formulation

$$H = \begin{pmatrix} E(1) + \varepsilon(2,3) & \Delta(3) & \Delta(2) \\ \Delta(3) & E(2) + \varepsilon(3,1) & \Delta(1) \\ \Delta(2) & \Delta(1) & E(3) + \varepsilon(1,2) \end{pmatrix} \tag{1}$$

with

$$E(i) = V^{(2)}_{[H_2,X]}(R_i)$$

$$\varepsilon(j,k) = \frac{1}{2}\left[V^{(2)}_{\left[H_2^+,X\right]}(R_j) + V^{(2)}_{\left[H_2^+,A\right]}(R_j) + V^{(2)}_{\left[H_2^+,X\right]}(R_k) \right.$$

$$\left. + V^{(2)}_{\left[H_2^+,A\right]}(R_k) \right] - 2E_H \tag{2}$$

$$\Delta(k) = \frac{1}{2}\left[V^{(2)}_{\left[H_2^+,X\right]}(R_k) - V^{(2)}_{\left[H_2^+,A\right]}(R_k) \right]$$

which is improved by addition of three-body terms to the matrix elements. Care must be taken not to destroy the permutation symmetry, $|\mathbf{H} - \lambda\mathbf{I}| = |\mathbf{PH_{ij}} - \lambda\mathbf{I}|$, and to keep the off-diagonal terms negative. Thus, the following substitutions have been made,

$$H_{ii} \rightarrow H_{ii} + V^{(3)}(\mathbf{R})\,, \quad H_{ij} \rightarrow H_{ij} - \left[\tilde{V}^{(3)}(\mathbf{R})\right]^2 \tag{3}$$

where $V^{(3)}(\mathbf{R})$ and $\tilde{V}^{(3)}(\mathbf{R})$ are totally symmetric and depend on the three internal coordinates.

1.1.1 Vibrational Resonance in the $B^1\Sigma^+$ State

The $B\,^1\Sigma^+$ state is characterised by six shallow minima only $57.07\,\text{cm}^{-1}$ below dissociation. They correspond to a hydrogen atom weekly bound to H_2^+, forming a linear complex. There is a weak rotational barrier of only $4.63\,\text{cm}^{-1}$. The barrier for proton exchange is $22{,}393.04\,\text{cm}^{-1}$, which is well above dissociation. Thus, proton exchange is not feasible. The question arising now is whether or not the $H_2^+ \cdots H$ complex is stable. For the isotopologue $^1H_3^+$ this is not the case. Furthermore, the hypothetical lowest vibrational state would have symmetry A_1' which would violate the Pauli principle. If, however, the protons are replaced by deuterons to form $D_2^+ \cdots D$, which is a boson system, such a state becomes symmetry-allowed and is stable by $20\,\mathrm{cm}^{-1}$ with respect to D_2^+ $(v = 0) + D$. We have not so far estimated its lifetime.

1.2   Triplet H_3^+

As outlined above, the potential energy surface of the $1\,^3E'$ state has two sheets that intersect at D_{3h} configurations. The system stabilizes on the lower sheet, $a^3\Sigma_u^+$. Already in 1974 Schaad and Hicks [15] predicted the existence of vibrational states for triplet H_3^+. This issue has since been investigated further by theoretical groups, but the first accurate vibrational [16] and rovibrational [17] energies have been reported only in 2001. Our surface has since been improved to yield accurate representations of the two sheets of the $1\,^3E'$ surface [18] using the double many-body expansion (DMBE) [19]

$$V_{u/l}(\mathbf{R}) = \sum_i V^{(1)} + \sum_i V_{u/l}^{(2)}(R_i) + V_{u/l}^{(3)}(\mathbf{R}) \tag{4}$$

The subscripts u/l refer to the upper and the lower sheet, respectively. A special form of the three-body term has been employed,

$$V_{u/l}^{(3)}(\mathbf{R}) = P_1(\mathbf{R}) \pm \Gamma_2 P_2(\mathbf{R}) \tag{5}$$

that not only ensures degeneracy at the intersection line ($\Gamma_2 = 0$) but also linear splitting with respect to the Jahn–Teller active coordinate, Γ_2, of the two sheets in the vicinity of this line.

The lower sheet, $a^3\Sigma_u^+$, is characterized by three equivalent minima of depth $2{,}947\,\mathrm{cm}^{-1}$, separated by saddle points of $2{,}598\,\mathrm{cm}^{-1}$. The rovibronic states of the homonuclear isotopologues may be described as symmetry-adapted superpositions of localized states such as to form a one-dimensional representation,

$$\left| \psi_A^{\pm} \right\rangle \sim \left| \psi_I^{\pm} \right\rangle + \left| \psi_{II}^{\pm} \right\rangle + \left| \psi_{III}^{\pm} \right\rangle \tag{6}$$

and a two-dimensional one

$$\left| \psi_{E,\xi}^{\pm} \right\rangle \sim \left| \psi_I^{\pm} \right\rangle + \omega \left| \psi_{II}^{\pm} \right\rangle + \omega^2 \left| \psi_{III}^{\pm} \right\rangle \tag{7}$$

$$\left| \psi_{E,\eta}^{\pm} \right\rangle \sim \left| \psi_I^{\pm} \right\rangle + \omega^2 \left| \psi_{II}^{\pm} \right\rangle + \omega \left| \psi_{III}^{\pm} \right\rangle \tag{8}$$

with $\omega = e^{\frac{2\pi i}{3}}$. Each of the localized functions $\left| \psi^{\pm} \right\rangle$, dropping now the localization index I, II or III, can be expanded approximately in terms of linear molecule basis functions as

$$|\psi^{\pm}\rangle = \frac{1}{\sqrt{2}} \left|v_1 v_2^{|\ell|} v_3\right\rangle (|N\ell m\rangle \pm |N - \ell m\rangle) \tag{9}$$

Such an expansion is useful from a qualitative point of view as it leads to a classification of the rovibrational states in terms of linear molecule quantum numbers and the symmetry index (A, E), which however do not hold rigorously. Therefore one cannot rely on the above expansion in order to actually solve the Schrödinger equation of the nuclear motion. We have used instead the accurate method of hyperspherical harmonics [20], in which all nuclear configurations are treated on equal footing to calculate the rovibronic states of the 19 lowest bands for $N \leq 10$, altogether 560 states [21]. Assignments in terms of approximate quantum numbers have been provided along with the exact symmetry quantum numbers.

The two symmetry components (A, E) of a rovibrational state are split in energy because of the potential barriers. In a first approach, this splitting should just depend on the energy. Our calculations [21] have shown that the splitting cannot be described in such simple terms, and recently we have investigated its nature [22]. The splitting can be understood in terms of hyperspherical motion of the nuclei, see Fig. 2. Motion along the hyperspherical angle ϕ, which is equivalent to the angle of pseudorotation, mainly accounts for the splitting and, since motion along orthogonal coordinates does not contribute, provides upper bounds. Near

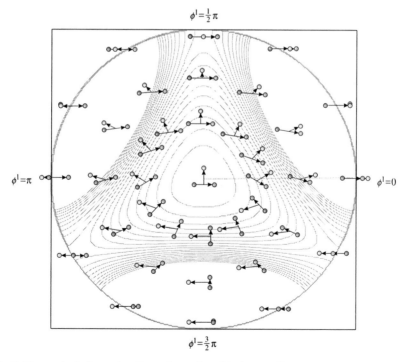

Fig. 2 Hyperspherical mapping: For a given value of the hyperradius, ρ, which controls the size of the triangle formed by the three nuclei, its shape depends on the polar angle, θ, and the azimuth, φ

the minima this hyperspherical motion can be related to the antisymmetric stretch quantum number v_3, such that within a family of states with symmetric stretch and bending quantum numbers v_1 and v_2 held fixed, the splitting increases with increasing v_3.

This is demonstrated in Fig. 3 for D_3^+. The left panel contains all splittings as a function of energy, while the right panel contains only those of three selected families of states, each of which now showing increase of the splitting with energy. In the case of the mixed isotopologue H_{2D}^+, hyperspherical motion links the two isomers HDH^+ and HHD^+ and might be responsible for perturbations found in the calculated spectra [23], see Fig. 4.

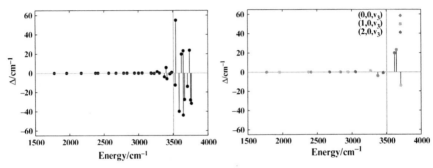

Fig. 3 Energy splittings $\Delta_E = E(A) - E(E)$ in D_3^+, on the left including all vibrational states, on the right for vibrational states of the first three families

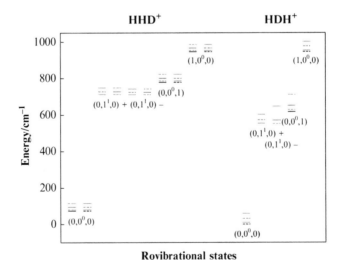

Fig. 4 Rovibrational states of H_{2D}^+

Table 1 Predicted rovibrational transition frequencies of triplet H_3^+ and some isotopologues

Molecule	Initial state	Γ_{rve} (w)	Final state	Γ_{rve} (w)	$\tilde{v}/cm^{-1}$	$\tilde{v}/GHz$	λ/nm
H_3^+	$(0, 0^0, 0)$	$A'_2(4)$	$(0, 1^1, 0)-$	$A''_2(4)$	665.89	19,963	15,017
		$E'(2)$	$(0, 1^1, 0)-$	$E''(2)$	665.89	19,963	15,017
H_3^+	$(0, 0^0, 0)$	$A'_2(4)$	$(0, 0^0, 1)$	$A''_2(4)$	749.71	22,476	13,339
		$E'(2)$	$(0, 0^0, 1)$	$E''(2)$	749.72	22,476	13,339
H_3^+	$(0, 0^0, 0)$	$E'(2)$	$(1, 0^0, 0)$	$E''(2)$	984.12	29,503	10,161
HDH^+	$(0, 0^0, 0)$	$B_2(3)$	$(0, 1^1, 0)-$	$B_1(3)$	551.46	16,532	18,134
HDH^+	$(0, 0^0, 0)$	$B_2(3)$	$(0, 0^0, 1)$	$B_1(3)$	628.32	18,837	15,915
HHD^+	$(0, 0^0, 0)$	$A_1(1)$	$(0, 1^1, 0)-$	$A_2(1)$	638.23	19,134	15,668
		$B_2(3)$	$(0, 0^0, 1)-$	$B_1(3)$	638.23	19,134	15,668
HHD^+	$(0, 0^0, 0)$	$A_1(1)$	$(0, 0^0, 1)$	$A_2(1)$	712.13	21,349	14,042
		$B_2(3)$	$(0, 0^0, 1)$	$B_1(3)$	712.13	21,349	14,042
HHD^+	$(0, 0^0, 0)$	$A_1(1)$	$(1, 0^0, 0)$	$A_2(1)$	879.64	26,371	11,368
		$B_2(3)$	$(1, 0^0, 0)$	$B_1(3)$	879.64	26,371	11,368

The rotational transition is $N = 1 \leftarrow N = 0$. The statistical weights of the states involved are given in parentheses.

1.2.1 Predicted Transition Frequencies

Neither H_3^+ nor its isotopologues have been observed so far. As a guidance for experimentalists, we have calculated the frequencies of transitions originating from the vibrational and rotational ground states. The selection rules are for H_3^+ and D_3^+: $A'_1 \leftrightarrow A''_1$, $A'_2 \leftrightarrow A''_2$, $E' \leftrightarrow E''$ while for H_2D^+ and D_2H^+ they are $A_1 \leftrightarrow A_2$, $B_1 \leftrightarrow B_2$. Furthermore, the angular momentum selection rule $\Delta N = 0, \pm 1$ applies. Table 1 indicates that lines are to be expected in the mid to far IR region.

2 H_4^+

H_3^+ is one of the most important species in astrochemical processes [24]. Acting as a strong proton donor, it catalyses the formation of a large variety of molecules such as water, amines or alcohols. It is formed by the very fast reaction $H_2 + H_2^+ \rightleftharpoons H_3^+ + H$ which takes place on the H_4^+ potential energy surface. H_4^+ itself is an interesting species in its own right, as it is the simplest weakly bound complex of the type $H_3^+ \cdots X$. It has been detected experimentally by mass spectroscopy [25] following theoretical predictions. Alvarez-Collado et al. [26] presented vibrational calculations on the H_4^+ cation based on a local potential energy surface obtained by the same authors, in which the H_3^+ moiety was kept frozen. Moyano et al. [27] reported a local potential energy surface and have localized a transition state which links two equivalent minimum structures. Jahn–Teller distortions and related structures have been described by Jiang et al. [28] and Jungwirth et al. [29].

It appears that some of the results by previous workers have been obtained at a modest level by today's computational standards and may be questionable. We have thus performed a systematic search for local minima and saddle points and performed benchmark calculations for those structures [30]. As an example we present in Table 2 our ab initio results for the minimum structure shown in Fig. 5. Varying systematically the number of active orbitals in the CASSCF/MRCI calculations we found that 16 active orbitals yield results of nearly full CI quality at a much lower cost. Hence this computational level lends itself to the extensive calculations needed for the construction of the potential energy surface. The minimum configuration of H_4^+ is a C_{2v} complex in which the H_3^+ core is deformed as compared to H_3^+ itself, stabilized with respect to dissociation into $H_3^+ + H$ by $1,952.75\,cm^{-1}$. Owing to permutational symmetry, there exist 12 such configurations which may be grouped into four sets of three. Within each set, there exist three low energy transition states, TS_1, only $1,062.48\,cm^{-1}$ above the minima. These transition states, obtained by moving the loosely bound hydrogen atom from the apex of the H_3^+ core to a side, have been discussed in an early paper by Poshusta and Zetik [31], but strangely their knowledge seems to have got lost. There are also transition states for proton exchange, TS_3, see Fig. 5, which are $3,127.29\,cm^{-1}$ above the minima and

Table 2 CASSCF/MRCI and FCI calculations at the minimum configuration of H_4^+

	Basis set						
n_a	VQZ	V5Z	AV5Z	V6Z	AV6Z	V7Z	AV7Z
4 – 1.851	871	2472	514	620	636	674	683
8	915	517	559	666	681		
12	919	520	563	669	685		
16	920	522	564	671	686	724	733
20	921	523	565	672	687		
All	922	524					

n_a denotes the number of active orbitals.

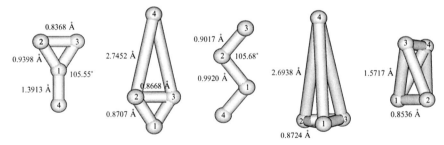

Fig. 5 H_4^+ structures: Minimum, first-order saddle points TS_1 and TS_3 and second-order saddle points TS_2 and TS_4 (from left to right)

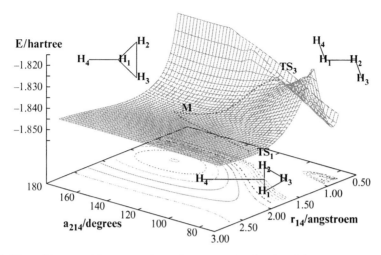

Fig. 6 View of the potential energy surface at low energy showing the minimum structure and two saddle points and connecting intrinsic reaction coordinate paths

thus located above dissociation. Some non-planar structures, which can be derived by symmetry arguments, are also shown in Fig. 5. The potential energy surface of H_4^+ is quite complex, consisting of 12 minimum structures and connecting transition states as well as conical intersections at square and tetrahedral configurations. Figure 6 offers a view of the low energy part of the potential energy surface up to dissociation. We have also examined the possibility of further true minima on the potential energy surface as suggested in the literature [28, 29] but could not find evidence for any other than the 12 lowest ones.

It is interesting to compare the structure of H_4^+ with that of the next higher mem-ber of the even H_n^+ clusters, H_6^+. As noted by Montgomery and Michels [32] and confirmed by Kurosaki and Takayanagi [33], H_6^+ exists in two isomeric structures, one with a H_3^+ core to which both a hydrogen atom and a hydrogen molecule are loosely bound, and the other with a H_2^+ core and two loosely bound hydrogen molecules. The latter isomer is more stable by $1,417\,cm^{-1}$ [33]. It is also more stable, by at least $2,085\,cm^{-1}$, than H_4^+ plus H_2. It is generally thought [32] that the odd-membered clusters H_n^+ are formed by adding H_2 units to an H_3^+ core while the even-membered ones consist of an H_2^+ core with added H_2 molecules. H_4^+ is special in that it doesn't fit into this scheme.

Acknowledgments A. A. would like to thank A. J. C. Varandas for hospitality and discussions and the Fundaçao para a Ciência e a Tecnologia for financial support. He is also grateful to the John von Neumann Institut fur Computing, Julich, for an allowance of super-computer time (Project EPG00).

References

1. See papers of a discussion meeting on the subject organized and edited by E. Herbst, S. Miller, T. Oka, and J. K. G. Watson, Philos. Trans. R. Soc. London A **358**, 2363 (2000).
2. See papers of a discussion meeting on the subject organized and edited by T. R. Geballe, D. Gerlich, J. Tennyson, and T. Oka, Philos. Trans. R. Soc. London A **364**, 2847 (2006).
3. R. Jaquet, W. Cencek, W. Kutzelnigg, and J. Rychlewski, J. Chem. Phys. **108**, 2837 (1998).
4. W. Cencek, J. Rychlewski, R. Jaquet, and W. Kutzelnigg, J. Chem. Phys. **108**, 2831 (1998).
5. O. Polyansky and J. Tennyson, J. Chem. Phys. **110**, 5056 (1999).
6. R. Jaquet, Spectrochim. Acta A **58**, 691 (2002).
7. P. Schiffels, A. Alijah, and J. Hinze, Mol. Phys. **101**, 175 (2003).
8. P. Schiffels, A. Alijah, and J. Hinze, Mol. Phys. **101**, 189 (2003).
9. R. Prosmiti, O. L. Polyansky, and J. Tennyson, Chem. Phys. Lett. **273**, 107 (1997).
10. O. L. Polyansky, R. Prosmiti, W. Klopper, and J. Tennyson, Mol. Phys. **98**, 261 (2000).
11. A. Aguado, O. Roncero, C. Tablero, C. Sanz, and M. Paniagua, J. Chem. Phys. **112**, 1240 (2000).
12. L. Velilla, B. Lepetit, A. Aguado, J. A. Beswick, and M. Paniagua, J. Chem. Phys. **129**, 084307 (2008).
13. L. P. Viegas, A. Alijah, and A. J. C. Varandas, J. Chem. Phys. **126**, 074309 (2007).
14. F. O. Ellison, J. Am. Chem. Soc. **85**, 3540 (1963).
15. L. J. Schaad and W. V. Hicks, J. Chem. Phys. **61**, (1974).
16. C. Sanz, O. Roncero, C. Tablero, A. Aguado, and M. Paniagua, J. Chem. Phys. **114**, 2182 (2001).
17. O. Friedrich, A. Alijah, Z. R. Xu, and A. J. C. Varandas, Phys. Rev. Lett. **86**, 1183 (2001).
18. A. J. C. Varandas, A. Alijah, and M. Cernei, Chem. Phys. **308**, 285 (2005).
19. A. J. C. Varandas, Adv. Chem. Phys. **74**, 255 (1988).
20. L. Wolniewicz, J. Chem. Phys. **90**, 371 (1988).
21. A. Alijah, L. P. Viegas, M. Cernei, and A. J. C. Varandas, J. Mol. Spectrosc. **221**, 163 (2003).
22. T. Mendes Ferreira, A. Alijah, and A. J. C. Varandas, J. Chem. Phys. **128**, 054301 (2008).
23. A. Alijah and A. J. C. Varandas, J. Phys. Chem. A **110**, 5499 (2006).
24. T. R. Geballe and T. Oka, Science **312**, 1610 (2006).
25. N. J. Kirchner, J. R. Gillbert, and M. T. Bowers, Chem. Phys. Lett. **106**, 7 (1984).
26. J. R. Alvarez-Collado, A. Aguado, and M. Paniagua, J. Chem. Phys. **102**, 5725 (1995).
27. G. E. Moyano, D. Pearson, and M. A. Collins, J. Chem. Phys. **121**, 12396 (2004).
28. G. Jiang, H. Y. Wang, and Z. H. Zhu, Chem. Phys. Lett. **284**, 267 (1998).
29. P. Jungwirth, P. Carsky, and T. Bally, Chem. Phys. Lett. **195**, 371 (1992).
30. A. Alijah and A. J. C. Varandas, J. Chem. Phys. **129**, 034303 (2008).
31. R. D. Poshusta and D. F. Zetik, J. Chem. Phys. **58**, 118 (1973).
32. J. A. Montgomery, Jr. and H. H. Michels, J. Chem. Phys. **87**, 771 (1987).
33. Y. Kurosaki and T. Takayanagi, J. Chem. Phys. **109**, 4327 (1998).

FDTD Studies of Metallic Nanoparticle Systems*

Ariel L. Atkinson, Jeffrey M. McMahon, and George C. Schatz

Abstract This paper provides an overview of the optical properties of plasmonic nanoparticles, using gold nanowires as a model system. The properties were calculated using classical electrodynamics methods with bulk metal dielectric constants, as these methods provide a nearly quantitative description of nanoparticle optical response that can be used for particles with dimensions of a few nanometers to many hundreds of nanometers. The nanowire calculations are based on the finite-difference time-domain (FDTD) method in two dimensions, and we specifically consider the transmission of light through nanowire arrays, as this provides a simple nanomaterial construct which still displays the richness of optical phenomena that is found for more general nanostructures. The calculations show a number of features that are known for other nanostructures, including the red-shifting of plasmon resonances as wire spacing is decreased, and as particle aspect ratio is increased. In addition, the influence of dielectric coatings on the wires is examined, including factors which determine dielectric sensitivity. These results provide insight into what structures will be most effective for index of refraction sensing applications.

Keywords Finite-difference time-domain · Extinction · Transmission · Plasmon · Hole-array · Nanowire array · Dielectric sensitivity · Gold · Nanoparticle

1 Introduction

The past few years have seen intense interest in the use of silver and gold particles in chemical and biological sensing applications using the intense absorption and scattering associated with plasmon resonance excitation as a reporter for the presence of molecules that are near to the particles [1–4]. Plasmon excitation involves the collective excitation of the conduction electrons in metals, leading to resonant wavelengths that are strongly dependent on nanoparticle size, shape, arrangement

A.L. Atkinson, J.M. McMahon, and G.C. Schatz (✉)
Department of Chemistry, Northwestern University, Evanston IL 60208-3113, USA
e-mail: schatz@chem.northwestern.edu

* Invited manuscript for inclusion in Proceedings of the NATO-ARW December 21, 2008

N. Russo et al. (eds.), *Self-Organization of Molecular Systems: From Molecules and Clusters to Nanotubes and Proteins*, NATO Science for Peace and Security Series A: Chemistry and Biology, © Springer Science+Business Media B.V. 2009

11

and dielectric environment. Because of the large size of the particles, and the fact that every conduction electron in the particle contributes to plasmon excitation, the extinction coefficients associated with these particles are enormous compared to the corresponding extinction coefficients in molecules, and as a result, sensing based on plasmon resonance excitation is superior to molecular fluorescence for some applications.

Plasmon-based sensing can be accomplished using a number of different spectroscopic methods. Extinction or scattering measurements can be used to determine the wavelength shift of the plasmon resonance as molecules bind to the particles (refractive index sensing) [5]. Alternatively, one can use extinction or scattering to detect the wavelength shifts that occur when molecular binding processes lead to aggregation of the particles (aggregation sensing) [2]. Holes in metal films are another popular platform for refractive index sensing where transmission or reflectivity are used [6–8]. It is also possible to detect molecules using surface enhanced Raman spectroscopy (SERS) in which polarization induced in the conduction electrons when plasmons are excited leads to enhanced electromagnetic fields around the nanoparticles [9, 10]. There are also a number of nonlinear optical techniques (hyper-Rayleigh, hyper-Raman) which can be enhanced by plasmon excitation, and which also have sensitivity to adsorbed molecules [11].

The experimental studies of sensing based on the optical properties of silver and gold particles have been accompanied by substantial theory work, mostly based on continuum electrodynamics methods for modeling optical response. In this one solves Maxwell's equations for the chosen nanoparticle structure and assumed values for the dielectric constants of the particles and the surrounding material [11–38]. This works quite well for particle sizes larger than 2 nm, especially for determining far-field properties such as extinction spectra. However for smaller particles, and even for bigger particles that have <2 nm substructures, there can be errors that arise from the use of a dielectric continuum approach, so there has been some activity using electronic structure theory as an alternative [39]. The focus of this work will, however, be on larger particles where classical electrodynamics is adequate.

Since most of the particle structures of recent interest are non-spherical, and often the particles are in arrays or aggregates, the calculation of optical properties using classical electrodynamics needs to be done using computational (rather than analytical) methods. Fortunately several computational electrodynamics methods have become available in the last few years (often derived from research in optical physics and electrical engineering that is unrelated to nanoscience) that are capable of describing particles with sizes up to a micron, and complicated dielectric environments, including coatings and substrates that in some cases include dye molecules or other nanoparticles [40].

The most popular methods for performing computational electrodynamics studies of silver and gold nanostructure optical properties include the discrete dipole approximation (DDA), the finite-difference time-domain (FDTD) method, and the Whitney form finite element (WF-FE) method [40]. These methods are based on fundamentally different concepts, but in all three cases the particles are represented in terms of discrete elements of some kind that are small compared to the particle size but large compared to the size of an atom (typical element size for metallic structures is around 1 nm). The DDA and FDTD methods use cubic elements. In DDA, the cubes are thought of as polarizable elements and the solution to Maxwell's equations involves determining the collective polarization in response to an applied field. The FDTD method uses a cubic grid to define finite difference approximations to Maxwell's equations. The WF-FE method typically uses tetrahedral elements, a vector-based polynomial representation of the field components within each element, and boundary condition matching on the surfaces of each element.

These three methods have good and bad features. DDA is only an approximate method due to errors in defining the polarizability of each cube from the dielectric function. As a result, there are errors in the extinction calculations (usually less than 10%) and there can be problems with getting high quality fields at positions near the particle surface. FDTD is an exact method, but it runs into convergence issues close to the nanoparticle surface due to the finite difference approximation. In addition it is computationally very demanding due to the large grids that are needed (larger than in DDA as grids are needed to represent empty space and to apply boundary conditions). WF-FE methods require a complicated gridding process, but are capable of providing much more accurate solutions than the other methods, especially for locations near the surfaces of the particles. Generally, high accuracy is not needed for the evaluation of far-field properties such as the extinction or scattering cross sections; however, accuracy is quite important to the evaluation of electric fields near the nanoparticle surfaces, such as is needed in the evaluation of SERS enhancement factors.

In the remainder of this manuscript, we first describe the FDTD method, and then we illustrate its use to describe the optical properties of metal nanostructures, including the dependence of plasmon properties on the size, shape, arrangement and dielectric environment of the nanoparticles. To provide a consistent set of results to illustrate these properties, we apply the FDTD calculations to determine the properties of gold nanowire arrays. These structures only require calculations on two-dimensional grids, which makes it relatively easy to study array structures that range from isolated particles to continuous metal films. In addition we consider particle sizes from a few to several hundred nanometers, with a range of shapes and spacings, all within a consistent framework, and all based on the use of transmission spectra (transmission versus wavelength) as the observed optical property.

2 Theory and Model

2.1 Finite-Difference Approach to Solving Maxwell's Equations

Finite-difference approaches to electrodynamics explicitly solve Maxwell–Ampere's law and Faraday's law in differential form, Eqs. (1) and (2), respectively:

$$\varepsilon \frac{\partial}{\partial t} \vec{E} = \nabla \times \vec{H} - \vec{J} \tag{1}$$

$$\mu \frac{\partial}{\partial t} \vec{H} = -\nabla \times \vec{E} \tag{2}$$

where $\vec{E}$ and $\vec{H}$ are the electric and magnetic fields, ε and μ are the permittivity and permeability of the medium, and $\vec{J}$ is the current density. To do this, the partial derivatives in the equations are approximated using Taylor expansions. For example, in one dimension the Taylor expansions of a function, $f(x)$, around the point x are given by

$$f(x + a) = f(x) + a \frac{\partial}{\partial x} f(x) + \dots \tag{3}$$

$$f(x - a) = f(x) - a \frac{\partial}{\partial x} f(x) + \dots \tag{4}$$

Equations (3) or (4) can be rearranged to get an expression for $\frac{\partial}{\partial x} f(x)$, leading to *forward* and *backward* finite-difference approximations, respectively, which would be first-order accurate (i.e. the truncation errors are of order a^2). A second-order accurate finite-difference approximation to $\frac{\partial}{\partial x} f(x)$ can be obtained by subtracting Eq. (4) from Eq. (3), known as a *central* difference,

$$\frac{\partial}{\partial x} f(x) = \frac{f(x + a) - f(x - a)}{2a}. \tag{5}$$

Central difference expressions are typically used in finite-difference algorithms because of the higher accuracy compared to forward and backward differences. The most straightforward way to use the finite-difference expressions in Eqs. (3)–(5) is to discretize the domain of $f(x)$ using a grid, where $f(x)$ is assumed to exist only at the discrete grid points. It is important to note that to obtain second-order accuracy in the temporal derivatives of Eqs. (1) and (2) $\vec{E}$ and $\vec{H}$ must be defined on time grids shifted by a. For example, in Eq. (1) the time derivative of $\vec{E}$ must "leap over" the spatial derivatives of $\vec{H}$, known as a *leap-frog algorithm* (see below).

Even though only Eqs. (1) and (2) are explicitly solved by finite-difference algorithms, special care must be taken in order to satisfy Gauss' laws,

$$\nabla \cdot \left(\varepsilon \vec{E} \right) = \rho \tag{6}$$

$$\nabla \cdot \left(\mu \vec{H} \right) = 0. \tag{7}$$

One way to satisfy Eqs. (6) and (7) is to shift all of the field components from each other, known as the Yee method, Fig. 1 [41]. For a proof of this result see Ref. [42].

Fig. 1 The Yee cell

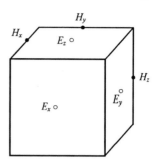

2.2 Finite-Difference Equations

2.2.1 Three-Dimensional Equations

Using the Yee cell (Fig. 1) and second-order accurate finite-differences (Eq. (5)) the finite-difference approximations to Eqs. (1) and (2) become

$$\varepsilon_{i+1/2,j,k} \frac{E_x^{n+1/2}{}_{i+1/2,j,k} - E_x^{n-1/2}{}_{i+1/2,j,k}}{\Delta t} =$$

$$\frac{H_z^n{}_{i+1/2,j+1/2,k} - H_z^n{}_{i-1/2,j+1/2,k}}{\Delta y} - \frac{H_y^n{}_{i+1/2,j,k+1/2} - H_y^n{}_{i+1/2,j,k-1/2}}{\Delta z} \tag{8}$$

$$\varepsilon_{i,j+1/2,k} \frac{E_y^{n+1/2}{}_{i,j+1/2,k} - E_y^{n-1/2}{}_{i,j+1/2,k}}{\Delta t} =$$

$$\frac{H_x^n{}_{i,j+1/2,k+1/2} - H_x^n{}_{i,j+1/2,k-1/2}}{\Delta z} - \frac{H_z^n{}_{i+1/2,j+1/2,k} - H_z^n{}_{i-1/2,j+1/2,k}}{\Delta x} \tag{9}$$

$$\varepsilon_{i,j,k+1/2} \frac{E_z^{n+1/2}{}_{i,j,k+1/2} - E_z^{n-1/2}{}_{i,j,k+1/2}}{\Delta t} =$$

$$\frac{H_y^n{}_{i+1/2,j,k+1/2} - H_y^n{}_{i-1/2,j,k+1/2}}{\Delta x} - \frac{H_x^n{}_{i,j+1/2,k+1/2} - H_x^n{}_{i,j-1/2,k+1/2}}{\Delta y} \tag{10}$$

$$\mu_{i,j+1/2,k+1/2} \frac{H_x^{n+1}{}_{i,j+1/2,k+1/2} - H_x^n{}_{i,j+1/2,k+1/2}}{\Delta t} =$$

$$\frac{E_y^{n+1/2}{}_{i,j+1/2,k+1} - E_y^{n+1/2}{}_{i,j+1/2,k}}{\Delta z} - \frac{E_z^{n+1/2}{}_{i,j+1,k+1/2} - E_z^{n+1/2}{}_{i,j,k+1/2}}{\Delta y}$$

$$\tag{11}$$

$$\mu_{i+1/2,j,k+1/2} \frac{H_y^{n+1}{}_{i+1/2,j,k+1/2} - H_y^n{}_{i+1/2,j,k+1/2}}{\Delta t} =$$

$$\frac{E_z^{n+1/2}{}_{i+1,j,k+1/2} - E_z^{n+1/2}{}_{i,j,k+1/2}}{\Delta x} - \frac{E_x^{n+1/2}{}_{i+1/2,j,k+1} - E_x^{n+1/2}{}_{i+1/2,j,k}}{\Delta z}$$

$$\tag{12}$$

$$\mu_{i+1/2,j+1/2,k} \frac{H_z^{n+1}{}_{i+1/2,j+1/2,k} - H_z^{n}{}_{i+1/2,j+1/2,k}}{\Delta t} =$$

$$\frac{E_x^{n+1/2}{}_{i+1/2,j+1,k} - E_x^{n+1/2}{}_{i+1/2,j,k}}{\Delta y} - \frac{E_y^{n+1/2}{}_{i+1,j+1/2,k} - E_y^{n+1/2}{}_{i,j+1/2,k}}{\Delta x}.$$

$$(13)$$

Equations (7)–(13) can be solved by discretizing a domain, specifying ε and μ at the discrete grid points, and sequentially solving Eqs. (7)–(9) and Eqs. (10)–(13).

2.2.2 Two-Dimensional Equations

If the system is invariant along one axis, which we will take to be the z-axis, then Eqs. (8)–(13) are greatly simplified.

TE$_z$ polarization

When $\vec{H}$ is oriented along the z-axis, such that the components of $\vec{E}$ are in the xy-plane, Eqs. (8)–(13) reduces to

$$\varepsilon_{i+1/2,j,k} \frac{E_x^{n+1/2}{}_{i+1/2,j,k} - E_x^{n-1/2}{}_{i+1/2,j,k}}{\Delta t} =$$

$$\frac{H_z^{n}{}_{i+1/2,j+1/2,k} - H_z^{n}{}_{i-1/2,j+1/2,k}}{\Delta y}$$

$$(14)$$

$$\varepsilon_{i,j+1/2,k} \frac{E_y^{n+1/2}{}_{i,j+1/2,k} - E_y^{n-1/2}{}_{i,j+1/2,k}}{\Delta t} =$$

$$-\frac{H_z^{n}{}_{i+1/2,j+1/2,k} - H_z^{n}{}_{i-1/2,j+1/2,k}}{\Delta x}$$

$$(15)$$

$$\mu_{i+1/2,j+1/2,k} \frac{H_z^{n+1}{}_{i+1/2,j+1/2,k} - H_z^{n}{}_{i+1/2,j+1/2,k}}{\Delta t} =$$

$$\frac{E_x^{n+1/2}{}_{i+1/2,j+1,k} - E_x^{n+1/2}{}_{i+1/2,j,k}}{\Delta y} - \frac{E_y^{n+1/2}{}_{i+1,j+1/2,k} - E_y^{n+1/2}{}_{i,j+1/2,k}}{\Delta x}.$$

$$(16)$$

TM$_z$ polarization

When $\vec{E}$ is oriented along the z-axis, such that the components of $\vec{H}$ are in the xy-plane, Eqs. (8)–(13) reduces to

$$\varepsilon_{i,j,k+1/2} \frac{E_z^{n+1/2}{}_{i,j,k+1/2} - E_z^{n-1/2}{}_{i,j,k+1/2}}{\Delta t} =$$

$$\frac{H_y^n{}_{i+1/2,j,k+1/2} - H_y^n{}_{i-1/2,j,k+1/2}}{\Delta x} - \frac{H_x^n{}_{i,j+1/2,k+1/2} - H_x^n{}_{i,j-1/2,k+1/2}}{\Delta y}$$

$$(17)$$

$$\mu_{i,j+1/2,k+1/2} \frac{H_x^{n+1}{}_{i,j+1/2,k+1/2} - H_x^n{}_{i,j+1/2,k+1/2}}{\Delta t} =$$

$$- \frac{E_z^{n+1/2}{}_{i,j+1,k+1/2} - E_z^{n+1/2}{}_{i,j,k+1/2}}{\Delta y}$$

$$(18)$$

$$\mu_{i+1/2,j,k+1/2} \frac{H_y^{n+1}{}_{i+1/2,j,k+1/2} - H_y^n{}_{i+1/2,j,k+1/2}}{\Delta t} =$$

$$\frac{E_z^{n+1/2}{}_{i+1,j,k+1/2} - E_z^{n+1/2}{}_{i,j,k+1/2}}{\Delta x}.$$

$$(19)$$

2.2.3 One-Dimensional Equations

If the system is further invariant along another axis, which we will take to be the y-axis, then Eqs. (14)–(19) reduce to the 1D scalar wave equations,

$$\varepsilon_{i,j+1/2,k} \frac{E_y^{n+1/2}{}_{i,j+1/2,k} - E_y^{n-1/2}{}_{i,j+1/2,k}}{\Delta t}$$

$$= - \frac{H_z^n{}_{i+1/2,j+1/2,k} - H_z^n{}_{i-1/2,j+1/2,k}}{\Delta x}$$

$$(20)$$

$$\mu_{i+1/2,j+1/2,k} \frac{H_z^{n+1}{}_{i+1/2,j+1/2,k} - H_z^n{}_{i+1/2,j+1/2,k}}{\Delta t}$$

$$= - \frac{E_y^{n+1/2}{}_{i+1,j+1/2,k} - E_y^{n+1/2}{}_{i,j+1/2,k}}{\Delta x}$$

$$(21)$$

or

$$\varepsilon_{i,j,k+1/2} \frac{E_z^{n+1/2}{}_{i,j,k+1/2} - E_z^{n-1/2}{}_{i,j,k+1/2}}{\Delta t}$$

$$= \frac{H_y^n{}_{i+1/2,j,k+1/2} - H_y^n{}_{i-1/2,j,k+1/2}}{\Delta x}$$

$$(22)$$

$$\mu_{i+1/2,j,k+1/2} \frac{H_y^{n+1}{}_{i+1/2,j,k+1/2} - H_y^n{}_{i+1/2,j,k+1/2}}{\Delta t}$$

$$= \frac{E_z^{n+1/2}{}_{i+1,j,k+1/2} - E_z^{n+1/2}{}_{i,j,k+1/2}}{\Delta x}.$$

$$(23)$$

2.3 Initial Conditions and Boundary Conditions

For practical use of the FDTD method, boundary conditions must be applied, and initial waves of arbitrary form must be easily introduced into the system.

2.3.1 Boundary Conditions

Even when simulating infinite domains, the computational domain must be truncated. However, many techniques have been developed to mimic open regions of space. One of the most successful techniques is to truncate the domain with artificial materials that absorb nearly all incident waves, called perfectly matched layers (PML) [43]. An efficient and accurate way of implementing PML with the Yee spatial lattice is to use convolutional PML (CPML) [44]. The implementation of CPML involves stretching the spatial derivatives and superimposing a time-dependent scalar function onto them. For a more complete discussion of PML see Ref. [42].

2.3.2 Initial Conditions

Given suitable initial conditions defined everywhere in the computational domain, the FDTD equations will properly evolve the fields according to Eqs. (1) and (2). However, defining computational domains for initial conditions with large spatial extent is often inefficient and unnecessary. A more efficient technique to introduce fields into the computational domain, particularly with the Yee spatial lattice, is to use the total field–scattered field (TF–SF) [45–47] technique. The implementation of this involves splitting the domain into two regions, an interior total field region and an exterior scattered field region. The FDTD method is applied directly in each region without modification. However, near the boundaries where the spatial derivatives extend into both regions, the fields are modified using the (known) incident field, so that all equations are consistent.

2.4 Dispersive Materials

Many materials have a frequency dependent dielectric response. For metals this response is often approximated using the Drude model. To model these materials using the FDTD method, auxiliary differential equations are used to link the material polarization and electric flux density [42, 48, 49]. These equations are updated self-consistently with the FDTD equations.

2.4.1 Auxiliary Differential Equation Method for the Drude Model

The Drude model approximates the frequency dependent relative permittivity of a material as

$$\varepsilon(\omega) = \varepsilon_\infty - \frac{\omega_D^2}{\omega^2 - i\omega\gamma_D} \tag{24}$$

where ε_∞ is the relative permittivity at infinite frequency, ω_D is the Drude pole frequency, and γ_D is the inverse of the pole relaxation time. To use Eq. (24) in the FDTD equations, the Maxwell–Ampere law is first expressed in the frequency domain,

$$i\omega\varepsilon\vec{E} = \nabla \times \vec{H} - \vec{J}. \tag{25}$$

Equation (24) is then inserted into Eq. (25), and rearranged such that the frequency dependence of the material is contained entirely in $\vec{J}$,

$$\vec{J} = -i\omega\varepsilon_0 \left(\frac{\omega_D^2}{\omega^2 - i\omega\gamma_D}\right) \vec{E}. \tag{26}$$

Fourier-transforming the results back to the time-domain and rearranging gives a modified set of FDTD equations for $\vec{E}$,

$$\vec{E}^{n+1} = \left(\frac{2\varepsilon_0\varepsilon_\infty - \Delta t\beta_D}{2\varepsilon_0\varepsilon_\infty + \Delta t\beta_D}\right) \vec{E}^n$$

$$+ \left(\frac{2\Delta t}{2\varepsilon_0\varepsilon_\infty + \Delta t\beta_D}\right) \cdot \left(\nabla \times \vec{H}^{n+1/2} - \frac{1}{2}(1 + k_D)\vec{J}_D^n\right) \tag{27}$$

where

$$k_D = \frac{1 - \gamma_D\Delta t/2}{1 + \gamma_D\Delta t/2} \tag{28}$$

$$\beta_D = \frac{\omega_D^2\varepsilon_0\Delta t/2}{1 + \gamma_D\Delta t/2} \tag{29}$$

and a corresponding update equation for $\vec{J}$,

$$\vec{J}_D^{n+1} = k_D\vec{J}_D^n + \beta_D\left(\vec{E}^{n+1} + \vec{E}^n\right). \tag{30}$$

3 Results and Discussion

3.1 Preliminary Work

3.1.1 Description of Model

In this work, the two-dimensional FDTD method is used to study a variety of metallic nanowire systems. As mentioned earlier, nanowires were chosen because they provide a computationally convenient structure (only requiring 2D grids) that can be varied to consider a broad range of problems that relate to the dependence of optical properties on particle size, shape and arrangement. To provide a consistent optical property for many different structures, our calculations will only refer to the transmission of light through nanowire arrays.

The layout of the FDTD computational area used for the nanowire calculations is depicted in Fig. 2. In these calculations, we take the x-direction to denote the film thickness, as well as the propagation direction of the light, while y measures distances along the film and is also the polarization direction. The grid is taken to be 230 nm in the x-direction while the y-direction is allowed be variable in size to account for different particle spacings and sizes. The "back" and "front" of the grid (at x = 0 and x = 230, respectively) have 15 nm of CPML on them to absorb scattered radiation. Periodic boundary conditions are defined in the y-direction. In cases

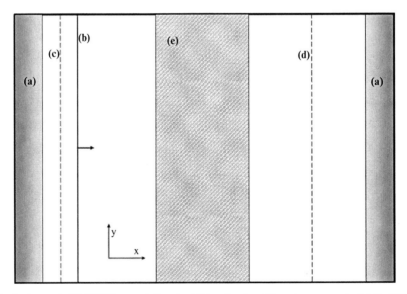

Fig. 2 A sample 2D FDTD computational grid showing the different components: (**a**) the CPML layers, (**b**) the initiation line for the plane wave, (**c**) and (**d**) the reflection and transmission spectra calculation points, respectively, and (**e**) the target area. In the convention used here, thickness is measured in the x-direction and length is measured in the y-direction

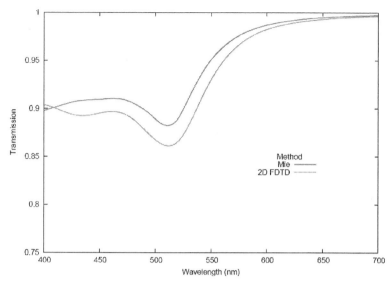

Fig. 3 Comparison between transmission cross-sections for a 30-nm diameter gold disk array by
(1) the 2D FDTD method and (2) Mie theory. In both arrays the disks have a spacing of 200 nm

where we study isolated particles, the y maximum is taken to be large enough relative to the particle size that there is no interaction between the particle and its
replicas. In most cases, 200 nm of spacing is more than sufficient. The section below entitled "Spacing Between Particles in an Array" covers this.

3.1.2 Comparison to Analytical Results

An important step in determining the reliability of any computational technique
is to compare the results with other known results. In this case, Mie theory (as
applied to circular wires) is used as the benchmark against which to measure the
2D-FDTD calculations. This was done for a 30 nm diameter gold wire, and we
see good agreement of the two results in Fig. 3. Note that the transmission has a
minimum at around 520 nm. This is the well-known extinction maximum associated
with gold nanoparticles. Note that 2D plasmon resonances are in general somewhat
blue-shifted compared to 3D resonances, but for gold these differences are small.

3.2 Effect of Particle Shape and Size

3.2.1 Equivalent Surface Area Shapes

A simple and rather significant variable that effects the interaction of light with a
particle is the particle's shape. In Fig. 4 we present a comparison of transmission

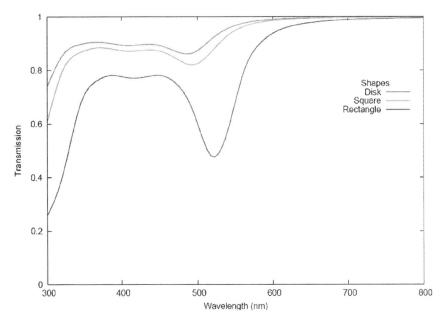

Fig. 4 Transmission spectra for three different structures, each with the same area, all calculated with 2D FDTD. The rectangle has an aspect ratio of 4:1, with the long axis oriented in the y-direction. The spacing between structures in each of these arrays was set at 200 nm

results for three different shapes: a square, a disk, and a rectangle. All three have the same surface area, and hence the differences in the transmission probability are based on the shapes themselves. The surface area of each is $100\,nm^2$, and the rectangle's aspect ratio [length (y-direction) to thickness (x-direction)] is 4:1, so the rectangle is 20 by 5 nm, the square's edge is 10 nm and the circle has a diameter of 11.3 nm. Notice that the rectangle gives a plasmon resonance that is red-shifted compared to the other particles. This is a well-known effect for anisotropic nanoparticles, but the red shift is small in this case as the particle is much smaller than the wavelength of light.

3.2.2 Particle Size

Another simple variable that has a strong influence on the optical properties of a particle or particle system is the size of the particles involved. In this work, particles with 2D cross sections of disks, squares, and rectangles were studied. The rectangular structures were scaled to keep the same aspect ratio of length to thickness (4:1), so that the effect of aspect ratio could be investigated separately. The results for these simulations are shown in Fig. 5(a)–(c).

Note that the smaller disk and square spectra are very similar to each other. The resonance peaks observed in the rectangle's spectrum are significantly different from the other two due to particle anisotropy. As all of the particles increase in size,

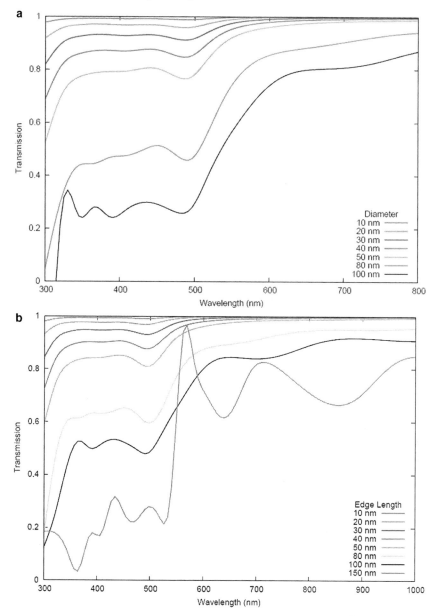

Fig. 5 (**a**) The effect of size on the transmission for an array of disks. The spacing in each array was 200 nm. (**b**) The effect of size on the transmission for an array of squares. The spacing in each array was 200 nm

the red-shifting and broadening of the plasmon peaks is clearly visible. These effects arise from two kinds of electrodynamic effects (as reviewed previously [15]): (1) the induced polarization is not all in-phase for plane wave excitation due to the

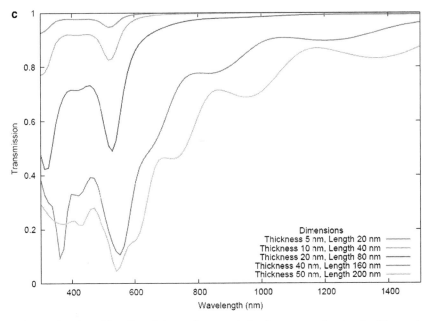

Fig. 5 (continued) (**c**) The effect of size on the transmission for an array of rectangles. The spacing in each array was 200 nm

finite wavelength, and (2) the plasmon is broadened due to radiative damping (i.e., the rate of emission by the particle is comparable to the rate of excitation). In addition, for all of the particles, the spectra show more plasmon peaks for the large particle sizes. This in part due to multipole resonance excitation, which is another effect that arises from the finite size of the particles compared to the wavelength.

3.2.3 Aspect Ratio

To study the effect of changing the aspect ratio of a rectangular structure, we keep the thickness of the particles constant at 10 nm, and vary the length in the range 10–150 nm (aspect ratios of 1:1 to 1:15). The resulting transmission spectra are presented in Fig. 6. Since the increasing length is parallel with the polarization of the incident light, the transmission is dramatically decreased as the aspect ratio is increased (corresponding to increased extinction), and the resonance peaks are red-shifted. These effects arise from a combination of the depolarization/damping effects that were mentioned in the previous section, and from electrodynamic boundary effects that relate to particle shape and the ability of the conduction electrons to oscillate relative to the positive background. These effects have previously been studied for spheroidally shaped gold particles [15], and they are also important for rods [50].

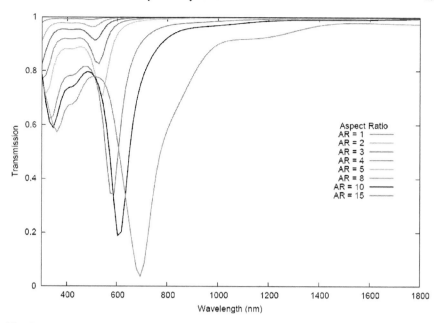

Fig. 6 The effect of the aspect ratio of gold particles on the optical signature. Here the thickness was kept constant at 10 nm while the length was varied from 10–150 nm. The spacing was kept constant at 200 nm

3.3 Particle Orientation and Arrangement

3.3.1 Spacing Between Particles in an Array

In all of the studies done so far, we modeled infinite arrays of metallic structures, but in most cases the structures/particles were spaced sufficiently far apart that no coupling or interactions between particles occur. In this section, the influence of particle spacing is considered for rectangles that are 80 nm in length and 10 nm thick for spacings of 4–200 nm. The results are presented in Fig. 7. For spacings of 100–200 nm, there is little interaction between the particles, and the dominant resonance is close to 600 nm. Here the transmission decreases as particle spacing is decreased simply due to changes in particle density. For smaller spacings, the plasmon resonances begin to couple and red-shift. This is again a known effect provided that the particle spacing is less than 100 nm, and a simple model for it based on dipole–dipole coupling was described by Zhao et al. [51].

Another feature of these spectra is that for the close spacings, the transmission drops close to 0%, a rather amazing result given that these particles are only 10 nm in thickness. In this limit the particle array is best thought of as a hole array in a metal film. Past work has demonstrated that hole array structures can efficiently convert incident plane wave excitation into surface plasmon polaritons (SPPs) which are

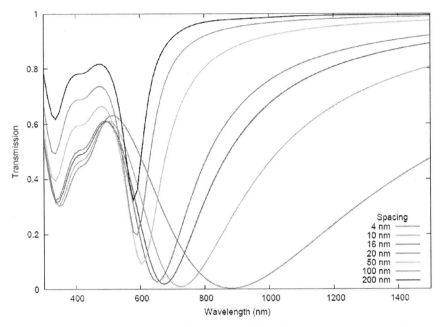

Fig. 7 Spectra of particle arrays each with particles of 10 nm thickness and 80 nm length but with varying spacings between them

propagating plasmons that can decay nonradiatively with reasonable efficiency [52]. In addition, it has been demonstrated that interference between SPP excitation and direct transmission through the film can lead to Fano profiles in transmission spectra in which the transmission drops to nearly zero over a limited wavelength range. This behavior was previously seen for 50 nm films [38], but the present application shows that very small transmission also arises for 10 nm films.

3.4 Environmental Dielectric Effects

3.4.1 Changing the Medium

A further well-known variable that affects the optical properties of nanostructured metallic systems is the refractive index of the surrounding medium in which the measurements are done. Here we use gold rectangular bars to study this effect, considering bars which are 80 nm in length and 10 nm in thickness to be embedded in a dielectric whose refractive index is varied from 1.0 to 2.0. The resulting transmission data are presented in Fig. 8. This shows that the plasmon resonance at 600 nm for n = 1.0 gradually red shifts as the index increases. This is a well-known effect [13] which typically leads to a roughly linear increase of wavelength with

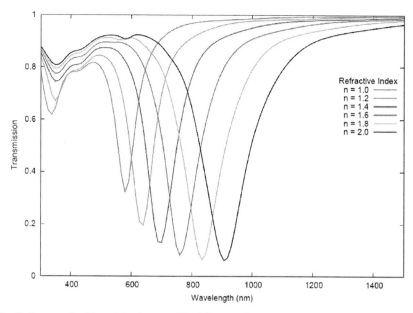

Fig. 8 Spectra of gold particles immersed in different refractive index media. These particles are 80 nm in length, 10 nm in thickness, and have a spacing of 200 nm

index. In the present study, there is a 330 nm increase in plasmon wavelength for a unit increase in index, which is a typical reactive index sensitivity for anisotropic nanoparticles [53].

Particle shape affecting refractive index sensitivity

An extension of the previous study is to examine the differing sensitivities of particles with different aspect ratios. In this case, spectra were computed for different dielectric environments for particle arrays with particles 10 nm thick and lengths ranging from 10 to 150 nm. Figure 9 plots plasmon wavelength versus refractive index, showing the linear dependence that was mentioned above. In addition, we see that the slope of these curves, which is the refractive index sensitivity, increases with increasing aspect ratio (particle length). Over the range of one refractive index unit, the small 10 nm by 10 nm square particles (1:1) have a shift of 65 nm whereas the much longer particles with an aspect ratio of 1:15 display a 500 nm shift. This increase in refractive index sensitivity is correlated with the larger plasmon wavelength associated with the longer particles. This effect that has been noticed in past work [53], although the present results are much more systematic, showing that there is a nearly linear dependence of slope (refractive index sensitivity) on aspect ratio.

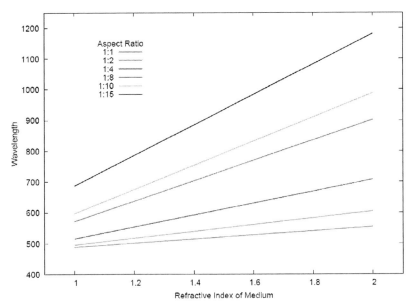

Fig. 9 Here the data from Fig. 8 was combined with other studies done with particles all 10 nm in thickness and with different lengths, varying from 10 to 150 nm. The plasmon peak positions were plotted against the refractive index of the media in which they were immersed

3.4.2 Dielectric Layer on the Surface

Different refractive indices

In many experiments the dielectric material is present as a layer of finite thickness rather than as a homogeneous material. To study this situation, we consider a layer of thickness 10 nm placed on the light-incident side of the metal rectangles. The thickness of this layer is taken to match the thickness of the metal (10 nm each). The refractive index of the coating layer is then varied from 1.0 to 2.0. The resulting transmission spectra are in Fig. 10. Note that though the larger refractive indices still shifts the resonance peak to the red, but the index sensitivity is greatly reduced (only about 40 nm shift per refractive index unit, compared to 330 nm in Fig. 8). This effect was explored previously for layers of molecules on triangular nanoparticles [5], and similar results were observed that are determined by the range of the plasmon enhanced electromagnetic field around the particle.

Different thicknesses

To continue this study of finite layers, we consider a similar layer structure as in Fig. 10 but now looking at the effect of varying layer thickness for a refractive index of 1.6. The gold structures are the same as was used in the previous study, except that the thickness has been increased to 20 nm. The results are shown in Fig. 11. This

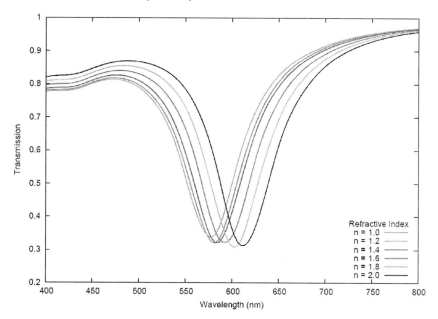

Fig. 10 Spectra of gold particles with a layer of a dielectric on the light-incident side while varying the refractive index of the layer. These particles are 80 nm in length, 10 nm in thickness, and have a spacing of 200 nm. The dielectric coating is 10 nm in thickness

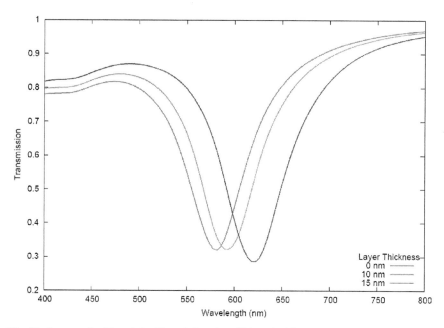

Fig. 11 Spectra of gold particles 80 nm in length and 20 nm in thickness with different thicknesses of a dielectric coating with a refractive index of 1.6 on the light-incident side. These particles have a spacing of 200 nm

shows that there is discernible red-shifting when changing the gold thickness from 0 to 20 nm, but at thicknesses of 25 and 30 nm the peak does not shift at all. This shows that the effect of layer thickness saturates at about 25 nm, which is similar to results seen in earlier work for particles in 3D. This saturation distance is similar to the range over which the electromagnetic field near the nanoparticles approaches its asymptotic form.

4 Conclusion

This article has provided an overview of the optical properties of metallic nano-structures with emphasis on the influence of plasmon excitation on these properties, and how these results can be used to make chemical and biological sensors. Our numerical results have explored the optical properties of gold nanowire structures using two-dimensional FDTD calculations. The ability to study a variety of parameters and systems with this simple model has provided a coherent picture concerning the sensitivity of plasmon resonance behavior to particle size, shape, spacing and dielectric environment. In most cases we were able to relate the properties observed to theories that were developed in earlier studies of particles in 3D that have proven useful for providing qualitative insight. In addition, the simple nanowire model provides opportunities for studying unusual effects, such as the transition from particle-like to film-like behavior as particle spacing is varied, revealing that even very thin (10 nm) films can have zero transmission over a narrow wavelength provided that the film has holes with the right spacing.

Acknowledgments This research was supported by the Northwestern Materials Research Center, sponsored by the National Science Foundation (DMR-0520513) and AFOSR/DARPA Project BAA07-61 (FA9550-08-1-0221).

References

1. Mirkin, C. A.; Letsinger, R. L.; Mucic, R. C.; Storhoff, J. J. *Nature* **1996**, *382*, 607–609.
2. Elghanian, R.; Storhoff, J. J.; Mucic, R. C.; Letsinger, R. L.; Mirkin, C. A. *Science* **1997**, *277*, 1078–1081.
3. Haynes, C. L.; Yonzon, C. R.; Zhang, X.; Walsh, J. T.; Duyne, R. P. V. *Anal. Chem.* **2004**, *76*, 78–85.
4. Haes, A. J.; Hall, W. P.; Chang, L.; Klein, W. L.; Duyne, R. P. V. *Nano Lett.* **2004**, *4*, 1029–1034.
5. Haes, A. J.; Zou, S.; Schatz, G. C.; Duyne, R. P. V. *J. Phys. Chem. B* **2004**, *108*, 109–116.
6. Reilly, T. H., III; Chang, S.-H.; Corbman, J. D.; Schatz, G. C.; Rowlen, K. L. *J. Phys. Chem. C* **2007**, *111*, 1689–1694.
7. Kwak, E.-S.; Henzie, J.; Chang, S.-H.; Gray, S. K.; Schatz, G. C.; Odom, T. W. *Nano Lett.* **2005**, *5*, 1963–1967.
8. Yin, L.; Vlasko-Vlasov, V. K.; Rydh, A.; Pearson, J.; Welp, U.; Chang, S. H.; Gray, S. K.; Schatz, G. C.; Brown, D. B.; Kimball, C. W. *Appl. Phys. Lett.* **2004**, *85*, 467–469.

9. Schatz, G. C.; Duyne, R. P. V. In *Handbook of Vibrational Spectroscopy*; Chalmers, J. M., Griffiths, P. R., Eds.; Wiley: New York, 2002; Vol. 1, pp. 759–774.

10. Qin, L.; Zou, S.; Xue, C.; Atkinson, A.; Schatz, G. C.; Mirkin, C. A. *Proc. Natl. Acad. Sci.* 2006, *103*, 13300–13303.

11. Hao, E.; Schatz, G. C.; Johnson, R. C.; Hupp, J. T. *J. Chem. Phys.* 2002, *117*, 5963–5965.

12. Jensen, T.; Kelly, K. L.; Lazarides, A.; Schatz, G. C. *J. Cluster Science* 1999, *10*, 295–317.

13. Jensen, T. R.; Duval, M. L.; Kelly, K. L.; Lazarides, A.; Schatz, G. C.; Duyne, R. P. V. *J. Phys. Chem.* 1999, *103*, 9846–9853.

14. Jensen, T. R.; Schatz, G. C.; Duyne, R. P. V. *J. Phys. Chem B* 1999, *103*, 2394–2401.

15. Kelly, K. L.; Coronado, E.; Zhao, L.; Schatz, G. C. *J. Phys. Chem. B* 2003, *107*, 668–677.

16. Kelly, K. L.; Jensen, T. R.; Lazarides, A. A.; Schatz, G. C. In *Metal Nanoparticles: Synthesis, Characterization and Applications*; Feldheim, D., Foss, C., Eds.; Marcel-Dekker: New York, 2002, pp. 89–118.

17. Kelly, K. L.; Lazarides, A. A.; Schatz, G. C. *Computing in Science & Engineering* 2001, *3*, 67–73.

18. Lazarides, A. A.; Kelly, K. L.; Jensen, T. R.; Schatz, G. C. *Theochem* 2000, *529*, 59–63.

19. Lazarides, A. A.; Kelly, K. L.; Schatz, G. C. *Mat. Res. Soc. Symp. Proc.* 2001, *635*, 1–10.

20. Lazarides, A. A.; Schatz, G. C. *J. Phys. Chem.* 2000, *104*, 460–467.

21. Lazarides, A. A.; Schatz, G. C. *J. Chem. Phys.* 2000, *112*, 2987–2993.

22. Malinsky, M. D.; Kelly, K. L.; Schatz, G. C.; Duyne, R. P. V. *J. Phys. Chem.* 2001, *105*, 2343–2350.

23. Malinsky, M. D.; Kelly, K. L.; Schatz, G. C.; Duyne, R. P. V. *J. Am. Chem. Soc.* 2001, *123*, 1471–1482.

24. Hao, E.; Bailey, R. C.; Schatz, G. C.; Hupp, J. T.; Li, S. *Nano Lett.* 2004, *4*, 327–330.

25. Hao, E.; Kelly, K. L.; Hupp, J. T.; Schatz, G. C. *J. Am. Chem. Soc.* 2002, *124*, 15182–15183.

26. Hao, E.; Li, S.; Bailey, R. C.; Zou, S.; Schatz, G. C.; Hupp, J. T. *J. Phys. Chem. B* 2004, *108*, 1224–1229.

27. Hao, E.; Schatz, G. C. *J. Chem. Phys.* 2004, *120*, 357–366.

28. Hao, E.; Schatz, G. C.; Hupp, J. T. *J. Fluoresc.* 2004, *14*, 331–341.

29. Zhao, L.; Kelly, K. L.; Schatz, G. C. *J. Phys. Chem. B* 2003, *107*, 7343–7350.

30. Zhao, L.; Schatz, G. C. SPIE Proceedings 2004, 5512 (*Plasmonics: Metallic Nanostructures and Their Optical Properties II*), 10–19.

31. Zou, S.; Janel, N.; Schatz, G. C. *J. Chem. Phys.* 2004, *120*, 10841–10875.

32. Zou, S.; Zhao, L.; Schatz, G. C. *SPIE Proceedings* 2003, *5221*(*Plasmonics: Metallic Nanostructures and Their Optical Properties*), 174–181.

33. Sherry, L. J.; Chang, S.-H.; Schatz, G. C.; Van Duyne, R. P.; Wiley, B. J.; Xia, Y. *Nano Lett.* 2005, *5*, 2034–2038.

34. Whitney, A. V.; Elam, J. W.; Zou, S.; Zinovev, A. V.; Stair, P. C.; Schatz, G. C.; Van Duyne, R. P. *J. Phys. Chem. B* 2005, *109*, 20522–20528.

35. Zou, S.; Schatz, G. C. *J. Chem. Phys.* 2005, *122*, 097102/1–097102/2.

36. Schatz, G. C.; Young, M. A.; Van Duyne, R. P. *Top. Appl. Phys.* 2006, *103*, 19–46.

37. Shuford, K. L.; Ratner, M. A.; Gray, S. K.; Schatz, G. C. *J. Comput. Theor. Nanosci.* 2007, *4*, 239–246.

38. McMahon, J. M.; Henzie, J.; Odom, T. W.; Schatz, G. C.; Gray, S. K. *Opt. Express* 2007, *15*, 18119–18129.

39. Jensen, L.; Aikens, C. M.; Schatz, G. C. *Chem. Soc. Rev.* 2008, *37*, 1061–1073.

40. Zhao, J.; Pinchuk, A. O.; McMahon, J. M.; Li, S.; Ausman, L. K.; Atkinson, A. L.; Schatz, G. C. *Accts. Chem. Res.* 2008, *41*, 1710–1720.

41. Yee, S. K. *IEEE Trans. Antennas Propagat.* 1966, *14*, 302–307.

42. Taflove, A.; Hagness, S. C. *Computational Electrodynamics: The Finite-Difference Time-Domain Method*; 3rd. ed.; Artech House:, Norwood, MA, 2005.

43. Berenger, J. P. *J. Comput. Phys* 1994, *114*, 185–200.

44. Roden, J. A.; Gedney, S. D. *Microwave Opt. Tech. Lett.* 2000, *27*, 334–339.

45. Merewether, D. E.; Fisher, R.; Smith, F. W. *IEEE Trans. Nucl. Sci.* 1980, *27*, 1829–1833.

46. Umashankar, K. R.; Taflove, A. *IEEE Trans. Electromagn. Compat.* **1982**, *24*, 397–405.
47. Mur, G. *IEEE Trans. Electromagn. Compat.* **1981**, *23*, 377–382.
48. Kashiwa, T.; Fukai, I. *Microwave Opt. Tech. Lett.* **1990**, *3*, 203–205.
49. Joseph, R. M.; Hagness, S. C.; Taflove, A. *Optics Lett.* **1991**, *16*, 1412–1414.
50. Payne, E. K.; Shuford, K. L.; Park, S.; Schatz, G. C.; Mirkin, C. A. *J. Phys. Chem. B* **2006**, *110*, 2150–2154.
51. Haynes, C. L.; McFarland, A. D.; Zhao, L.; Schatz, G. C.; Duyne, R. P. V.; Gunnarsson, L.; Prikulis, J.; Kasemo, B.; Käll, M. *J. Phys. Chem B* **2003**, *107*, 7337–7342.
52. Chang, S.-H.; Gray, S. K.; Schatz, G. C. *Opt. Express* **2005**, *13*, 3150–3165.
53. Sherry, L. J.; Jin, R.; Mirkin, C. A.; Schatz, G. C.; Van Duyne, R. P. *Nano Lett.* **2006**, *6*, 2060–2065.

Exploring the Nitric Oxide Detoxification Mechanism of *Mycobacterium tuberculosis* Truncated Haemoglobin N

A. Bidon-Chanal, M.A. Martí, D.A. Estrín, and F.J. Luque

Abstract Mycobacterium tuberculosis, the causative agent of human tuberculosis, encodes a haemoprotein named Truncated Haemoglobin N (trHbN), which in its active site transforms nitric oxide (NO) to nitrate anion (NO_3^-). The NO-dioxygenase activity of trHbN seems to be crucial for the bacillus, which can survive under the nitrosative stress conditions that occur upon infection of the host. As a defense mechanism against the copious amounts of NO produced by macrophages upon infection, the protein must achieve a high level of NO-dioxygenase activity to eliminate NO, but this is modulated by its efficiency in capturing O_2 and NO. Migration of small diatomic ligands through the protein matrix is related to the presence of a doubly branched tunnel system connecting the surface and the haem cavity site. In this work, we have studied the mechanism that controls ligand diffusion and product egression with state-of-the-art molecular dynamics simulations. The results support a dual path mechanism for migration of O_2 and NO through distinct branches of the tunnel, where migration of NO is facilitated upon binding of O_2 to the haem group. Finally, egression of NO_3^- is preceded by the entrance of water to the haem cavity and occurs through a different pathway. Overall, the results highlight the intimate relationship between structure, dynamical behavior and biological function of trHbN.

Keywords Truncated haemoglobin N · Mycobacterium tuberculosis · Ligand migration · NO detoxification · Molecular simulations

A. Bidon-Chanal (✉) and F.J. Luque
Departament de Fisicoquímica and Institut de Biomedicina (IBUB), Facultat de Farmàcia, Universitat de Barcelona, AV. Diagonal 643, 08028, Barcelona, Spain
e-mail: fjluque@ub.edu

M.A. Martí and D.A Estrín
Departamento de Química Inorgánica, Analítica y Química Física/INQUIMAE-CONICET, Facultad de Ciencias Exactas y Naturales, Universidad de Buenos Aires. Ciudad Universitaria, Pabellón II, Buenos Aires (C1428EHA), Argentina
e-mail: dario@qi.fcen.uba.ar

N. Russo et al. (eds.), *Self-Organization of Molecular Systems: From Molecules and Clusters to Nanotubes and Proteins*, NATO Science for Peace and Security Series A: Chemistry and Biology, © Springer Science+Business Media B.V. 2009

1 Introduction

Tuberculosis has been declared as a global emergency by the World Health Organization (WHO) and represents one of the main open threads in health sciences. The disease is caused by *Mycobacterium tuberculosis*, which infects about one third of the human population and causes about two million deaths per year over the world [1]. In the early stages of the infection, the immune system reacts by increasing the production of nitric oxide (NO) in the macrophages, a process that should contribute to reduce the replication rate of the bacilli and eliminate the pathogen from the host by inhibiting key enzymes such as the terminal respiratory oxidases and iron-sulfur centers of key enzymes [2, 3]. However, *M. tuberculosis* can resist the hazardous environments of its intramolecular niche and eventually enter an induced dormancy state and rest in latency. The ability of the microorganism to evade the toxic effects of NO and nitrosative stress appears to be associated with the NO-dioxygenase activity of the bacillus [4–6], which seems to be crucial for the survival and pathogenicity of *M. tuberculosis*. The detoxification activity is related to a small hemoprotein called Truncated Hemoglobin-N (trHbN), where the haem-bound O_2 reacts with NO to yield the harmless nitrate anion.

Truncated hemoglobins (trHb), also called 2/2Hb [7], are small hemoproteins found in bacteria, unicellular eukaryotes and higher plants [8, 9]. They form a distinct group in the globin super-family and a phylogenetic analysis shows that they can be divided into three subgroups, named I, II and III (also known as N, O and P, respectively), with less than 20% identity between members of the different classes [9]. The main differences with mammalian globins are a shorter primary sequence that lacks 20–40 residues, and a 2-on-2 a-helical fold instead of the 3-on-3 a-helical classical globin fold found in myoglobin [7, 10–13]. This shortening in sequence results in the conservation of only a-helices B, E, G and H, while helices C and D are merged together to form the C-D loop and helix F is replaced by a long, flexible loop that connects helices E and G (see Fig. 1).

Although a comprehensive knowledge of the functional role of truncated hemoglobins remains to be elucidated, several studies have hypothesized some physiological functions. Scavenging of nitrogen and oxygen species, O_2 transport and uptake, cellular respiration and (pseudo-)enzymatic reactions seem to be the most plausible ones [14]. For the particular case of trHbN, several evidences indicate that the oxygenated form of trHbN can convert NO to nitrate anion either *in vitro* or *in vivo* [4–6]. The lack of this protein in *M. bovis* results in a decrease of respiration activity upon exposure to NO, and expression of *M. tuberculosis* trHbN in *M. Smegmatis* and flavohaemoglobin lacking *Escherichia coli* and *Salmonella enterica Typhimurium*, also enhances the survival capacity of the organisms under oxidative stress in the presence of NO [6, 15]. Other truncated hemoglobins also present this dioxygenase activity, like trHbO of *M. tuberculosis* and trHbN of *M. smegmatis*, but with lower reaction rates that do not confer a notable increase in the survival of the organism.

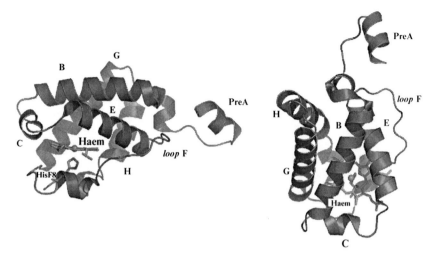

Fig. 1 Side view (*left*) and top view (*right*) of the truncated haemoglobin fold. The haem group and the coordination residue HisF8 are represented in sticks

2 *Mycobacterium tuberculosis* trHbN

The crystal structure of the oxy form of trHbN is an homodimer where the two chains present minimal structural differences between them, as noted by a root-mean square deviation (rmsd) between chains of only 0.7 Å. It exhibits a 2-on-2 a-helical sandwich fold where helices B and G along with CD an F loops provide a cage that accommodates the haem group. Like 100% of the globins known at the present moment, the Fe atom of the haem group is coordinated to the NE2 atom of the HisF8 residue, which acts as anchoring group for the porphyrin ring in the proximal site. The cavity containing the haem is mainly built up of hydrophobic residues in well-conserved topological positions within group I truncated haemoglobins (C6, C7, CD1, E14, F4, FG3, G8, H11) which in trHbN correspond to Leu(42), Phe(45), Phe(46), Phe(61), Met(77), Arg(84), Val(94), Ile(119). These residues contribute to haem stabilisation through van der Waals contacts. Furthermore, haem stabilisation is also achieved through hydrogen bonds with residues ThrE2 and TyrEF6, and also through salt bridges between the propionate groups and residues LysE10, ArgE6 and ArgFG3.

The distal site of the haem cavity is mainly packed with the side chains of apolar (PheB9, ValB13, PheCD1, LeuE7, PheE14, PheE15, ValG8) residues. In fact, only two polar residues, TyrB10 and GlnE11, are found in the haem binding pocket. Nevertheless, ligand binding to the haem is modulated by the interactions formed by these polar residues. Thus, TyrB10 forms a hydrogen bond with the oxygen molecule that strongly stabilises it upon binding to the Fe atom, and GlnE11 contributes to ligand stabilisation by forming a hydrogen bond with the hydroxyl group of TyrB10. This hydrogen-bond network largely contributes to the decrease in the

k_{off} experimentally measured for O_2 ($2.0 \times 10^{-1}\,M^{-1}$), which is ~2 orders of magnitude lower than that determined for Sperm whale myoglobin ($1.2 \times 10^1\,M^{-1}$), thus increasing the possibility that the active site is loaded with oxygen when NO reaches the cavity.

Inspection of the three dimensional structure of the protein also highlights the presence of a two-branched apolar tunnel that connects the solvent with the protein core, with a total volume of ~345 Å^3. The long tunnel branch lies between helices B and E and has a length of ~20 Å, while the short branch is ~8 Å long, nearly orthogonal to the long branch of the tunnel, and is placed between helices G and H (see Fig. 2). Soaking of the crystal structure with Xe atoms shows the presence of five different cavities where atoms are trapped with high residence times: two are found in the short tunnel branch, two sites are identified in the long branch, and the fifth site is located in the region where both tunnel branches converge. Of these

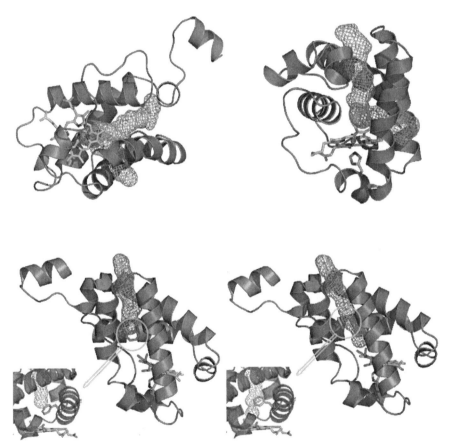

Fig. 2 Bottom (*top-left*) and frontal (*top-right*) views of the doubly branched tunnel found in the crystal structure of truncated haemoglobin N (PDB ID: 1IDR). Side view of the tunnel long branch showing PheE15 in the closed conformation (*bottom-left*) and in the open conformation (*bottom-right*)

Table 1 Kinetics and equilibrium constants for the reactions of ferrous HbN with oxygen and nitric oxide compared to that from sperm whale myoglobin

Enzyme	O_2 binding to ferrous Hb		NO oxidation by oxy Hb	
	k_{on} (M^{-1} s^{-1})	k_{off} (s^{-1})	K (M)	k ($M^{-1}s^{-1}$)
trHbN [4, 5]	2.5×10^7	2.0×10^{-1}	8.0×10^9	7.5×10^8
Myoglobin [17, 18]	1.4×10^7	1.2×10^1	8.6×10^7	3.4×10^7

five positions, the one that shows the highest occupancy is placed at the middle of the long branch of the tunnel, next to PheE15, whose side chain protrudes into the tunnel. Migration of a small diatomic ligand through the long branch is thus sterically limited by PheE15, and Milani et al. proposed that this residue could act as a gate for the migration of diatomic ligands to the haem cavity [16]. Remarkably, in the X-ray crystal structure the side chain of PheE15 is found in two different conformations interchangeable by a rotation of 100° around the Ca–Cb bond. The rotation changes the position of the phenyl ring, which protrudes into the tunnel (*closed state*) and then lies parallel to the tunnel axis (*open state*), thus reducing the steric hindrance that impedes access through the tunnel long branch [16].

Kinetic data presented in Table 1 shows that the conversion of Fe(II)–O_2 + NO to Fe(III) + NO_3^- in trHbN occurs with a rate constant 20 times greater than in Mb, but the rate constant measured for the entrance of O_2 (k_{on}) is of the same order of magnitude in both cases and only two times greater in trHbN with respect to Mb. These findings point out the possibility of a diffusion controlled mechanism where NO migration to the haem would be the limiting step of the process.

3 Ligand Migration

As stated before, for the reaction to occur the protein must contain oxygen bound to the haem Fe before NO reaches the active site. Thus, the first question that arises is how O_2 diffuses from the solvent to the active site of trHbN and, furthermore, which changes occur upon its binding. Experimental studies indicate that ligand diffusion could proceed through the double branched tunnel, that is, O_2 has two possible pathways to reach the haem cavity. However, access through the tunnel long branch may be blocked by PheE15, as hypothesized by Milani et al. and therefore diffusion through the tunnel short branch appears to be the most feasible pathway. To study the behavior of PheE15 and the accessibility to the active site in the oxy and deoxy forms of trHbN, a series of 0.1 μs molecular dynamics simulations were performed with the deoxygenated and oxygenated forms of the protein to study its dynamics, both at global and residue levels, along with steered molecular dynamics simulations to investigate the free energy profiles associated with the diffusion process [19].

3.1 O_2 Diffusion

Analysis of the deoxy-trHbN MD simulation points out that the tunnel long branch is completely blocked by PheE15 during the whole trajectory and rotation of the PheE15 side chain around the Ca-Cb bond leading to conversion between *closed* and *open* orientations found in the X-ray crystal structure is not observed. When the side chain of PheE15 protrudes into the tunnel long branch, molecular interaction potential energy maps computed for a diatomic ligand show clearly a discontinuity in the energy isocontour at the position occupied by PheE15, thus reflecting the steric hindrance exerted by the benzene ring. A more quantitative analysis comes from the free energy profile associated with the diffusion of a diatomic ligand (O_2 or NO) through the tunnel as determined from steered molecular dynamic calculations (see Fig. 3). When the side chain of PheE15 is buried into the tunnel (*closed state*), the free energy increases steadily as the ligand is forced to become closer to the haem cavity.

The preceding findings suggest that ligand diffusion to the active site in the deoxy state of trHbN must occur through the short branch of the tunnel. To explore the feasibility of this hypothesis, free energy calculations where performed for the migration of O_2 through this pathway. The results reveal that migration must overcome a barrier of ~2 kcal/mol to reach a flat minimum at a position that corresponds to one of the binding sites in the Xe soaking X-ray crystallographic experiments. From this position, located between residues PheG5, AlaG9 an IleH11, the ligand can access the Fe atom surpassing a ~4 kcal/mol barrier, which mainly reflects a narrow passage between ValG8 and the haem group (see Fig. 3).

3.2 NO Diffusion

To investigate the possibility of PheE15 to act as a gate to ligand migration through the tunnel long branch, the fluctuations of the H_α–C_α–C_β–C_γ dihedral angle along the trajectory of the oxy form were examined. Surprisingly, several transitions between *open* and *closed* states are detected, and in fact one of those transitions permitted the migration of a water molecule through the tunnel long branch reaching the haem active site. Moreover, steered molecular dynamics calculations show that when the phenyl ring of PheE15 is placed parallel to the tunnel axis (*open* state), the diffusion process becomes nearly barrierless and access to the haem binding pocket is favored by 3–4 kcal/mol (see Fig. 4). In contrast, migration through the short tunnel branch is not energetically accessible as the free energy barrier increases as the ligand becomes closer to the haem cavity (see Fig. 4).

These results sustain the hypothesis of a dual migration mechanism where diffusion of O_2 and NO would occur through different pathways. In the deoxy form, oxygen can reach the haem group through the tunnel short branch, while the long branch remains completely blocked by PheE15. However, upon binding of O_2 to the haem, diffusion of NO through the tunnel short branch is impeded and access to the

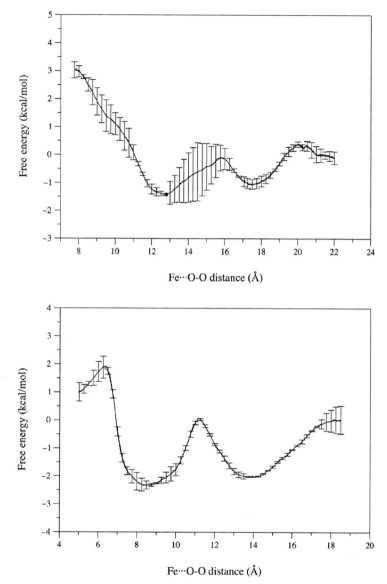

Fig. 3 Free energy profiles obtained for the diffusion of an oxygen molecule through the long tunnel branch when the PheE15 residue adopts a closed conformation (*top*) and through the short tunnel branch in the deoxy state (*bottom*)

haem cavity takes place through the long branch. This mechanism is operative due to the very high oxygen affinity of the protein (O_2 binding affinity to ferrous trHbN of 8.0×10^{-9}), which guarantees that it is mainly loaded with O_2. It also benefits from the high hydrophobic character of the ligand diffusion tunnel, which can act

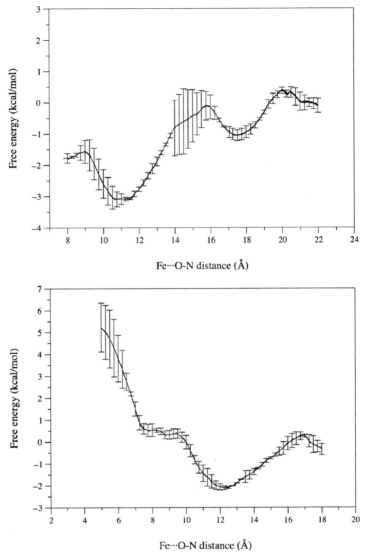

Fig. 4 Free energy profiles obtained for the diffusion of a nitric oxide molecule through the long tunnel branch when PheE15 adopts an open conformation (*top*) and through the short tunnel branch when O_2 is bound to the haem (*bottom*)

as a potential reservoir to concentrate nonpolar diatomic ligands, and at the same time limit the rate at which water can occupy the heme pocket when vacated by the dissociated ligand. These factors would then contribute to enhance the efficiency of NO detoxification, while minimizing the fraction of NO that directly binds to the deoxy-heme, inhibiting the detoxification process.

3.3 Molecular Basis of the Dual Migration Mechanism

The mechanism of ligand migration outlined above raises the question of the relationship between O_2 binding and opening of the long branch of the tunnel, thus enabling access of NO to the O_2-bound haem. To answer this question, a close look to the structural and dynamical features of trHbN in deoxy and oxy states is necessary.

X-ray crystallographic and spectroscopic data [16, 20] highlighted the presence of a particular hydrogen-bond network in the active site of the oxy form of trHbN, where the hydroxyl group of TyrB10 interacts with haem-bound O_2, while the amine group of GlnE11 is hydrogen bonded to it. Inspection of the molecular dynamics simulations reveals that such a hydrogen-bond network is stable and maintained along the trajectory, as noted in the lack of significant disruptions in the hydrogen-bond contacts. In turn, such a hydrogen-bond pattern imposes restrains to the conformational flexibility of the side chains of TyrB10 and GlnE11. The hydrogen bond formed between TyrB10 and the oxygen molecule restricts the movement of the residue side chain, as the hydroxyl group is acting as hydrogen-bond donor. In turn, the terminal amido unit of GlnE11 acts as hydrogen-bond donor to the TyrB10 hydroxyl group, which forces the side chain of GlnE11 to adopt a staggered conformation (Fig. 5).

This situation is very different from the structural preferences detected in the deoxy state of trHbN. In this latter state, the terminal amido moiety of GlnE11 is found to act as hydrogen-bond donor to or hydrogen-bond acceptor from the hydroxyl unit of TyrB10, which plays the reverse role in those interactions. However, since the position of TyrB10 is no longer restrained by the interaction with O_2 in the deoxy state, thee chain of GlnE11 adopts an *all-trans* extended conformation. Thus, the conformational change experienced by GlnE11 is related to the modification in the hydrogen-bond patterns reflected in dynamical fluctuation

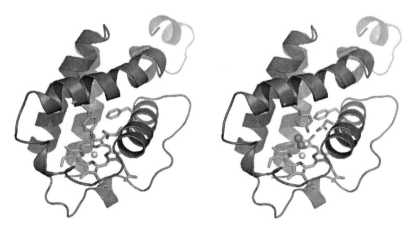

Fig. 5 TyrB10 and GlnE11 conformations in the deoxy (*left*) and oxy (*right*) states. PheE15 and the haem group are shown as sticks, and oxygen is shown in spheres

between (TyrB10)O–H$\cdots$O $=$ C(GlnE11) and (TyrB10)O$\cdots$H–N(GlnE11) hydrogen bonds, which are constantly exchanged during the whole simulation.

PheE15 is located in helix E just one turn further away from GlnE11 and the staggered conformation adopted by the side chain of GlnE11 in the oxy stateplaces the side chains of PheE15 and GlnE11 at a distance $\sim$1.5 Å closer than in the deoxy form. In fact, the average distance between side chains in oxy-trHbN is 3.9 Å, (i.e., the van der Waals contact between –CH_2– and >CH groups according to Pauling's radii) and a remarkable fraction of structures (around 45%) present distances below this number and sometimes as short as 3.1 Å. As a consequence GlnE11 exerts a *mechanical pressure* over PheE15, pushing its side chain and promoting a change in the orientation of the benzene ring that would facilitate the opening of the tunnel long branch, which eventually would allow ligand diffusion. On the other hand, in the deoxy form the distance between the two side chains is larger, and the *mechanical pressure* exerted by the side chain of GlnE11 over PheE15 is lower, which favors the population of the *closed* state that impedes the accessibility through the tunnel long branch. In fact, the transition barrier between the *closed* and *open* states of PheE15, which was estimated to be around 6 kcal/mol in the deoxy state, was predicted to be reduced to 3 kcal/mol in the oxy-trHbN, which reflects a facilitated transition from the *closed* state to the *open* one.

To asses the role of both residues in the diffusion mechanism control, molecular dynamics simulations of the TyrB10$^{(\rightarrow)}$Phe and the GlnE11$^{(\rightarrow)}$Ala oxy-trHbN mutants were performed [21]. These mutations were chosen because they disrupt the hydrogen-bond network formed by the haem-bound O_2 molecule. To the best of our knowledge, kinetic data are available only for the TyrB10$^{(\rightarrow)}$Phe mutant, but they are conclusive in stating that, although the mutated protein can bind O_2, there is a drastic decrease in the NO consumption activity [5].

Previous studies of the chemical reaction by Crespo et al. [22] demonstrated that this activity decrease cannot be attributed to a loss in the catalytic efficiency of the mutant, as the energetics of the reaction was not significantly perturbed with the mutation. Accordingly, it can be speculated that the loss of activity of the TyrB10$^{(\rightarrow)}$Phe mutant arises from the difficulty of NO to migrate to the active site as the molecular mechanism that assists opening of the tunnel long branch is disrupted. At this point, inspection of the PheE15 H_α–C_α–C_β–C_γ dihedral angle in molecular dynamics simulations of the TyrB10$^{(\rightarrow)}$Phe mutant clearly states that the side chain populates the *closed* conformation during the whole trajectory, thus blocking ligand access to the active site. The disruption of the hydrogen-bond network through residue mutation has the same effect found in the deoxy-trHbN simulation, where the conformation adopted by the side chain of GlnE11 increases its distance from the benzene ring of PheE15. In the TyrB10$^{(\rightarrow)}$Phe mutant, the (Tyr)OH$\cdots$(Gln)NH_2 hydrogen bond cannot be formed and the GlnE11 residue forms a hydrogen-bond with the haem-bound O_2 mole. To do this, the side chain of the residue adopts an *all-trans* extended conformation similar to that found in the deoxy-trHbN and the distance to the PheE15 side chain is increased on average to 5.2 Å. In the GlnE11$^{(\rightarrow)}$Ala mutant, the fixed conformation of PheE15 is directly related to the small volume of the Ala side chain, which prevents any significant interaction with the side chain of GlnE11.

Overall, both TyrB10 and Glne11 are very important for the protein to carry out the conversion of NO to nitrate anion. Noteworthy, this mechanism is modulated by O_2 binding to the haem Fe, as it fixes the conformation of TyrB10 and forces GlnE11 to adopt a *staggered* conformation, which promotes the conformational change of PheE15 side chain and opening of the tunnel long branch.

4   Protein Dynamics and Ligand Migration

At first sight, there is no major structural difference between the average structures of the protein skeleton in the oxy and deoxy states of trHbN. However, this finding does not necessarily mean that the peptide backbone is playing no role in modulating the NO-dioxygenase activity of the protein. Rather, a careful analysis of the dynamical behavior of the protein reveals an unexpected, but crucial contribution to the migration of ligand through the protein matrix.

Inspection of the main global motions of the protein backbone determined from essential dynamics demonstrates that a drastic alteration in the "breathing" of the protein takes place upon binding of O_2 to the haem Fe atom. Thus, the essential dynamics modes for the deoxy state reveals that the major motion involves the displacement of helices G and H, which mainly define the short branch of the tunnel, and the F loop. However, in the oxy state, the major motion affects the relative displacement of helices B and E, which are the structural elements that contribute to delineate the walls of tunnel long branch. Accordingly, the increased flexibility observed in distinct structural elements in the deoxy and oxy states should facilitate the migration of small ligands through the short and long branches of the tunnel. In particular, the reduced friction between helices B and E in the oxy state should contribute to explain the decrease in the barrier for the conformational transition between *closed* and *open* states of the PheE15 gate.

Interestingly, it turns out that the major structural fluctuations detected in TyrB10$^{(\rightarrow)}$Phe and GlnE11$^{(\rightarrow)}$Ala mutants comes from the displacements of the helices C, G, H and the loop F, which are the same structural elements identified in the deoxy state of trHbN. Again, this finding agrees with the experimentally observed decrease in the detoxification activity of the mutated TyrB10$^{(\rightarrow)}$Phe protein [5].

These results evidence that the binding of O_2 to trHbN has not only a local effect of increasing the *mechanical pressure* exerted by GlnE11 onto the benzene ring of PheE15, but also triggers a global effect, which affects the dynamical behavior of the whole protein by increasing the mobility of helices B and E. Overall, it might be concluded that the simultaneous effect of both local and global changes triggered upon O_2 binding facilitate the access of NO to the haem cavity and thus contribute to the efficiency of trHbN as a defense mechanism for the survival of the microorganism.

5   Egression of Nitrate Anion

Under nitrosative stress, the protein must be able to recover its initial state as quick as possible to achieve a safe NO detoxification regime and guarantee the survival of the bacillus. Accordingly, it is not only necessary to accomplish the NO-dioxygenase activity, but also to release efficiently the nitrate anion formed in the active site cavity to the solvent, thus enabling the protein to start a new enzymatic cycle.

Clearly, the driving force for the release of the nitrate anion to the bulk aqueous solution is the hydration of the negative charge. However, this process faces two problems. First, the increased size of the nitrate anion relative to the reactants (O_2, NO) implies that migration through the protein tunnel should be more difficult. Second, the hydrophobic nature of the tunnel does not support the migration of a negatively charged chemical species. Finally, an additional problem to be considered is the breaking of the bond between the haem Fe and the nitrate anion, which in turn is expected to interact with GlnE11 and TyrB10 through hydrogen bonds in the haem binding pocket. Accordingly, to reach the aqueous solvent, the Fe–ONOO bond must be broken, the hydrogen bond interactions lost, and the anion must diffuse through the protein matrix.

Molecular dynamics simulations of the haem-bound nitrate anion of trHbN have been valuable to elucidate the puzzle associated with egression of the product [23]. It was found that the presence of the nitrate anion bound to the Fe atom promotes a sizable distortion of the cavity walls. The hydrophobic residues lining the cavity (ValB13, LeuB14, PheCt, PheCD1, MetE4 and LeuE7) are pushed up to 1.5 Å far away from the Fe atom relative to their position in the oxy trHbN, thus increasing the cavity volume by a factor of ca. 2. Furthermore, the conformational rearrangement of these residues reduces the compactness of the cavity and facilitates the entrance of water molecules to the active site through two non-simultaneous channels (see Fig. 6). The first one is a consequence of the fast relocation of MetE4 side chain that creates a pore in the cavity as the residue moves away from the nitrate anion. The path is located between residues TyrB10, LeuB14, MetE4 and LysE8 and water molecules diffuse through it and solvate the nitrate anion. Later on, a new pathway

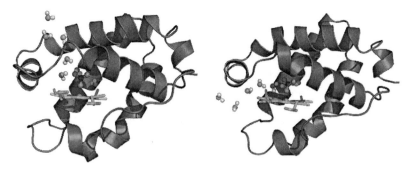

Fig. 6 Side view of the two different water pathways that appear in the molecular dynamics simulation of the nitrate-bound truncated hemoglobin N

appears due to the rearrangement of residues PheE7 and PheCD1, which is maintained along the rest of the simulation. As a result, the nitrate anion is effectively solvated by a few water molecules during the whole simulation.

To investigate the role played by hydration on the breaking of the bond between haem Fe and nitrate anion, a series of QM and QM/MM calculations were performed [23]. The results showed a crucial role of water in lowering the barrier of the Fe–ONOO bond breaking. For a model system derived from a representative snapshot of the MD simulation in which waters around the nitrate anion were removed, the barrier associated to bond breaking amounts to ∼18 kcal/mol. On the other hand, when water molecules are kept and included in the QM subsystem, the barrier is lowered to ∼4 kcal/mol showing that hydration is essential to efficiently break the bond.

Once the hydrated nitrate anion is released from the haem Fe atom, molecular dynamics simulations showed the existence of two possible egression pathways. In one case the nitrate anion escapes through the upper part of the haem cavity within 3 ns, while in the other it passes below the CD loop and escapes in ∼4.5 ns. In both cases, the anion egression is preceded by the formation of a hydrogen bond with ThrE2, which is placed at ∼9 Å away from the Fe atom. These results indicate that the anion can be rapidly released to the bulk solvent once the Fe–ONOO bond is broken and point to the entry of water as the crucial event in triggering the release of the nitrate anion.

This mechanism is supported by the free energy profiles obtained from steered molecular dynamics simulations (see Fig. 7) [23]. The anion must overcome a small barrier of less than 1 kcal/mol to reach the ThrE2 interaction site, which can be seen in the energy profile as a minimum that provides a 1 kcal/mol stabilization. Finally, it can escape from the haem cavity surpassing a second barrier of ∼2 kcal/mol.

Overall, these results indicate that the pathway by which the nitrate anion escapes from the haem cavity is completely different from those used by O_2 and NO to reach

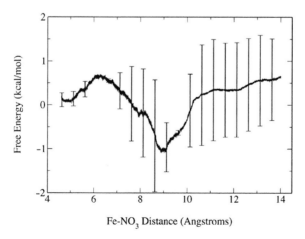

Fig. 7 Free energy profile for the egression of NO_3^- from the active site

the binding pocket. Furthermore, egression of the nitrate anion is largely determined by the entrance of water molecules into the active site, which lower the energy necessary to break the bond with the haem Fe atom and stabilize the unbound nitrate anion in the cavity.

6 Conclusion

The results point out the intimate nature between structure, dynamics and function in trHbN. Clearly, a comprehensive understanding of the biological role played by trHbN for the survival of *M. tuberculosis* cannot be achieved without an integrated knowledge of the protein structure and dynamics.

The analysis of the information collected from different computational tools permit us to reconcile the migration of the ligands O_2 and NO through the highly hydrophobic branches of the protein tunnel with the larger size and charged nature of the nitrate anion, as the product of the reaction leaves the enzyme using a different pathway. Taken together, this information permits to identify certain residues that play a crucial role in the NO-dioxygenase activity of trHbN, such as PheE15, which acts as the gate of the long tunnel branch, the pair TyrB10–GlnE11, which not only modulates the O_2 binding affinity and the correct positioning of NO in the heme-bound O_2 cavity but also contributes to facilitating the opening of the gate by combining both local and global conformational changes, and ThrE2, which assists the nitrate anion along the regression pathway.

The knowledge gained from the detailed analysis of these results should be valuable to suggest possible mutations, which should affect the efficiency of the NO-detoxification mechanism by trHbN, to explain the differences in activity between related truncated hemoglobins and eventually to provide a basis for the design of a pharmacological strategy against tuberculosis based on the definition of trHbN as a potential therapeutic target.

Acknowledgments Authors would like to thank the Spanish Ministerio de Ciencia e Innovación (SAF2008–05595-C02–01 and PCI2006-A7–0688), Agencia Nacional de Promoción Científica y Tecnológica (PICT 25667), CONICET (PIP 5218), Universidad de Buenos Aires, the John Simon Guggenheim Foundation, and the EU FP7 program (project NOstress) for providing financial support for carrying out this work. Calculations were performed in the Marenostrum Supercomputer at the Barcelona Supercomputer Center.

References

1. Hussian T (2007) Leprosy and tuberculosis: an insight-review. Crit Rev Microbiol 33:15–66.
2. MacMicking JD, North RJ, LaCourse R, Mudgett JS, Shah SK, Nathan CF (1997) Identification of nitric oxide synthase as a protective locus against tuberculosis. Proc Natl Acad Sci USA 94:5243–5248.
3. Shiloh MU, Nathan CF (2000) Reactive nitrogen intermediates and the pathogenesis of *Salmonella and Mycobacteria*. Curr Opin Microbiol 3:35–42.

4. Couture M, Yeh SR, Wittenberg BA, Wittenberg JB, Ouellet Y, Rousseau DL, Guertin M (1999) A cooperative oxygen-binding hemoglobin from *Mycobacterium tuberculosis*. Proc Natl Acad Sci USA 96:11223–11228.

5. Ouellet H, Ouellet Y, Richard C, Labarre M, Wittenberg B, Wittenberg J, Guertin M (2002) Truncated hemoglobin HbN protects *Mycobacterium bovis* from nitric oxide. Proc Natl Acad Sci USA 99:5902–5907.

6. Pathania R, Navani NK, Gardner AM, Gardner PR, Dikshit KL (2002) Nitric oxide scavenging and detoxification by the *Mycobacterium tuberculosis* haemoglobin, HbN in *Escherichia coli*. Mol Microbiol, 45:1303–1314.

7. Nardini M, Pesce A, Milani M, Bolognesi M (2007) Protein fold and structure in the truncated (2/2) globin family. Gene 398:2–11.

8. Wittenberg JB, Bolognesi M, Wittenberg BA, Guertin M (2002) Truncated hemoglobins: a new family of hemoglobins widely distributed in bacteria, unicellular eukaryotes, and plants. J Biol Chem 277:871–874.

9. Vuletich DA, Lecomte JT (2006) A phylogenetic and structural analysis of truncated hemoglobins. J Mol Evol 62:196–210.

10. Pesce A, Couture M, Dewilde S, Guertin M, Yamauchi K, Ascenzi P, Moens L, Bolognesi M (2000) A novel two-over-two α-helical sandwich fold is characteristic of the truncated hemoglobin family. EMBO J. 19:2424–2434.

11. Milani M, Pesce A, Bolognesi M, Bocedi A, Ascenzi P (2003) Substrate channelling: molecular bases. Biochem Mol Biol Edu 31:228–233.

12. Vinogradov SN, Hoogewijs D, Bailly X, Arredondo-Peter R, Gough J, Dewilde S, Moens L, Vanfleteren JR (2006) A phylogenomic profile of globins. BMC Evol Biol 6:31–47.

13. Perutz MF (1979) Regulation of oxygen affinity of hemoglobin: influence of structure of the globin on the heme iron. Annu Rev Biochem 48:327–386.

14. Ascenzi A, Bolognesi M, Milani M, Guertin M, Visca P (2007) Mycobacterial truncated hemoglobins: from genes to functions. Gene 398:42–51.

15. Pawaria S, Rajamohan G, Gambhir V, Lama A, Varshney GC, Dikshit KL (2007) Intracellular growth and survival of *Salmonella enterica* serovar Typhimurium carrying truncated hemoglobins of *Mycobacterium tuberculosis*. Microb Pathog 42:119–128.

16. Milani M, Pesce A, Ouellet Y, Ascenzi P, Guertin M, Bolognesi M (2001) *Mycobacterium tuberculosis* hemoglobin N displays a protein tunnel for O_2 diffusion to the heme. EMBO J 20:3902–3909.

17. Springer BA, Egeberg KD, Sligar SG, Rohlfs RJ, Mathews AJ, Olson JS (1989) Discrimination between oxygen and carbon monoxide and inhibition of autooxidation by myoglobin. Site-directed mutagenesis of the distal histidine. J Biol Chem 264:3057–3060.

18. Eich RF, Li T, Lemon DD, Doherty DH, Curry SR, Aitken JF, Mathews AJ, Johnson KA, Smith RD, Phillips GN Jr., Olson JS (1996) Mechanism of NO-induced oxidation of myoglobin and hemoglobin. Biochemistry-US 35:6976–6983.

19. Bidon-Chanal A, Martí MA, Crespo A, Milani M, Orozco M, Bolognesi M, Luque FJ, Estrin DA (2006) Ligand-induced dynamical regulation of NO conversion in *Mycobacterium tuberculosis* truncated hemoglobin-N. Proteins 64:457–464.

20. Ouellet Y, Milani M, Couture M, Bolognesi M, Guertin M (2006) Ligand interactions in the distal heme pocket of *Mycobacterium tuberculosis* truncated hemoglobin N: roles of TyrB10 and GlnE11 residues. Biochemistry-US 2006 45:8770–8781.

21. Bidon-Chanal A, Martí MA, Estrin DA, Luque FJ (2007) Dynamical regulation of ligand migration by a gate-opening molecular switch in truncated hemoglobin-N from *Mycobacterium tuberculosis*. J Am Chem Soc 129:6782–6788.

22. Crespo A, Martí MA, Kalko SG, Morreale A, Orozco M, Gelpi JL, Luque FJ, Estrin DA (2005) Theoretical study of the truncated hemoglobin HbN: exploring the molecular basis of the NO detoxification mechanism. J Am Chem Soc 127:4433–4444.

23. Martí MA, Bidon-Chanal A, Crespo A, Yeh SR, Guallar V, Luque FJ, Estrin DA (2008) Mechanism of product release in NO detoxification from *Mycobacterium tuberculosis* truncated hemoglobin N. J Am Chem Soc 130:1688–1693.

Complex Symmetry, Jordan Blocks and Microscopic Self-organization

An Examination of the Limits of Quantum Theory Based on Nonself-adjoint Extensions with Illustrations from Chemistry and Physics

Erkki J. Brändas

Abstract The basis and motivation for extending quantum mechanics beyond its traditional domain are recognized and examined. The mathematical details are briefly discussed and a convenient compact complex symmetric representation derived. An original formula is proved and demonstrated to incorporate general Jordan block configurations characterized by Segrè characteristics larger than one. It is verified that these triangular forms can portray realistic evolutions via maps established both within fundamental quantum mechanics as well as within a generalized thermodynamic formulation displaying features that are reminiscent of self-organization on a microscopic level. Various applications of these so-called coherent dissipative structures in physics and chemistry are pointed out, and discussed with possible inferences also made to the biological domain.

Keywords Aqueous solutions · Analytic continuation · Complex symmetry · Resonances · Jordan blocks · Segrè characteristics · Condensed matter · Coherent-dissipative structures · ODLRO · Universal relation · High-T_C · Molten salts · Special- and general theory of relativity

1 Introduction

Everybody working in the theoretical fields of rigorous quantum mechanics as applied to micro-, meso- and macroscopic systems know the fundamental contributions of Alexander S. Davydov, whether it concerns the excellent exposition of quantum mechanics [1] or the development of the so-called Davydov soliton [2]. To pay honor to this outstanding performance the present NATO Advanced

E.J. Brändas
Department of Physical and Analytical Chemistry, Quantum Chemistry,
Uppsala University, SE-751 20 Uppsala, Sweden
e-mail: erkki.brandas@kvac.uu.se

N. Russo et al. (eds.), *Self-Organization of Molecular Systems: From Molecules and Clusters to Nanotubes and Proteins*, NATO Science for Peace and Security Series A: Chemistry and Biology, © Springer Science+Business Media B.V. 2009

Workshop aimed at unveiling the principles that govern the countless processes in micro-, meso- and macroscopic phenomena, where in particular molecular aggregation at various levels of organization play a crucial role. Recent advances to understand the disparity between physics, chemistry and biology contends that this gap cannot be explained or reduced to physics alone. Rather than engaging in a discussion of reductionism versus a constructionist viewpoint, see e.g. the work of S. Kauffman [3], we will advocate below a somewhat different view-point that is microscopic yet emphases a holistic conception.

To embark on the trails of the NATO Research Workshop program, we will provide an examination of systems where attributes like coherence/decoherence as well as various levels of dissipation are conjured. Not only will we consider the limits of quantum theory in probing non-selfadjoint extensions of standard quantum mechanics, but we will also examine the properties of emerging thermal- and quantum correlations. Under certain circumstances, e.g. at interfering timescales, the mixing of emerging correlation structures will show evidence of persisting long-lived conformation patterns, which through their constructive character descend to the concept of microscopic self-organization.

The consideration given here is based on traditional linear algebra quantum mechanics [4], but augmented with an important generalization, namely the proviso to extend quantum concepts, like resolvents or Green's functions, into the complex plane as well as their associated Fourier/Laplace transformed propagators into apposite semi-groups with potentially well-established contraction properties. As a result it will be possible to include dynamical characteristics such as time-, length- and temperature scales into the theory [5–7]. A simple scaling argument leads further to a complex symmetric formalism, which imparts a wider set of possibly broken symmetry solutions. Hence there might emerge structures, recognized as Jordan blocks, i.e. block diagonal triangular units with Segrè characteristics larger than one. Nevertheless it is important to remember that proper invariance laws, e.g. gauge invariance, unitarity and time reversibility are appropriately embedded in the formulation and accessible when necessary.

The outline of this review will be as follows. The subsequent sections entail in a few words the mathematical background and the relevant details for the appreciation of the extension programmed above. We will make the prerequisites for the so-called resonance picture of unstable states including a quantum statistical description based on explicit quantum theoretical density matrices and associated reduced system operators. As a prerequisite for the applications we will define the concept of a dissipative system and particularly introduce a new type of time irreversible building blocks denoted as a coherent dissipative structure. Appraisals of situations where alike configurations subsist are briefly examined and various conclusions regarding associated emergent properties compared. In particular we will consider anomalies of proton transport in water and ionic conductance of molten salts, conjectures regarding long-range proton correlations in DNA and further, quantum correlation effects in high-T_C cuprates. As a final point a surprising derivation of Einstein's laws of relativity is provided.

To close up the introductory section we call attention to two statements detailed by A. S. Davydov's in the preface to the 2nd edition of his *Quantum Mechanics* [1]. The first one concerns the remark that "we show the inapplicability of the concept of an essentially relativistic motion of a single particle". Further below it is followed up by "we do not consider in this book special methodological problems, the exposition is based upon dialectical materialism". As will be evident from our theoretical development, and of course leaving out political implications, the laws of opposites, negation and transformation has a certain appeal as regards the emergence of Jordan blocks, the transformation properties of the preferred physical localized basis, the principle of microscopic self-organization and the materialization of new dissipative structures defining new levels of organization.

2 Mathematical Details

We will here give a brief mathematical exposition and demonstration regarding the extensions necessary for the present description. The objective and motivation for our endeavor is primarily to find new realistic and rigorous time evolutions that goes beyond traditional time dependence based on the unitary time-evolution operator. At the same time we need to find new complex structures on the so-called unphysical Riemann sheet that build up the resolvent or the Green's function improving the analysis of associated auto-correlation functions and their Fourier–Laplace transforms. In short, we will look for a more compact way of examining the signal determining the spectrum and vice versa, and this may give new insights into the physical and chemical properties of the system.

First we will illustrate an inherent way to do appropriate analytic continuations both in connection with the numerical integration of the actual differential equation as well as in reference to solving standard secular equations. It is further shown that a complex symmetric representation becomes the natural ansatz. To prove that this choice does not impose any restrictions on the model, an explicit formula for the general Jordan canonical form is established. The techniques ascertain the resonance picture of unstable states, including solutions with broken symmetries, emergence of timescales and microscopic self-organization.

The theory summarized here incorporates the Schrödinger—as well as the Liouville equation. The former, inserted in a non-isolated system framework, will be shown to provide a basis, not only for a mathematical rigorous development of scattering theory, but also for an account of the theory of relativity. The Liouville superoperator formulation, extended to include thermalization via the Bloch equation and referring directly to the density matrix, will be shown to produce explicit relationships between the temperature, the size and the timescale of the alleged time irreversible dissipative structure. From the general definition of complex open systems we achieve a thorough description of the system-environment split and the precise characterization of a coherent-dissipative system.

2.1 The Differential Equation

We will here discuss the various consequences in the mathematical development caused by the generalizations introduced in the introduction. To avoid misunderstandings we need to understand some of the fine points in the development of quantum mechanics. For this reason it is important to say a few words on the "classical" Schrödinger equation (8).

As everybody studying quantum mechanics know, Schrödinger described in his famous treatise, posing quantization as an eigenvalue problem, how to acquiesce physically suitable solutions of the boundary value problem of the (Schrödinger) differential equation. He imposed the requirements of continuity, single valued- and finiteness. The quandary or complication, though well understood and worked out today yet sometimes misread, is the relation between these demands and the ones of square integrability, i.e. that the solutions of interest should belong to a suitably defined Hilbert space. For instance the so-called continuum wave-functions present a normalization problem, since the characteristic aspect of such oscillatory wavefunctions (rather than exponentials) are that their orthonormality is expressed by the use of the Dirac delta function. One might say that square integrability is sufficient for bound-state problems where the potential is bounded from below and that continuum wave-functions never needed quantization. It is nevertheless interesting to note the way in which eigenfunctions of the differential equation operator may be used to form an eigenfunction expansion of the Hilbert space without ever being required to belong to Hilbert space themselves. In summary one might say that Schrödinger was able to formulate quantization as an eigenvalue problem precisely because of his familiarity with the developments of Hilbert and Weyl [9]. To rigorously develop any useful generalizations one needs either to pose precise boundary conditions for the physically interesting solutions or to find the relevant topology for the actual spaces.

Let us start with the boundary condition problem. The first one to study these aspects for second order ordinary differential equations in a realistic and useful way was Herman Weyl [10]. The theory, which preceded both the development of Hilbert Space as well as Schrödinger's original papers, was extended to treat perturbations, which make the spectrum continuous, by Titchmarsh [11]. In fact it can be shown that the Weyl-Titchmarsh m-function through its intimate connection with the spectral density forms a key quantity in the solution of the boundary value problem, i.e.

$$\rho(E_\beta) - \rho(E_\alpha) = \lim_{\varepsilon \to 0+0} \frac{1}{\pi} \int_{E_\alpha}^{E_\beta} \mathrm{Im}[m(E + i\varepsilon)]dE \qquad (1)$$

where ρ is the spectral function of the differential operator. Furthermore, m is the uniquely given energy dependent coefficient obtained from requesting that a suitable combination of the two linearly independent initial value solutions of the 2nd-order ordinary differential equation should be $\mathcal{L}^2$ or square integrable. Note that $\mathrm{Im}[m]$ should be evaluated at the limit when the imaginary part of the complex energy parameter $E + i\varepsilon$ goes to zero. The asymptotic boundary condition is handled by another limit that gives the unique $\mathcal{L}^2$ solution at infinity via the so-called limit-point classification, a feature dependent on the properties of the potential. Although this categorization is different from the conditions required in standard scattering theory, see e.g. Newton [12], one can e.g. prove that the m-function is entirely dependent on the logarithmic derivatives of the physical solution [13, 14], and that its imaginary part can be expressed, when appropriate, in the standard form, as [14]

$$\pi \left(\frac{d\rho}{d\omega} \right)_{\omega=E} = \frac{k^{2l+1}}{|F_l(k)|^2} \tag{2}$$

where $k = \sqrt{E}, l \neq 0$ is the angular momentum quantum number and F_l is the well-known Jost function [12]. For some key applications referring to the continuous spectrum as well as to the problem of extending the limit-point limit-circle theory to the complex plane, see particularly references [13–16]. Without going into more details we will emphasize the possibility to extend the techniques to a set of coupled ordinary second order differential Eq. (17). Although one might integrate the partial differential equation directly along a suitable path in the complex plane we will instead resort to another different but very convenient approach below.

2.2 The Complex Symmetric Form

In order to set the stage for a more general development, we will board on a theorem derived and supported by Balslev and Combes [18]. Using rigorous mathematical characteristics of so-called dilatation analytic interactions they proved important spectral properties of many-body Schrödinger operators [18]. The possibility to "move" or rotate the continuous spectrum was immediately subject to successful applications in a variety of quantum theoretical applications in both quantum chemistry and nuclear physics [17]. The key to this development can be grasped from a simple scaling argument [19]. Before explaining how to generate the dilatation transformation we will make a simple but significant conclusion. The basic idea is mainly to make a suitable change of coordinates in every matrix element building up the secular equation. Consider for instance the general operator $\Omega(r)$, where we write $r = r_1, r_2, \ldots r_N$; assuming $3N$ fermionic degrees of freedom. By simply carrying out the scaling $r' = \eta^{3N} r$; $\eta = e^{i\vartheta}$, where $\vartheta < \vartheta_0$ for some ϑ_0 that generally depends on the potential, one finds trivially

$$\int \varphi^*(r)\Omega(r)\phi(r)dr = \int \varphi^*(r'^*)\Omega(r')\phi(r')dr' \tag{3}$$

It is assumed that the operator $\Omega(r)$ as well as $\varphi(r)$ and $\phi(r)$ are appropriately defined for the scaling process to be meaningful, and that the interval of the radial components of r generally is $(0, \infty)$; we will return to this issue further below.

The important conclusion from Eq. (3) is the requirements of analyticity with reference to the parameter η, hence the complex conjugate in $\varphi(\eta^* r)$, and this is the main reason why most complex scaling applications in quantum chemistry are implemented via complex symmetric representations, see e.g. Ref. [20].

Returning briefly to the domain issue. Following Balslev and Combes [18] we introduce the N-body Hamiltonian as $H = T + V$, where T is the kinetic energy operator and V is the (dilatation analytic [18]) interaction potential. We clearly need to restrict the domain of H, since it is an unbounded operator, although bounded from below (compare, e.g. the distinction with the Stark Hamiltonian treated above with Weyl's theory [13, 14]). Without going into unnecessary detail we write the domain of H as (below H is the well-known Hilbert space, see e.g. [5, 18] for further details)

$$D(H) = \{\Phi \in \mathcal{H}, H\Phi \in \mathcal{H}\} \tag{4}$$

The condition on the dilatation analytic potential is essentially that the (pair) potential does not dominate the kinetic energy and hence the unboundedness is due to the latter or $\mathcal{D}(H) = \mathcal{D}(T)$. With these preliminaries we can introduce the scaling operator $U(\vartheta) = \exp(iA\theta)$ where the generator A is given by

$$A = \frac{1}{2} \sum_{k=1}^{k=N} [\vec{p}_k \vec{x}_k + \vec{x}_k \vec{p}_k] \tag{5}$$

Note that the parameter ϑ is here real, but will be made complex in what follows commensurate with what is said above. Also $\vec{x}_k$ and $\vec{p}_k$ are the coordinate and momentum vectors of the particle k. As a result we obtain

$$U(\vartheta)\Phi(r) = e^{\frac{3N\vartheta}{2}} \Phi(e^{\vartheta} r) \tag{6}$$

The form of the generator, defined by Eqs. (5) and (6), can be simply derived by direct differentiation of both sides of Eq. (6) with respect to the parameter ϑ. A more interesting route is to work with the Mellin transformation

$$v(\lambda) = (2\pi)^{-\frac{3}{2}} \frac{1}{2} \int_{\mathbf{R}^3} \tilde{v}(\mathbf{r}) \mathbf{r}^{-i\lambda - \frac{3}{2}} d^3 \mathbf{r} \tag{7}$$

with its eigenfunction satisfying (with A here restricted to $N = 1$ for simplicity)

$$A|\mathbf{r}|^{i\lambda - \frac{3}{2}} = \lambda |\mathbf{r}|^{i\lambda - \frac{3}{2}} \tag{8}$$

For the centrally symmetric case ($|\mathbf{r}| = r$), cf. the possibility to connect the Mellin transform with the Riemann zeta function, one can write down the eigenfunction of the generator A as

$$\psi_\lambda = \frac{1}{\sqrt{2\pi}} r^{i\lambda - \frac{1}{2}} \tag{9}$$

This leads to the following inversion formulas

$$S(\lambda) = \int_0^\infty \psi_\lambda^*(r)\varphi(r)dr; \quad \varphi(r) = \int_{-\infty-\frac{1}{2}i}^{+\infty-\frac{1}{2}i} \psi_\lambda(r)S(\lambda)d\lambda \qquad (10)$$

and the eigenvalue relation displaying an alternative way to carry out the scaling using the relation

$$e^{iA\vartheta}\psi_\lambda(r) = e^{\frac{1}{2}\vartheta}\psi_\lambda(e^\vartheta r) = e^{i\lambda\vartheta}\psi_\lambda(r) \qquad (11)$$

By turning $\vartheta \to i\vartheta$ the one-parameter unitary dilation group, defined in Eq. (6), becomes an unbounded similitude. This affects not only the operator domains, but it alters the spectrum significantly, i.e. the absolutely continuous spectrum (there is no singularly continuous spectrum for dilatation analytic operators [18]) will be "rotated down" in the complex plane with a rotation angle of 2ϑ. As we will se further below this possibility to make the scaling parameter ϑ complex will contract ψ_λ accordingly and lead to novel spectral representations in terms of resonances, deflated rotated cuts including Jordan blocks, where the corresponding transformed canonical vectors appear partitioned into all possible factorizations of any natural number.

Summarizing: We have considered the general scaling parameter $\eta = |\eta|e^{i\vartheta}$ (for some $0 \leq \vartheta < \vartheta_0$ as mentioned above). What is more, provided that the potential V is a sum of two-body, Δ-compact (bounded by the kinetic energy operator) potentials, we have restricted the Hilbert space to a dense subspace on which the unbounded complex scaling operation is defined. Upon closing the subset to the whole Hilbert space one obtained the simple form of the deformed spectrum, defining the dilation analytic family of Hamilton operators depending explicitly on the analytic parameter η. The development leads naturally to complex symmetric representations, in which the standard self-adjoint formulation is embedded. Although there exists other tools to investigate the "unphysical sheet" the present approach has been very convenient for atomic and molecular physicists and chemists since the dilatation analytic Hamiltonian is the perfect tool, save complications related to the Born-Oppenheimer approximation, see Appendix 1, which naturally opens up the complex plane via spectral rotation. Analogous analytic continuation in the Liouville case is not trivial but can be done [5], see e.g. the developments in the next section.

As we will, see there will be a price to pay for this extension. To examine this complication we will first show that the present complex symmetric representation invokes no restriction on the formulation. This will lead to a simple and useful complex symmetric form of the most general finite matrix description [21, 22], and this will be of immediate use in what follows.

2.3   The Emergence of the Jordan Block

To display the emergence of the Jordan block structure we will give a physical argument that leads directly to the forms that we can identify as non-diagonalizable triangular units in disguise. Although Gantmacher proved [23] that every square matrix is similar to a complex symmetric matrix and that every symmetric matrix is orthogonally similar to an explicitly given normal form, we have found, in our scientific work, a particularly suggestive and elegant representation [21, 22] which, contrary to previous existence proofs produce a very simple, explicit form. We will return to some illustrative applications of this formulation in coming sections [5, 6].

To begin with, we will give a brief account of the quantum statistical framework to be examined. Not only do we need to consider how to treat the system–environment partition, but we also need to incorporate Liouville-like master equations in the evolution dynamics. Although this sounds a bit too general to be practical, we will see some surprising simplifications as well as make some useful definitions. It is important to remember that one wants to start at the most basic level of formulation in order to justify the extensions made previously. We will henceforth focus on the density matrix ρ (not to be confused with the spectral function of Section 2.1) subject to the Liouville equation ($\hbar$ suppressed)

$$i\frac{\partial\rho}{\partial t} = \hat{L}\rho \tag{12}$$

where

$$\hat{L} = H\rho - \rho H^\dagger \tag{13}$$

Note that we have kept the Hermitean conjugate in the second term above for the possibility to do dilatation analytic extensions [5, 26]. The use of reduced density matrices and their associated, representability properties for solving the Schrödinger equation has a long and interesting history [24] and technical progress is still developing. Although we will revisit the problem again, it is necessary to bring up and stress the fundamental importance of Yang's celebrated Off-Diagonal Long-Range Order, ODLRO [25], and its profound theoretical and physical consequences. For later use we will also introduce the definition of the N-particle (and its p-reduced companions) representable density matrix $\Gamma^{(p)}$ as follows

$$\Gamma^{(p)}(x_1 \ldots x_p | x'_1 \ldots x'_p) = \tag{14}$$

$$\binom{N}{p} \int \Psi^*(x_1 \ldots x_p, x_{p+1} \ldots x_N)\Psi(x'_1 \ldots x'_p, x_{p+1} \ldots x_N)dx_{p+1} \ldots dx_N$$

where the (normalized) wave function $\Psi(x_1 \ldots x_N)$ represents a fermion many-body quantum mechanical system. One usually considers two types of particles, bosons and fermions, but since quantum chemical applications concern the electronic structure of atoms and molecules it is common practice to focus on the characteristics of reduced fermionic density matrices, particularly with $p = 1$ or

$p = 2$. Note also that the normalization is set to the number of pairings, but other choices are frequently used. For the present purpose an oversimplified model should be sufficient. It will be rigorously rationalized later on.

We will consider $M = N/2$ bosons (or N fermions) described by set of $m \geq N/2$ localized pair functions or geminals $\boldsymbol{h} = (h_1, h_2, \ldots h_m)$ obtained from appropriate pairing of one-particle basis spin functions (m should not be confused with the m-coefficient of previous sections). For the moment we will take no notice of the fermionic level. Although this model is somewhat primitive (it can easily be improved [42]) it will allow the portrayal of interesting phenomena via the density operator

$$\Gamma^{(2)} = \rho = \sum_{k,l}^{m} |h_k\rangle \rho_{kl} \langle h_l|; \ Tr\{\rho\} = \frac{N}{2} \tag{15}$$

to be subsequently examined in what follows. Note the inconsistency in the normalization, but the reader should not worry, as it will be shown to have a natural explanation. A straightforward statistical argument will describe the model and its quantum content. The matrix elements ρ_{kl} define probabilities for finding particles at site k and transition probabilities for "particles to go" from site k to l. Hence the matrix ρ has the elements (a connection with the structure of fermion density matrices [38] will be made further below and also in the coming sections)

$$\rho_{kk} = p; \ \rho_{kl} = p(1 - p); \ k \neq l; \ p = \frac{N}{2m} \tag{16}$$

The associated secular equation reveals a non-degenerate large eigenvalue $\lambda_L = mp - (m - 1)p^2$ and a small $(m - 1)$-degenerate $\lambda_S = p^2$. Note that for large m, $\lambda_L \approx N/2$. Consequently the density operator becomes

$$\Gamma^{(2)} = \rho = \lambda_L |g_1\rangle \langle g_1| + \lambda_S \sum_{k,l=1}^{m} |h_k\rangle (\delta_{kl} - \frac{1}{m}) \langle h_l| \tag{17}$$

Using the transformation $|\boldsymbol{h}\rangle \boldsymbol{B} = |\boldsymbol{g}\rangle = |g_1, g_2, \ldots g_m\rangle$, see Appendix 2 for the origin of this transformation

$$\boldsymbol{B} = \frac{1}{\sqrt{m}} \begin{pmatrix} 1 & \omega & \omega^2 & \cdot & \omega^{m-1} \\ 1 & \omega^3 & \omega^6 & \cdot & \omega^{3(m-1)} \\ \cdot & \cdot & \cdot & \cdot & \cdot \\ \cdot & \cdot & \cdot & \cdot & \cdot \\ 1 & \omega^{2m-1} & \omega^{2(2m-1)} & \cdot & \omega^{(m-1)(2m-1)} \end{pmatrix}; \ \omega = e^{\frac{i\pi}{m}} \tag{18}$$

a compact diagonal representation for the degenerate part obtains

$$\Gamma^{(2)} = \rho = \lambda_L |g_1\rangle \langle g_1| + \lambda_S \sum_{k=2}^{m} |g_k\rangle \langle g_k| \tag{19}$$

Rationalizations of this result will be discussed, see e.g. Eq. (30), and studied further below.

To fully integrate thermal- and quantum fluctuations into the theoretical formulation an intensive quantity like the temperature must be consistently incorporated. This is fundamentally a difficult problem since we are dealing with systems out of equilibrium, yet we will show how to introduce the "temperature" in a quasi-equilibrium context taking into consideration the constructive interaction between the thermal inputs from the environment on the open system. With an open or dissipative system (we will here use the term inter-changeably) we mean: *a system in which there exists a flow of entropy due to exchange of energy or matter with the environment*, see e.g. the wonderful discourse "From being to becoming" [27]. Additionally we will append specifications for so-called *coherent-dissipative structures*. In passing we should also make clear that we do not apply the thermodynamic limit, unless specifically implemented. Continuing to build on these traits, we will carry out the very well known mathematical trick of invoking the temperature by making time imaginary i.e.

$$t \rightarrow t - i\hbar\beta;$$

where

$$\beta = \frac{1}{k_B T}$$

Here k_B is Boltzmann's constant and $\hbar$ is Planck's constant divided by 2π. We also supply a word of warning that the extended formulation for time and β must be done separately [5, 22]. The usual thermalization procedure resorts to the Bloch equation using, instead of the Liouvillian, the energy superoperator $\hat{L}_B$, i.e.

$$-\frac{\partial\rho}{\partial\beta} = \hat{L}_B\rho \tag{20}$$

with

$$\hat{L}_B = \frac{1}{2}\{H|\ \rangle\langle*|\ +\ |\ \rangle\langle*|H\} \tag{21}$$

appropriately extended to a biorthogonal complex symmetric representation. In particular we note the complex conjugate in the bra-position, which is in contrast to the extended Liouville time generator in Eq. (13). Since we may use the mathematical machinery presented above, we can assign to our model a complex energy $\mathcal{E}_k = E_k - i\varepsilon_k$ to every site described by the basis function h_k. Note also that the total energy expression is given by

$$\mathcal{E} = Tr\{H_2\Gamma^{(2)}\} \tag{22}$$

where H_2 is the reduced Hamiltonian of the ensemble. The total energy for the many-body Hamiltonian, involving at most two body terms, can alternatively be decomposed into a sum of pair energies, see e.g. [5] for more details. We will give the exact form for this expression in Section 3.1. Using Eqs. (20) and (21) we can write the formal solution of the Bloch equation

$$\Gamma_T^{(2)} = e^{-\beta \hat{L}_B} \Gamma^{(2)}$$

using the standard factorization property of the exponential superoperator

$$e^{-\beta \hat{L}_B} \Gamma^{(2)} = \lambda_L \sum_{k,l}^{m} |h_k\rangle e^{i\beta \frac{1}{2}(\varepsilon_k + \varepsilon_l)} \langle h_l| + \lambda_S \sum_{k,l}^{m} |h_k\rangle e^{i\beta \frac{1}{2}(\varepsilon_k + \varepsilon_l)} \left(\delta_{kl} - \frac{1}{m} \right) \langle h_l| \tag{23}$$

In the example above we have assumed that the real part of the energies $\mathcal{E}_k$ for each site is independent of k and hence without restriction can be set equal to zero. To avoid confusion, we have not explicitly indicated that the analytic picture behind the thermalization, and the corresponding transformation, generates a complex symmetric bi-orthogonal representation. In particular a complex conjugate should appear in the bra-position, although the choice of a real localized basis set $\{h_k\}; k = 1, 2, \ldots m$, would make this temporarily unnecessary.

As usual we assign the relation between the imaginary part of the energy, the half-width and the timescale as

$$\varepsilon_k = \frac{\Gamma_k}{2} = \frac{\hbar}{2\tau_k} \tag{24}$$

In previous derivations we have as a rule established the following "quantization condition", see Eq. (25) below, based on our request that the thermalized density matrix in Eq. (23) should produce a Jordan block and that this feature presumes a special physical property of the dissipative system e.g. provides a prolonged survival time or structural timescale. In the appendix we have examined a simple thermal scattering process and derived basically the same quantization condition between timescales and the "temperature" of the coherent dissipative structure, the difference being that the quantization condition here follows from a physical argument rather than from a mathematical coincidence. Applying this condition, see Appendix 1

$$\beta \varepsilon_l = 2\pi \frac{l-1}{m}; l = 1, 2, \ldots m \tag{25}$$

We obtain directly

$$\Gamma_T^{(2)} = \lambda_L \sum_{k,l}^{m} |h_k\rangle e^{i\frac{\pi}{m}(k+l-2)} \langle h_l| + \lambda_S \sum_{k,l}^{m} |h_k\rangle e^{i\frac{\pi}{m}(k+l-2)} \left(\delta_{kl} - \frac{1}{m} \right) \langle h_l| \tag{26}$$

By inspection we conclude that the first "large" part of Eq. (26) is proportional to a Jordan block of order 2 and the "small" part to one of order m. This insight comes from the simple fact that

$$
J = \begin{pmatrix} 0 & 1 & 0 & . & 0 \\ 0 & 0 & 1 & . & . \\ . & . & . & . & 0 \\ . & . & . & . & 1 \\ 0 & . & . & . & 0 \end{pmatrix} \tag{27}
$$

and

$$
Q_{kl} = (\delta_{kl} - \frac{1}{m})e^{i\frac{\pi}{m}(k+l-2)}; \ k,l = 1,2,\ldots m \tag{28}
$$

are similar. Using the same transformation, as defined in Eq. (18), the statement is easily confirmed [21, 22], see Appendix 2, i.e.

$$
Q = B^{-1}JB \tag{29}
$$

The transformation B^{-1} defines the basis $|f\rangle$ that represents the canonical basis for the Jordan block, i.e.

$$
|h\rangle B^{-1} = |f\rangle = |f_1, f_2, \ldots f_m\rangle \tag{30}
$$

As a consequence the Dunford formula for the thermalized superoperator in Eq. (23) becomes (to avoid extra notation, we have only distinguished the operator from its matrix representation in Eq. (3) by a calligraphic J)

$$
\Gamma_T^{(2)} = \lambda_L \mathcal{J}^{(m-1)} + \lambda_S \mathcal{J} \tag{31}
$$

$$
\mathcal{J} = \sum_{k=1}^{m-1} |f_k\rangle \langle f_{k+1}|
$$

Note that $\mathcal{J}^{(m-1)} = |f_1\rangle \langle f_m|$ is a Jordan block of order 2. Since $\Gamma_T^{(2)}$ is traceless the energy, see Eq. (22), becomes

$$
E = Tr\left\{H_2\Gamma_T^{(2)}\right\} = 0 \tag{32}
$$

As a final point, we remark that the change of basis due to the thermalization procedure is neither a unitary- nor a similarity transformation.

Before leaving this section one should make the following distinction. The Jordan blocks may appear in many situations in quantum theory. Hence it is important to point out that we do not consider traditional use of nilpotent operators, i.e. step operators in angular momentum algebra or creation- and annihilation operators in second quantization. Instead we re-stress that our focus is put on the dynamics of our open dissipative systems and as a result a multiplicity in the related operator representation, with Segrè characteristics larger than one, may arise (i) in the density matrix (e.g. studies of proton transfer processes in water and aqueous solutions [29]) (ii) in the Hamiltonian (the complex symmetric ansatz of the Klein Gordon Equation (Eq. (30)) (iii) and at the Liouville level (spontaneous and stimulated emission of radiation in masers [31]). It will be shown, that the results obtained in case (i), are consistent with the evolution properties of a dynamical process of type (iii).

With the reasoning above as a background, we will define, as proposed in Section 2.3 the concept of a *coherent dissipative system* by requiring additionally that *(a) they are created or destroyed by integrated quantum- and thermal correlations ($T \neq 0$), (b) they exchange energy with an (partially) entangled environment and (c) they can not have a size smaller than a critical one.* Unequivocal dynamical evolution on such systems will lead to non-exponential decay and the law of microscopic self-organization.

In the sections to follow, we will analyze the significance and the relationship of this new phase organization onto the microscopic level. The mathematical reasons for this interpretation will be given in Appendix 3.

3 Quantum Technology

After the preliminary mathematical details given in the first two sections we will continue with some specific applications to non-isolated systems where realistic maps of the dynamical behavior can be portrayed by the theory developed here.

There has been, already since the millennium shift, a dramatic theoretical and experimental development, both fundamental and innovative, of modern physics and chemistry not to mention biology. The tools of state-of-the-art science and technology extend into many new areas of industrial significance with quantum physics and chemistry becoming essential both for miniaturized objects on the nanoscale— as well as on the higher complexity levels, and for advanced computer and data communications. On the biological side we are understanding and learning how to operate the hierarchy of the levels of complexity from the lowest quantum molecular level all the way up to large-scale systems consisting of all components, controlled by mechanisms at the lower level, linked into networks that makes the dissipative system robust and flexible. These features lead to novel advanced technologies depending heavily on fundamental principles. We have termed those original developments characterized by such traits *quantum technology*. To convey the idea, we have indicated in Table 1, what we mean, by listing some recent topics and techniques starting from pure quantum systems going towards more "mixed-dirty"

Table 1 Quantum technology

Application Area	Mechanism/Technique
Quantum information	Nonclassical
Condensed matter	Broken symmetry
Ultracold matter	Dissipation-dispersion
Interferometry	Coherence–decoherence
Coherent dissipative systems	Quantum–thermal correlations
Stochastic non-linear dynamical systems	Stochastic resonances

fields of applications, for some more details on this development, see e.g. the proceedings from the Nobel Symposium Nr. 104 [28].

Below we will adopt examples from high T_C superconducting cuprates, anomalous conductance of H^+ and OH^- in aqueous solutions, electric conductivity of molten salts and finally some observations of interesting mathematical structures within a biological frame. Under a new heading we will present a surprising application to Einstein's laws of relativity. The last topic concerns "Relativistic innovations" under *Application Area* and "Quantum mechanical superposition principle" under *Mechanism/Technique*.

3.1 High-T_C Cuprates—General Features

We will start the sections on applications by attending to the development of ODLRO [25] and the phases of high-T_C cuprate superconductors. The field has for a long time been one of high controversy as to what constitutes the fundamental mechanism behind the Cooper pairing, not to mention the semantic difficulties causing misunderstandings between quantum chemists and condensed matter physicists. For instance, while the resonating valence bond state [32] predicted many unusual properties of the early oxide super-conductors indicating a common unique mechanism, there appeared many different interpretations like the nature of the carriers and the symmetry of the gap function. Although the book *The Theory of Superconductivity in the High-T_C Cuprate Superconductors* [33] by P. W. Anderson, containing a full presentation of the experimental and theoretical material up to about 10 years ago, including *Central Dogmas* and selected re- and preprints of work made by the author and his collaborator's, full consensus in the scientific community has not yet been achieved. For instance, there is a growing belief in the quantum chemistry community that new theoretical approaches together with the utilization of the enormous advance in computer capacity will lead to first principles based predictions of real materials in general and to physical phenomena involving high temperature superconductors in particular. Recent DFT studies supports a chiral plaquette polaron theory [34] of cuprate superconductivity, see also related developments in terms of d-wave polaronic condensates within an extended Hubbard Hamiltonian model [35–37].

One of the crucial differences between solid-state theoreticians, with a quantum chemical orientation, and condensed matter physicists are the latter's predilection to work within the thermodynamic limit. To keep away from possible misunderstandings, we will not presuppose this limit from the beginning. The main objective will therefore be to set up the conditions of ODLRO [25] or rather use Coleman's extreme type configuration [38] as a precursor for the onset of the superconducting phase.

We have already mentioned the fundamental importance of Yang' concept of ODLRO [25]. Related to this development is what Coleman calls wave functions of the extreme type [38]. This is a configuration of strongest possible correlations between the fermions of the system. Assume that our system consists of N fermions and that we have selected a finite basis of m spatial orbitals, from which we construct an equal amount of α and β spin orbitals. All in all we have $2m$ orthogonal spin orbitals, which we can pair in such a way that we obtain m singlet spin pair determinants $|k, k + m\rangle = |h_k\rangle$ for $k = 1, 2, \ldots m$. The nature of the basis depends naturally on the problem at hand. For simplicity one might think of the basis element as being localized at a particular site k in the system, but symmetry considerations may also give rise to more than one basis geminal per site. One might suspect that the totally symmetric combination, cf. Section 2.3, will play a special role. The function or geminal $|g\rangle$, defined by

$$|g\rangle = \sum_{k=1}^{m} c_k \, |h_k\rangle \tag{33}$$

plays the key role in the construction of a particularly interesting class of functions, i.e. the explicitly constructed many-body wavefunction termed an antisymmetrized geminal power $\left|g^{N/2}\right\rangle$. It is defined as the normalized combination of wedge products, see below

$$\left|g^{N/2}\right\rangle \propto |g \wedge g \wedge \ldots \wedge g\rangle \tag{34}$$

The coefficients of the geminal will in the extreme case be given by

$$c_k = \frac{1}{\sqrt{m}}; k = 1, 2 \ldots m$$

We can now state Coleman's fundamental observation and result.

Theorem 1. *The geminal $|g\rangle$ is an eigenfunction of $\Gamma^{(2)}(g^{N/2})$ with a non-vanishing eigenvalue iff $|g\rangle$ is of extreme type, i.e. the eigenvalues of $\Gamma^{(1)}(g)$ are all equal.*

Above we have indicated the dependence on the wavefunction, i.e. the one that has been used to obtain the pth reduce density matrix. Hence $\Gamma^{(1)}(g)$ refers to the first order reduced density matrix, which is obtained from the geminal (2-particle wavefunction) g. It is surprising that the form of $\Gamma^{(2)}(g^{N/2})$ in form coincides with formula (17,19) ($g = g_1$) of Section 2.3, with two minor differences (actually the differences can be avoided by a more rigorous statistical ansatz).

$$\Gamma^{(2)}(g^{N/2}) = \lambda_L \, |g_1\rangle \, \langle g_1| + \lambda_S \sum_{k=2}^{m} |g_k\rangle \, \langle g_k| + \Gamma_T^{(2)} \tag{35}$$

First there appears, in addition to the "large" and "small" part a so-called "tail" component consisting of all unpaired spin orbital contributions with a degenerate eigenvalue equal to λ_S. Furthermore the eigenvalues are slightly different although they carry the same macroscopic features, i.e.

$$\lambda_L = \frac{N}{2} - (m-1)\lambda_S; \ \lambda_S = \frac{N(N-2)}{4m(m-1)} \tag{36}$$

The appearance of $\Gamma_T^{(2)}$ in Eq. (35) explains why the full normalization of $\Gamma^{(2)}$, set to the number of pairings, see Eq. (14), is not in conflict with Eq. (15), that is the normalization over the $m \times m$ matrix, historically denoted [24] "the box", constituting only the large and small parts. From the factorized density matrix (in the limit $m \to \infty$), we will relate the superconducting gap with the following energy expression, cf. Section 2.3,

$$\langle H_{tot} \rangle_{Av} = Tr \{ H_2 \Gamma^{(2)} \} \tag{37}$$

where H_{tot} is the total Hamiltonian and the reduced Hamiltonian H_2 given by

$$H_2 = \frac{1}{(N-1)} (h_1 + h_2) + h_{12} \tag{38}$$

containing the one body part h_i and the two body part h_{ij}, see e.g. Coleman [24] and others [5] for more details. We note that in the limit $m \to \infty$ only the large component contributes to the energy expression (37). However, as will be easily seen, this may also occur for finite m. Associating the superconducting energy gap 2Δ with the relevant two-body operator V_{12}, accounting for the, as yet, unknown mechanism, then it follows that an enormous energy stabilization occur provided all matrix elements

$$\langle h_k | V_{12} | h_l \rangle = \upsilon_g; \ k,l = 1,2\ldots m \tag{39}$$

i.e. are independent on k and l. It follows further from Eq. (39) that the contribution from the small part becomes rigorously zero, since the phases in the eigenvectors corresponding to $|g_k\rangle; k = 2,3\ldots,m$, see the definition of the basis $|g\rangle$ from Eq. (18), "wash" out the sum to zero. Hence we obtain

$$2\Delta = \lambda_L |Tr \{V_{12} |g\rangle \langle g|\}| = \left[\frac{N}{2} - \frac{N(N-2)}{4m} \right] |\langle g| V_{12} |g\rangle| \tag{40}$$

which yields with the use of Eq. (39)

$$2\Delta = \lambda_L |\langle g| V_{12} |g\rangle| = \lambda_L m |\upsilon_g| = \frac{N}{2} \bar{m} |\upsilon_g|; \ (m-1) \leq \bar{m} \leq m \tag{41}$$

Note that the present derivation does not require taking the thermodynamic limit and that it is enough for $\bar{m}\upsilon_g \to \gamma < \infty$, i.e. to stay finite in the limit. As such our derivation acts as a precursor to the onset of ODLRO.

Returning to the many anomalies displayed by the high-T_C cuprates, we will show that the present setting provides a good portrayal and explanation of why we see so many different properties in comparison to the traditional low-temperature superconductivity, e.g., exceedingly high T_C values, strong anisotropies, anomalous energy gaps, d-wave character of the gap function, the universal linear relation between T_C and carrier concentration, saturation effects, not to mention general thermal behavior, doping dependence of condensed electrons, heat capacity, Knight shifts, and existence of both a condensate- and a spin gap. Although many of these properties have been examined separately in the light of various different mechanisms, we believe that the present description give ample explanations and non-contradictory justification, even before the exact nature of the pairing has been determined.

One of the most striking features of Eq. (49) is the quadratic behaviour of λ_L as a function of the number of fermions. In addition to the universal linear relation and the pronounced saturation effect, it also predicts a sudden break-down of the superconducting state at higher carrier concentrations, the break-down corresponding to $2m = N$. Note also that the energy gap 2Δ is not due to simple pair breaking, as in the BSC case, but rather to a more complex breakage of coherent sub-clusters of various sizes [6]. In analogy with the discussion in Appendix 1 it is possible to compute the gap [6] determining successive cluster break-ups, destroying islands of various sizes in the precursor [37], until the condensate is destroyed. Original reflectivity data indicated gaps around

$$\frac{2\Delta}{k_B T_C} \approx (6.3 - 9.4) \tag{42}$$

where T_C is the critical temperature. However, the situation turned out to be more involved [35–37] and the consensus today is based on a more complex structure, i.e. a superconducting- and a pseudo gap. Combining the apposite gap parameter with Eq. (40) normalizing appropriate quantities to carrier- (here holes, p-type) and virtual particle concentrations per effective mass, it is possible to determine the maximum critical temperature, $T_C^{\max}$ and to obtain the universal relation between the normalized T_C and the hole content x, i.e.

$$\frac{T_C}{T_C^{\max}} = 4(x - x^2) \tag{43}$$

This feature is reflected by muon-spin-relaxation experiments [39] displaying a universal linear relation between T_C and the muon spin relaxation rate (proportional to the carriers per effective mass) with increasing carrier doping suggesting also a saturation effect for each class of superconductors. The latter is further in agreement with the negative quadratic deviation exhibited in the formulas above.

3.2 Microscopic Mechanisms

It is an interesting fact that the over all features of Cu-O based super-conductors can be described without evoking an explicit microscopic mechanism. Although there exists a plethora of different propositions, we will only mention and discuss a few representative situations. Effectively we will illustrate our point of view by bringing up the following ones: (i) spontaneous break-up of the superconducting state by quantum-thermal correlations [37], (ii) repulsive electronic interactions/correlations [36], (iii) non-adiabatic molecular electron-vibrational theory [40], and (iv) short-lived magnetic clusters with domain wall motion [41].

In the first case [37] an attempt was made to extract general physical information independent from the precise nature of any pair formation mechanism. As a result the energy gap is not due to simple pair break-up processes, but rather with a gradual restriction of pair mobility through successive disturbances of the actual coherence. In this way the superconducting gap, the universal linear relation as well as current experiments on muon-spin-relaxation, infrared-reflectivity and uv-photoemission could be rationalized and quantified.

Introducing the interactions between the electrons, via a short-range screened Coulomb repulsion, adopting an extended Hubbard Hamiltonian [36], one is lead, in a straightforward way, to an adequate description of the condensate and much of the essential behavior of these highly correlated materials. Note that the extreme configuration can be derived from statistical arguments [42] similar to the discussions in Section 2.2. It is furthermore possible to devise an approach, where a fraction of the electrons are condensed, while the remainder is unpaired. The associated wave function can all the same be written as a geminal power, accounting for a pair of Wannier functions, localized on every lattice point at the centre of the unit cell. The Cu atom is in an almost D_{4h} environment, but small puckerings of the cuprate layer and external ions lower the symmetry to C_{4v}. Hence localized E-functions in this group are symmetry adapted combinations of Cu e-orbitals and inplane oxygen p_z-orbitals. Apical oxygen p_x and p_y orbitals also have such symmetry and combining all these orbitals yield pairs of E-basis functions on every lattice cell. The thermal average of the eigenvector of the second order reduced density matrix thus exhibits a d-wave symmetry, transforming as the B_1 irreducible representation of the C_{4v} group.

The theory [35, 36] accounts for (i) short coherence lengths, (ii) transition temperature dependences, (iii) unusual heat capacity behavior, (iv) Knight shift temperature dependence. At the same time the two-fluid model accounts for the normal spin gap although the latter is not essentially related to superconducting pairing.

It is interesting to note that the general theory, advocated here, has intimate relations with non-adiabatic electron-vibration coupling [40]. Although not mentioned here, there is a non-stated supposition resting in the definition of the hierarchy of reduced density matrices. Most applications invoke a parametric dependence on the inter-nuclear coordinates while a fully reduced electronic density matrix should incorporate suitable averages over these, see a recent discussion [43] on this topic. Leaving the details of this question aside here, we proceed by asking what would happen if the large eigenvalue in Eq. (35) would emerge corresponding to some g_k, $k \neq 1$ (equivalent to assigning a particular predetermined phase to each site). Since energy supported at the level of the small part of $\Gamma^{(2)}$ does wash out for all eigenvectors, except the one with all elements equal to $1/\sqrt{m}$, the resultant energy gap is still unimportant as $\lambda_S \rightarrow 0$ for large m. As a consequence it follows that the two-body contribution, see Eqs. (30–41) all together vanishes and the one-body part becomes crucially important for energy stabilization to emerge. This is in a sense a restricted one-body effect, since it is dependent on the two-body phase controlled interaction through ODLRO. Nevertheless it is to some extent dependent on electron transfer rates through non-adiabatic electron-vibration couplings, yielding gap formation and Jahn–Teller like nuclear arrangements, which are all characteristics of the same fermionic ground state energy, cf. electron transfer phenomena related to conical intersections. The present mechanism models accurately the less pronounced phonon dependence compared to the standard BCS situation.

Finally we want to mention the possibility to link the high-T_C phenomenon to the interplay between quantum correlations and short-lived magnetic clusters [41]. Note that we are using the more strict definition of *a coherent dissipative structure* in that we require the criteria that (a) the correlations are created by quantum and thermal correlations ($T \neq 0$), (b) they exchange energy with an entangled surroundings and (c) they can not have a size smaller than a critical one, *viz.* they must intermix quantum and thermal correlations in accordance with the density matrix formulation described above. Introducing the key quantity for this integration, i.e. the thermal de Broglie wave length,

$$\Lambda_{db} = \sqrt{\frac{h^2}{2\pi M k_B T}} \tag{44}$$

where M is the effective mass, to be identified with the "magnetic wall effective mass" [41]. Combining Eqs. (44) and A(5) with the criteria that our coherent dissipative structure exhibits a spatial dimension, d_{min}, depending linearly on Λ_{db} and the number of degrees of freedom $s_{min} = m$, for some minimum value of m, i.e.

$$d_{min} = F(H) \cdot s_{min} \cdot \Lambda_{db} \tag{45}$$

where $F(H)$ depends on the Hamiltonian H and on some thermodynamic variables as well as external parameters [42], one obtains

$$d_{min} = \left[\frac{32\pi^3}{(F)^{-2}M} \right]^{\frac{1}{2}} \cdot \sqrt{k_B T} \cdot \tau_{rel} \tag{46}$$

Eqs. (45–46) can be interpreted either as (a) s_{min} times F multiplied by the correlation length ζ, where ζ is the de Broglie wave length based on the effective mass or (b) as the de Broglie wave length based on the reduced mass $M' = (s_{min})^{-2}M_{phen}$, where $M' = F^2 M_{phen}$, giving necessary input for a quantitative analysis. Hence the underlying magnetic mechanism, based on Eq. (46), leads to consistent estimates of correlation lengths, effective masses and relaxation times as well as giving good qualitative figures for the maximum current density limited by the frustrated spins following the charge carriers [41].

Before finishing this section, we reemphasize that this examination of some of the possible mechanisms of high-T_C superconducting cuprates is not complete in any sense, for instance one might mention that accurate band theory calculations for the YBaCuO's yield consistent results including a d-wave gap function [45]. Instead, the purpose has been to demonstrate that many different mechanisms, supported by the general coherent dissipative picture presented above, promote a very detailed as well as a qualitative over all picture of high-T_C cuprates. It goes without saying that this picture should be applicable to other systems under similar but also quite different conditions.

3.3   Coherent Dissipative Structures

We will here discuss briefly coherent dissipative structures in amorphous condensed systems [6, 29, 42, 46–48]. The first case concerns the evaluation of the rate constants characterizing the processes (k_i and E_i are the reaction rates and the activation energies respectively)

$$
\begin{aligned}
H_3O^+ + H_2O &\xrightarrow{k_1,E_1} H_2O + H_3O^+ \\
OH^- + H_2O &\xrightarrow{k_2,E_2} H_2O + OH^-
\end{aligned}
\tag{47}
$$

This is perhaps the most fundamental process in existence and it plays an important role in many chemical systems not to mention in the biological organism. For the pioneering work of Meiboom, using NMR spectroscopic measurements, through the classic work of Eigen to the detailed analysis of Hertz we refer to previous discussions [6, 29, 48] and references therein. The motivation compelling Hertz to reconsider the so-called H^+-particle (or the H_3O^+ ion) as a dynamic entity in aqueous solutions, and not representing a particle in the conventional sense, came from the classical predictions of the connections between (i) the proton transfer rates, k_i, of the reactions Eq. (47) and the excess conductivities of H^+ and OH^-. The latter can be defined from the experimentally measured ionic conductances, λ_X of the ion X in water, i.e.

$$
\lambda^e_{H+} = \lambda_{H+} - \lambda_{X+}
$$

with $X^+ = K^+$ or Na^+ and

$$
\lambda^e_{OH^-} = \lambda_{OH^-} - \lambda_{Cl^-}
$$

From well-known theories by Nernst and Einstein using the standard Arrhenius type ansatz, the classical relation becomes

$$
\frac{\lambda^e_{H+}}{\lambda^e_{OH^-}} = \frac{k_1}{k_2}
\tag{48}
$$

and

$$
\log\left(\frac{\lambda^e_{H+}}{\lambda^e_{OH^-}}\right) = C - \frac{E_1 - E_2}{RT}
\tag{49}
$$

where C is a temperature independent constant and R the gas constant. The classically predicted values, obtained from Eqs. (47–49) above, are $k_1/k_2 \approx 2.35$ at $T = 25°C$ and $E_1 - E_2 \approx -2.0$ kJ/mol for $T = 15°C, \ldots, 55°C$. As we will se this is in complete disagreement with the predictions made from the theory of coherent-dissipative structures [29] and accurate measurements by means of advanced [1] H-NMR spin-echo techniques [49].

To prove our point we will employ the thermal de Broglie wave length Λ_{db}, see Eq. (44) for the definition, and the size estimate of the coherent-dissipative structure, Eq. (45). Since Λ_{db} for a "quasi-free" proton is about 1 Å at room temperatures, one might find "water protons" within a distance of order Λ_{db} around each H^+. The coherent-dissipative structure thus consists of H^+ constituents forming positive ions indistinguishable from those belonging to the water molecule. The hypothesis, i.e. the assumption that thermal–quantum correlations between the protons belonging to the ions and those belonging to the water molecules provide the pair correlated entities of the system. From this over-all conjecture of quantum correlated pairs of fermionic entities H^+ (and similarly for OH^-) follows two important predictions. Assuming similar correlations between the positive and the negative ions one can derive the following modifications of Eqs. (48–49), i.e. and

$$\frac{\lambda^e_{H^+}}{\lambda^e_{OH^-}} = \sqrt{\frac{m_{OH^-}}{m_{H^+}} \frac{k_2}{k_1}} \tag{50}$$

$$\log\left(\frac{\lambda^e_{H^+}}{\lambda^e_{OH^-}}\right) = C' + \frac{E_1 - E_2}{RT} \tag{51}$$

Comparing with the classical derivation one obtains that $k_1/k_2 \approx 1.75$ at $T = 25°C$ and $E_1 - E_2 \approx +2.1$ kJ/mol (for $X^+ = K^+$) for $T = 15°C, \ldots, 55°C$. The value for the difference in activation energies, using $X^+ = Na^+$, is slightly smaller. All in all these predictions are in good agreement with the previously quoted experiments that was carried out in the laboratory of H. G. Hertz [49]. In fact the present theory did promote another prediction [29, 50], viz. an anomalous decrease of H^+ conductance in H_2O/D_2O mixtures pointing at a fundamental quantum statistical difference between H^+ and D^+ entities, i.e. portraying fermionic- versus bosonic properties.

A closely related phenomenon is the high electric conductivity of molten alkali chlorides [47]. Using essentially the same arguments. Defining the molar conductivity

$$\Xi_{XCl} = \lambda_{X^+} + \lambda_{Cl^-}$$

where $X = Li, Na, K, Rb$ a similar reasoning that lead up to Eq. (45), or equivalently that a property like the conductance should (in the first approximation) be linearly dependent on the size on the coherent-dissipative structure, i.e.

$$\Xi_{XCl} = c \cdot F \cdot \left(\Lambda_{db}(X^+) + \Lambda_{db}(Cl^-)\right) \cdot \frac{4\pi k_B T}{\hbar} \cdot \tau_{XCl} \tag{52}$$

$$\tau_{XCl} = \tau_{X^+} + \tau_{Cl^-}$$

In Eq. (52), τ_{XCl} is the relaxation time for each melt, a quantity that goes into the equation for s_{min}, see Eqs. A(5) and (45), and $\Lambda_{db}(ion)$ is the relevant thermal de Broglie wave length. Note that the theory assumes that the coefficients c and F have a general dependence on thermodynamic parameters, but are independent of the specific melt under consideration. Since the ionic masses appears above explicitly through the de Broglie wave length Λ_{db} it is natural to define the scaled molecular conductivity as

$$\Xi^*_{XCl} = \Xi_{XCl}/\left[(m_{X^+})^{-1/2} + (m_{Cl^-})^{-1/2}\right] \propto \tau_{XCl} \tag{53}$$

It is a striking result that the scaled molecular conductivities, obtained from standard conductivity and density data, confirm the simple relationship of Eq. (53). For instance at $T = 800°C$, one obtains for the chlorides of Li, Na, K, and Rb: $\Xi^*_{Cl} = 350 \pm 3\%$ and at $T = 900°C$, Ξ^*_{Cl} $398 \pm 2\%$. Note that the quoted values have a variation of about 4%, the same order of magnitude as the experimental accuracy and reading errors for molecular conductivity data, see the original reference for details [37]. It is interesting to note that the scaled molecular conductivity of CsCl does not obey the present constancy, a 13% deviation at $T = 800°C$ and equal to 22% at $T = 900°C$. This rests most likely on the fact that the crystalline phase of CsCl is bcc in contrast to the other four alkali chlorides that form fcc lattices. One would expect that the microstructure of the melts, determining the relevant Hamiltonian have to influence the factor F in Eq. (52), and hence this should be reflected in the scaled molecular conductivity data.

In summary: We have, from the theory of coherent-dissipative structures, analyzed long-range correlation effects based on charge fluctuations. Anomalous features of proton transfer processes in water and aqueous solutions have been predicted and confirmed by experiments. We have also examined the general microdynamical behavior of Coulombic interactions dominating inter-ionic long-range correlations in molten salts. One concludes that the relaxation times of the charge fluctuations, spontaneously appearing around each ion, are independent of the masses of the ions, indicating that the microscopic relaxation process is intrinsically connected with the emergence of an irreducible coherent-dissipative structure.

4 Theory of Relativity

A somewhat surprising application of the present formulation comes from the theory of relativity. We remind the readers of Davydov's remark, quoted in the introduction, namely that "we show the inapplicability of the concept of an essentially relativistic motion of a single particle". It is nevertheless a well-known fact that the Klein–Gordon equation can be written formally as a standard self-adjoint secular problem based on the simple Hamiltonian matrix (in mass units)

$$H = \begin{pmatrix} m_0 & p/c \\ p/c & -m_0 \end{pmatrix}$$

where m_0 is the rest-mass, $p = mv$ is the momentum of the particle and c the velocity of light—note also that the entities above are operators and the velocity v of the particle(s) is relative a system in rest, wherever the rest masses of the particles involved are m_0 and $-m_0$ respectively. This formulation, generalized to fermions via the Dirac equation, accounts for most of the very accurate predictions of microscopic phenomena by relativistic quantum theory with powerful techniques extended to the general many-body picture [52].

As shown elsewhere [7, 30, 51] an analogous formulation can be made by a complex symmetric ansatz. The latter corresponds to the philosophy of this article, i.e. of embedding the formulation in a more general framework of (coherent) dissipative systems and this is also in agreement with the citation made by the man we honor. This principle permits an important generalization here, since it will simultaneously allow the introduction of time- and length scales as well as mimic the non-positive definiteness of the Minkowski metric.

In this section we will briefly review the model, starting with the special theory of relativity. We will then proceed to discuss the extension to the general theory. The end result of this endeavour will be to connect with some of the most well known facts of the laws of special and general relativity, i.e. the contraction of length- and timescales, Einstein's law of light deflection in a gravitational field and the compatibility with the Schwarzschild gauge in the minimal two-component metric. The conclusions imply that "Einstein laws of relativity" are founded on the quantum mechanical superposition principle and hence a characteristic quantum effect.

4.1 The Special Theory

With the backing of the present generalized description we will promptly set up a simple 2×2 complex symmetric matrix that (without interaction) displays perfect symmetry between the states of the particle and its antiparticle image.

$$H = \begin{pmatrix} m & -iv \\ -iv & -m \end{pmatrix} \tag{54}$$

In Eq. (54) the diagonal elements are the energies associated with a particle with mass m in a state with its wave vector denoted by $|m\rangle$ and the antiparticle state, assigned a negative energy $-m$ with the state vector $|\bar{m}\rangle$. The extension to the Dirac equation is essentially straightforward, but will be left out here [30]. Above, $-iv$ is the complex symmetric interaction, to be defined below; the minus sign is by convention. For zero interaction the diagonal elements are $\pm m_0$. Note also that the vectors $|m_0\rangle$ and $|\bar{m}_0\rangle$ by choice are orthonormal, while $|m\rangle$ and $|\bar{m}\rangle$ as a rule are bi-orthogonal.

Solving the secular equation corresponding to the ansatz (Eq. (54)) we obtain directly the roots $\lambda_\pm = \pm m_0$ from $\lambda^2 = m_0^2 = m^2 - v^2$. Defining the kinematic perturbation as $v = p/c$ we identify the familiar relation $m^2 c^4 = m_0^2 c^4 + p^2 c^2$. In passing we observe that $p = mv$, with appropriate modifications for a particle in an electromagnetic [43] (or other field), is in general an operator, which in its extended form may not be self-adjoint. Returning to the Klein–Gordon type equation, we obtain the associated "eigensolutions"

$$|m_0\rangle = c_1 |m\rangle + c_2 |\bar{m}\rangle \,; \lambda_+ = m_0; \qquad |m\rangle = c_1 |m_0\rangle - c_2 |\bar{m}_0\rangle \,;$$
$$|\bar{m}_0\rangle = -c_2 |m\rangle + c_1 |\bar{m}\rangle \,; \ \lambda_- = -m_0; \ |\bar{m}\rangle = c_2 |m_0\rangle + c_1 |\bar{m}_0\rangle \,; \tag{55}$$

with

$$c_1 = \sqrt{\frac{1+X}{2X}} \,; \ c_2 = -i \sqrt{\frac{1-X}{2X}} \,; \ m = \frac{m_0}{X} \,; \ c_1^2 + c_2^2 = 1$$
$$X = \sqrt{1-\beta^2} \,; \ \beta = p/mc.$$

For "classical particles" we recover the familiar β factor (not to be mixed with absolute temperature parameter), e.g. $p/mc = v/c$. In general we must remember to keep the order of the entities appearing in the operator secular equation. The present formulation is somewhat unspecified for simplicity, but the procedure need to be checked, se below. Since we respect complex symmetry our model admits, under suitable environmental interactions and/or correlations primary complex resonance energies commensurate with rigorous mathematics and precise boundary conditions, see sections above. Hence we find that

$$m_0 c^2 \to m_0 c^2 - i\frac{\Gamma_0}{2}; \ \tau_0 = \frac{\hbar}{\Gamma_0}; \ mc^2 \to mc^2 - i\frac{\Gamma}{2}; \ \tau = \frac{\hbar}{\Gamma}, \tag{56}$$

where Γ, τ and Γ_0, τ_0 are the half widths and lifetimes of the state respectively. Inserting the modifications of Eq. (56) into our complex symmetric secular equation and separating real and imaginary parts one gets the contractions

$$\Gamma_0 = \Gamma \sqrt{1-\beta^2}; \ \tau = \tau_0 \sqrt{1-\beta^2}. \tag{57}$$

By comparing times in the two scales, enforcing Lorentz-invariance for the length l, one concludes that

$$l = \frac{l_0}{\sqrt{1-\beta^2}}; \ t = \frac{t_0}{\sqrt{1-\beta^2}}; \ m = \frac{m_0}{\sqrt{1-\beta^2}}. \tag{58}$$

From this analysis we deduce that the laws of special relativity appears as a simple consequence of the quantum mechanical superposition principle. Next we will extend the consideration to gravitational interactions.

4.2 General Theory

We will first extend the model by including the scalar gravitational interaction by augmenting the present development in the basis $|m, \bar{m}\rangle$:

$$H = \begin{pmatrix} m(1 - \kappa(r)) & -iv \\ -iv & -m(1 - \kappa(r)) \end{pmatrix}$$

$$\lambda^2 = m^2(1 - \kappa(r))^2 - p^2/c^2; \; \lambda = m_0(1 - \kappa(r)); \; v = p/c$$

(59)

with

$$\kappa(r) = \mu/r; \; \mu = \frac{G \cdot M}{c^2}$$

Here μ is the gravitational radius, G the gravitational constant, M a "classical mass" (which does not change sign when $m \to -m$) and $v = p/c$ as before. Also $\kappa(r) \geq 0$ depends on the coordinate r of the particle m, with origin at the center of mass of M. The coordinate r (and t) refers to a flat Euclidean space and the emerging scales define the curved space-time. Eq. (59) leads to the eigenvalues $\lambda_\pm$

$$m_0^2 = m^2 - p^2/(1 - \kappa(r))^2 c^2;$$

(60)

$$\lambda_\pm/(1 - \kappa(r)) = \pm m_0 = \pm\sqrt{m^2 - p^2/(1 - \kappa(r))^2 c^2}$$

$$m = m_0/\sqrt{1 - \beta'^2}; \; \beta' \leq 1; \; 1 > \kappa(r); \; \beta' = p/mc(1 - \kappa(r)) = v/c(1 - \kappa(r)).$$

(61)

Utilizing the simple fact (in this model) that the angular momentum, mvr, for a particle under the influence of a central force is a constant of motion we obtain (here $m = \hat{m}_{op}$ has the eigenvalue m_0) the relation

$$m_0 v r = m_0 c \mu; \; v = \kappa(r)c = \mu c/r.$$

(62)

where the constant have been evaluated at the limiting velocity c and the limiting distance, the gravitational radius. It follows, for a particle with a non-zero restmass, that a degeneracy (Jordan block) occurs at the Schwarzschild radius, $r = R_{LS}$, provided the mass M is entirely localized inside the sphere, i.e.

$$\frac{1}{2}m = mv/c = m\kappa(r); \; r = R_{LS} = 2\mu$$

$(61')$

Here we need to distinguish two cases (in addition to $m_0 = 0$) in approaching the singularity:

1. Either $m \to \infty$ adiabatically (or on-adiabatically from e.g. an electro-magnetic fluctuation) with m_0 finite
2. m is finite with $m_0 \to 0$ adiabatically (or non-adiabatically).

For instance in the limit of a finite m one obtains from Eq. (59) at $r = R_{LS}$

$$H_{deg} = \frac{1}{2} \begin{pmatrix} m & -im \\ -im & -m \end{pmatrix} \to H_{deg} = \begin{pmatrix} 0 & m \\ 0 & 0 \end{pmatrix}$$

$$|0\rangle = \frac{1}{\sqrt{2}} |m\rangle - i \frac{1}{\sqrt{2}} |\bar{m}\rangle \tag{62}$$

$$|\bar{0}\rangle = \frac{1}{\sqrt{2}} |m\rangle + i \frac{1}{\sqrt{2}} |\bar{m}\rangle$$

explicitly displaying a Jordan block singularity at the Schwarzschild radius.

Thus, as a first conclusion, we have established that the present scalar model specify that a quantum particle will occupy one of two possible states. The identification of these states occurs through the interaction v and the emergence of the length and timescale contractions. For zero rest-mass particles we further find, using $\lambda_0 = m_0 = 0$ and the requirement that our equations should be consistent with the singularity, at $r = R_{LS}$, that

$$m^2 c^4 (1 - \kappa_0(r))^2 = p^2 c^2 \tag{63}$$

$$\kappa_0(r) = \frac{2GM}{c^2 r} = 2\kappa(r)$$

Relation (63) means that zero rest-mass particles like, e.g. photons obey the law commensurate with the effect of light deflection in a gravitational field.

As a final point we will indicate through a non-scalar description that Eq. (63) is compatible with the Jebsen–Birkoff stationary, spherically symmetric solution [53]. We will set up and generalize the following familiar formalism

$$\langle \mathbf{r} \mid \mathbf{p} \rangle = (2\pi\hbar)^{-3/2} e^{i/\hbar \bar{\mathbf{r}} \cdot \mathbf{p}} \tag{64}$$

where

$$\mathbf{r} = \begin{pmatrix} x \\ y \\ z \end{pmatrix} ; \ \mathbf{p} = \begin{pmatrix} p_x \\ p_y \\ p_z \end{pmatrix}$$

to four dimensions obtaining

$$\left\langle \mathbf{r}, -ict \left| \mathbf{p}, \frac{iE}{c} \right. \right\rangle = (2\pi\hbar)^{-2} e^{i/\hbar(\bar{\mathbf{r}} \cdot \mathbf{p} - Et)}$$

or

$$\langle x^* \mid \Pi \rangle = (2\pi\hbar)^{-2} e^{i/\hbar(\tilde{\mathbf{x}} \cdot \Pi)} \tag{65}$$

with the apparent definition

$$\mathbf{x} = \begin{pmatrix} x \\ y \\ z \\ ict \end{pmatrix} ; \ \Pi = \begin{pmatrix} p_x \\ p_y \\ p_y \\ iE/c \end{pmatrix} \tag{66}$$

Note the complex conjugate in the bra-position, invoked to guarantee complex symmetric constructions. Rewriting our equations, displaying usual operator identifications

$$\vec{p} = -i\hbar\nabla; E = i\hbar\frac{\partial}{\partial t}; \vec{\Pi} = -i\hbar\left(\nabla, i/c\frac{\partial}{\partial t}\right)$$

we obtain next the operator secular equation (identity operator suppressed)

$$\lambda^2 = (E^2 - p^2 c^2) = -c^2 \vec{\Pi} \cdot \vec{\Pi} = -c^2 \tilde{\Pi} \cdot \Pi = -c^2 \Pi^2 = m_0^2 c^4 \qquad (67)$$

This relation can be utilized to define the restmass, i.e. from

$$\Pi^2 = -\hbar^2\left(\Delta - \frac{1}{c^2}\frac{\partial^2}{\partial t^2}\right)$$

and

$$\langle\mathbf{x}^*|-\Pi^2|\,\Pi\rangle = m_0^2 c^2\,\langle\mathbf{x}^*\mid\Pi\rangle \qquad (68)$$

From Eqs. (59–61) follows the relation

$$m = \frac{m_0(1 - \kappa(r))}{\sqrt{1 - 2\kappa(r)}} = \frac{\lambda_0}{\sqrt{1 - 2\kappa(r)}} \qquad (69)$$

where λ_0 is the positive (real part) eigenvalue of the secular equation of the matrix

$$m\begin{pmatrix}(1 - i\kappa(r)) & \kappa(r)\\ \kappa(r) & -(1 - i\kappa(r))\end{pmatrix} \qquad (70)$$

Since the eigenvalues of the matrix above may be non-real in our complex symmetric setting, i.e.

$$m = m_r - i\Gamma; \lambda_0 = \lambda_r - i\Gamma_0 \qquad (71)$$

we find by projecting out the real and imaginary parts of Eqs. (70–71) that

$$\tau = \tau_0\sqrt{1 - 2\kappa(r)}; \Gamma = \frac{\hbar}{\tau}; \Gamma_0 = \frac{\hbar}{\tau_0}$$

or for differential "times"

$$d\tau^2 = d\tau_0^2(1 - 2\kappa(r)) \qquad (72)$$

Eqs. (59) and (72) implies that we can introduce the gravitational interaction as follows in the present case of $m_0 \to 0$ (static, spherically symmetric case), i.e.

$$\Pi^2 \to \Pi^2_{\text{grav}} = \left(1 - \frac{2Gm}{c^2 r}\right)^{-1} p_r^2 - \left(1 - \frac{2Gm}{c^2 r}\right) \frac{E^2}{c^2} \tag{73}$$

or in its appropriate symmetrized forms

$$\Pi^2 \to \Pi^2_{\text{grav}} = \left(1 - \frac{2Gm}{c^2 r}\right)^{-1/2} p_r^2 \left(1 - \frac{2Gm}{c^2 r}\right)^{-1/2}$$
$$- \left(1 - \frac{2Gm}{c^2 r}\right)^{1/2} \frac{E^2}{c^2} \left(1 - \frac{2Gm}{c^2 r}\right)^{1/2} \tag{73$'$}$$

and/or

$$\Pi^2 \to \Pi^2_{\text{grav}} = p_r \left(1 - \frac{2Gm}{c^2 r}\right)^{-1} p_r - \frac{E}{c} \left(1 - \frac{2Gm}{c^2 r}\right) \frac{E}{c} \tag{73$''$}$$

Eqs. (73) correspond to a change of the coordinate-, and recipocal coordinate system: $\mathbf{x}' = \alpha^{-1}\mathbf{x}$, $\Pi' = \alpha\Pi$, where α is a 4×4 similarity transformation, but here restricted to two dimensions in the basis (r, ict) and $(p_r, iE/c)$ with corresponding modifications in Eqs. (65–68)

$$\alpha = \begin{pmatrix} \left(1 - \dfrac{2Gm}{c^2 r}\right)^{-1/2} & 0 \\ 0 & \left(1 - \dfrac{2Gm}{c^2 r}\right)^{1/2} \end{pmatrix} \tag{74}$$

Rather than using the traditional covariant formalism, we have conformed to the present complex symmetric picture using simple matrix algebra [4] to analyze the consequences of the transformations. From Eqs. (73) follow directly

$$\langle (r, ict)^* | \Pi^2 | p_r, iE/c \rangle \to \langle (r', ict')^* | \Pi'^2 | p'_r, iE'/c \rangle =$$
$$\langle (r', ict')^* | \Pi^2_{\text{grav}} | p'_r, iE'/c \rangle = \langle (r, ict)^* | \Pi^2_{\text{grav}} | p_r, iE/c \rangle$$

Hence the coordinate transformation α, which in general should be commensurate with the appropriate boundary conditions for the quantum system under study, appears in the present formulation in such a way that the surrounding gravitational field develops as an effect of the geometry characterizing the source of the interaction. In addition, it follows directly from Eqs. (63) and (72–74), that we have established the sought after compatibility with the Schwarzschild metric in the spherically symmetric, static vacuum.

5 Concluding Remarks

We will begin the conclusions with some specific conjectures regarding processes within the biological domain. Since the theoretical development centers around the transformation matrix B, see Eq. (18) and Appendix 2, we will consider its properties further. One notice directly that it lends itself to an interesting factor-structure, cf. also discussions in connection with Mellin transform in Section 2.2. The factoring property implies that certain groups of sites will act coherently and/or strongly correlated [5, 6, 22, 54]. A simple example will explicitly display what we mean, below [55].

Let us briefly introduce the following notation for the cyclic structure shown in B. Denoting the simple column $(\omega^*, \omega^{*3}, \omega^{*5}, \ldots, \omega^{*2n-1})^\dagger$, for any arbitrary n, with the symbol $(n)^\dagger$ where $n \leq m$, it is easy to see how the structure of the general matrix will develop. After the trivial organization of the first two rows, the next one will be partitioned into two, if m is even. In general m will be partitioned into the factors of m in ascending order. For instance choosing $m = 12$ we can write for $\sqrt{12}\,B$ the symbolic form

$$
\begin{array}{c}
(1)\\(1)\\(1)\\(1)\\(1)\\(1)\\(1)\\(1)\\(1)\\(1)\\(1)\\(1)
\end{array}\Bigg\}12
\;\;
\left\{\begin{array}{c}6\\ \\4\\ \\ \\4\\ \\ \\6\\ \\4\end{array}\right\}
\left\{\begin{array}{c}4\\ \\3\\ \\ \\3\\ \\ \\3\\ \\3\end{array}\right\}
\Bigg\}12
\;
\left\{\begin{array}{c}2\\2\\2\\2\\2\\2\\2\\2\\2\\2\end{array}\right\}12
\;
\left\{\begin{array}{c}3\\ \\3\\ \\ \\3\\ \\ \\3\\ \\3\end{array}\right\}
\left\{\begin{array}{c}4\\ \\6\\ \\ \\4\\ \\ \\6\\ \\4\end{array}\right\}12
$$

(75)

For more details on the symmetries involving both rows and columns, we refer to a recent study [55]. One might e.g. speculate what would be the consequences for the appearance of a large prime number $m = p$. This means that no partitions will emerge in Eq. (75). Whether this irreducibility infers any physico-biological significance would of course be an interesting possibility. The present structure, interpreted within our coherent-dissipative ensemble, would further suggest possible interpretations in the biological field, e.g. proton correlations in DNA, the origin of the screw like symmetry of the double helix and possible long-term correlations of the smallest microscopic self-organizing units co-operating *in vivo* systems [54]. To examine the latter, we will rely on the following pair entropy

$$
S_{\text{pair}}^{(2)} = k_B \frac{\ln(N-1)}{N-1} \tag{76}
$$

referring to the weight of physical pairs in all possible pairings of N fermionic particles. It is well-known that a double helix of identical $C - G$ base pairs, e.g. Cytosine and Guanine forms a left-handed screw with a $30\,^\circ$ angle between two consecutive "stairs" forming a full rotation in 12 base-pair units. As the base pairs are held together via H-bond patterns of the Watson Crick type exhibiting a specific phase stability we will conceive of a simple quantum correlated model of tunneling protons. The smallest case corresponds to $N = 4$. It follows that the number of pairings and the relevant probability are

$$r = \binom{N}{2} = \binom{4}{2} = 6; \ \lambda^{(1)} = \frac{N}{2r} = \frac{1}{3} \tag{77}$$

where $\lambda^{(1)} = p$, see Eq. (16) is the probability of finding a proton pair in a particular correlated state. The entropy (Eq. 76)) for the ($N = 4$) pairing system is hence given by

$$S_{\text{pair}}^{(2)} = -k_B \frac{1}{3} \ln \left(\frac{1}{3} \right) \tag{78}$$

noting that every unit, see Eq. (75), is in contact with an environment of other similar or identical units. This result is consistent with $m = 6$, which implies that we need six correlated pairs (or base pairs) to "get around" 180° in the vector space. Hence 12 pairs would be needed for a full rotation but since we "do not return" to the origin here we have instead a pseudo-cyclic "screw-like" symmetry reflected in Eq. (18), i.e. the appropriate column of Eq. (75). Note further that the Gibb's partition of the micro entropy above is not maximum for a near equilibrium situation. This follows from the fact that the function $-x \ln x; 0 \leq x \leq 1$, has a maximum for $x = 1/e$. Assuming that our correlation model in a more realistic dissipative system of varying occurrence of base pairs would "prefer" a "close to equilibrium" entropy, we find that a recalculation of our parameters above yields the result

$$N - 1 \to e; r = \binom{N}{2} \to \binom{e+1}{2} \approx 5.05$$

indicating a less orderly double helix with about 10–11 base pairs in a full turn as well as a possible change in directions of the screw. One might further imagine that a coherent situation linking together very many units would cause a sudden drop in the pair entropy, corresponding to a large N in Eq. (76), resulting in macroscopic selforganization of all participating base pairs, thereby extending the characteristic lifetime for the relevant units of protonic pairs.

As a final conclusion we propose that the general characteristics of the present extended theoretical picture, and supported from the illustrative study cases given here, are generic manifestations of thermally activated quantum correlations of a coherent-dissipative structure. We have shown that a small but crucial extension of standard quantum mechanics leads to a formulation that allows for general broken symmetry solutions, yet have conventional quantum mechanics embedded. The complex symmetric framework provides a natural approach to Hamiltonian and Liouvillian isometric and contractive evolution and a consistent organization of energy superoperator thermalization based on the Bloch equation. The over-all description is not explicitly dependent or susceptible to precise physico-chemical mechanisms as the latter can be successively employed separately to derive qualitative conclusions and quantitative data.

Not only do we have a general organization of coherent-dissipative structures effectively representing quantum correlation effects in condensed and soft condensed matter, but we have also, as a result of the present account, portrayed Einstein's laws of relativity as a quantum effect. Notwithstanding this statement, one should be aware of the remarkable fact that the current formulation of gravitational interactions is essentially "classical" outside the domain boundary characterized by the Schwarzschild radius. Yet the laws of relativity derive here from the quantum mechanical superposition principle. Except from the occurrence of a general Jordan block singularity, the equations appear mostly to be of classical-orthodox character. Note, however, that a theoretical structure exists inside the singularity, a situation, which has not been satisfactorily determined by all classical theories. The present black-hole-like configuration develops via the degeneracy condition, see Eqs. (61'–63), acceding the latter interactions, or correlations, to condense or unify [7, 30] according to Yang's ODLRO [25]. In a sense "the law of the opposite, negation and transformation" has here been repeatedly applied in harmony with the statement Alexander S. Davydov.

6 Appendices

6.1 Quantum Diffusion and Thermal Scattering

Here we will examine particular aspects of a process that we have named thermally induced quantum diffusion using the language of standard scattering theory. The outcome leads directly to the derivation of the formula used in Section 2.3. The precursor is the following thermal scattering model.

Consider m degrees of freedom (they could be bosonic or paired fermionic, the quantum statistical analysis can be carried out separately) and assume that they are correlated on a relaxation timescale given by τ_{rel}. The latter corresponds to the average lifetime of the building blocks of the system, and depends in general on the type of particles or properties of the localized units being represented, cf. the Einstein relation which combines transport displacements with the diffusion constant D. To complete the picture we will define an area or region of the correlations, corresponding to a spherically averaged total scattering cross section denoted by σ_{tot}. This area should be consistent with the physical parameters of the model so that on average we will detect one particle or degree of freedom in the differential solid-angle element $d\Omega$ during the limit timescale τ_{lim} given by Heisenberg's uncertainty relation, i.e.

$$\tau_{lim} = \frac{\hbar}{kT} \tag{A1}$$

Here as usual k_B is Boltzmann's constant and T the absolute temperature. We restress that our goal is not to determine cross section data or evaluating lifetimes, reaction rates etc. Instead our aim is to find consistent relations between temperatures, the size of the dissipative structure, various lifetimes, reaction rates etc. and to use this information as input to our quantum statistical equations.

With the ingredients introduced above it is a simple matter to set up the consistency relations between incident and scattered fluxes. The incident flux, N_{inc} of number of particles/degrees of freedom per unit area and time is

$$N_{inc} = \frac{m}{\sigma_{tot} \tau_{rel}} \tag{A2}$$

Since the number, $N_s d\Omega$ of particles scattered into $d\Omega$ per unit time is

$$N_s d\Omega = \frac{d\Omega}{\tau_{lim}} = \frac{k_B T}{\hbar} d\Omega \tag{A3}$$

we obtain straightforwardly

$$\sigma_{tot} = \int \sigma(\Omega) d\Omega = \int \frac{N_s}{N_{inc}} d\Omega \tag{A4}$$

From which we obtain the first relation between our experimental parameters

$$m = \frac{4\pi k_B T}{\hbar} \tau_{rel} \tag{A5}$$

The relation (A5) (note that σ_{tot} cancels out in Eq. (A4)) tells us that the number m of the model's correlated degrees of freedom depends uniquely on the temperature and the relaxation time.

The next step is to use this information as input to the thermalization formula of Section 2.3, i.e. we want to study the behavior of the matrix element $\hat{\rho}_{kl} = |h_k\rangle \langle h_l^*|$ which yields, (the degenerate energy value is set to zero)

$$e^{-\frac{\beta}{2} H_2} \hat{\rho}_{kl} e^{-\frac{\beta}{2} H_2} = e^{i\frac{\beta}{2}(\varepsilon_k + \varepsilon_l)} \hat{\rho}_{kl} \tag{A6}$$

Recognizing the temperature dependence from our oscillating building blocks of our dissipative system we make the simplifying assumption that we can organize the excitation spectrum of the collective cluster of particles/degrees of freedom harmonically, i.e. with the distance between the levels being $\hbar \tau_{rel}^{-1}$ (except the zero point vibration). Hence the $(l-1)$th level is characterized by the angular frequency τ_l^{-1} which is uniquely determined by the harmonic spectrum. Hence it follows that $\tau_{rel} = (l-1)\tau_l$; $l = 2, 3, \ldots m$, with $\tau_1 = \infty$; $l = 1$, corresponding to the zero reference energy. From this follows

$$\frac{1}{2}\beta(\varepsilon_k + \varepsilon_l) = \frac{\hbar}{4k_B T \tau_{rel}} \left\{ \frac{1}{\tau_k} - \frac{1}{\tau_l} \right\} = \frac{\hbar}{4k_B T \tau_{rel}} \{k + l - 2\} = \frac{\pi}{m}(k + l - 2) \tag{A7}$$

where we have used Eq. (A5) in the last equality above. Together with Eq. (A6) we have the result needed in Section 2.3.

6.2 The Complex Symmetric Jordan Form

To make this review self-contained we will make a straightforward proof establishing a complex symmetric representation of the classical canonical Jordan form. We will assume that the reader is familiar with standard linear algebra theories and understand why the present complex symmetric representation of a non-normal operator cannot be diagonal, see e.g. Löwdin [4] for a very clear account of the necessary background knowledge. For simplicity we will not discuss the fundamental theories behind the construction [21] or prove the more general formula for the various powers of the nilpotent operator [22]. The latter follows along the same lines, but would make the appendix unnecessary convoluted.

We will hence show by a straight forward computation, $n > 1$ that

$$Q_n = B_n^{-1} J_n B_n \qquad (A8)$$

where

$$[Q_n]_{ik} = \omega^{(i+k-2)}(\delta_{ik} - \frac{1}{n}); \omega = e^{\frac{i\pi}{n}}; [J_n]_{ik} = \delta_{i+1,k}; [B]_{ik} = \omega^{(2i-1)(k-1)}$$

Note that the indices i and k runs from 1 to n and that there will be no problems if an index should become larger than n, since the corresponding matrix element will be zero anyway not contributing to the summations in the proof. Because $J_n^n = 0; J_n^{n-1} \neq 0$ its Segrè characteristic is n.

Since $\det(B_n) \neq 0$ (note the slight difference between B_n and the unitary transformation matrix in Eq. (18)) we can define the inverse as

$$\alpha = B_n^{-1} = \frac{1}{n} B_n^\dagger \qquad (A9)$$

Direct evaluation of the right hand side of Eq. (A8) gives

$$[Q_n]_{ik} = \sum_{j=1}^{n-1} \sum_{l=1}^{n} \alpha_{ij} [J_n]_{jk} \omega^{(2l-1)(k-1)} = \sum_{j=1}^{n-1} \alpha_{ij} \omega^{(2j+1)(k-1)}$$

$$= \left\{ \sum_{j=1}^{n-1} \alpha_{ij} \omega^{(2j-1)(k-1)} \right\} \omega^{2(k-1)}$$

The last expression can be simplified by the use of Eq. (A9) not forgetting to subtract the missing term corresponding to $j = n$ above. The result yields

$$[Q_n]_{ik} = \omega^{2(k-1)} \{ \delta_{ik} - \alpha_{in} \omega^{(2n-1)(k-1)} \} = \omega^{2(k-1)} \left\{ \delta_{ik} - \frac{1}{n} \omega^{i-k} \right\}$$

In the last expression above we have used Eq. (A9) and the fact that $\omega^{2n} = 1$. Furthermore since trivially $\delta_{ik} = \delta_{ik}\omega^{i-k}$ the proof follows, i.e.

$$[\boldsymbol{Q_n}]_{ik} = \omega^{2(k-1)}\left\{\delta_{ik}\omega^{i-k} - \frac{1}{n}\omega^{i-k}\right\} = \omega^{(k+i-2)}\left\{\delta_{ik} - \frac{1}{n}\right\}$$

Analogously one can prove the more general formula

$$[\boldsymbol{Q_n^r}]_{kl} = \omega^{r(k+l-2)}[\delta_{kl} - [\boldsymbol{R_n^r}]_{kl}]; k,l = 1, 2, \ldots m \qquad (A10)$$

$$[\boldsymbol{R_n^r}]_{kl} = \begin{cases} \frac{1}{m}\frac{\sin\left(\frac{\pi r(l-k)}{m}\right)}{\sin\left(\frac{\pi(l-k)}{m}\right)} & k \neq l \\ \frac{r}{m} & k = l \end{cases} \qquad (A11)$$

using the same strategy.

6.3 Microscopic Self-Organization

One of the many surprising consequences of the extended quantum formulation, developed above, is the appearance of Jordan blocks and its physical consequences. As mentioned earlier, these structures may appear directly in the Hamiltonian [5–7, 30], see also scattering matrix formulations based on the Gelfand triple [56], or in the Liouvillian [5, 6]. It is a complicated matter to analyze the effects of complex scaling on the time evolution, since the sought after contractive semigroup properties of the dilated propagator cannot be rigorously proven for, e.g. the Coulomb potential [57]. In the latter case one needs to remove the subspace associated with the bound states, and in general, the space and the Hamiltonian have to be carefully chosen such that the dilatation operator is able to convert isometric evolution of unscaled wave functions to contractive evolution of scaled ones [56]. To integrate the case of Jordan blocks into the picture we need specifically to look at the consequences of the associated evolution, restricted to the subspace of degenerate eigenvectors [57].

To simplify things we will formalize the description in two ways. First, it is necessary to work in a retarded-advanced formulation, i.e. we define the propagator and resolvent as

$$\mathcal{G}^{\pm}(t) = \mp i\theta(\pm t)e^{-iLt}$$
$$\mathcal{G}(z) = (zI - L)^{-1} \qquad (A12)$$

where $\theta(x) = 1; x \geq 0$ and $\theta(x) = 0; x < 0$. It is assumed that we have a well-defined mathematical structure allowing analytic continuation across the real axis into the complex plane. Although the extension of complex scaling to the Liouville case is not trivial [26], the analytical structure is here essentially the same whether the operator, or generator of the evolution, L, is the Hamiltonian, evolving wave functions, or the Liouvillian, evolving density matrices ($\hbar$ is suppressed for simplicity). With this in mind we can write down the connecting Fourier–Laplace transforms, i.e.

$$\mathcal{G}^{\pm}(t) = \frac{1}{2\pi} \int\limits_{C^{\pm}} \mathcal{G}(z) e^{-izt} dz \tag{A13}$$

$$\mathcal{G}(z) = \int\limits_{C^{\pm}} \mathcal{G}^{\pm}(t) e^{izt} dz \tag{A14}$$

where the contour $C^{\pm}$ runs in the upper $(+)$ and the lower $(-)$ complex half plane respectively, from $-\infty$ to $+\infty$. With the initial condition and the retarded-advanced evolution one gets with $\rho(t) = \rho_0; t = 0; \rho^{\pm}(t) = \pm i \, \mathcal{G}^{\pm}(t)\rho_0$, the inhomogeneous equations

$$\left(i\frac{\partial}{\partial t} - L \right) \mathcal{G}^{\pm}(t) = \delta(t) \tag{A15}$$

and

$$\rho^{\pm}(z) = \pm i \mathcal{G}(z)\rho_0 \tag{A16}$$

In Eq. (16) the $+$ sign indicates that $t > 0$ and z is in the upper half plane while the $-$ sign indicates that $t < 0$ and z is in the lower half plane.

The second formalization will be introduced by examining the case of an emerging Jordan block of order m in the Hamilton/Liouvillian formulation. By way of choice we will directly work with frequencies and timescales [6, 54] instead of complex energy differences. Thus we define ($t > 0$, z in the upper half plane—note no appearance of Planck's constant)

$$\mathcal{P} = (\omega_0\tau - i)\mathcal{I} + \mathcal{J} \tag{A17}$$

and

$$\mathcal{G}(t) = e^{-i\mathcal{P}\frac{t}{\tau}}; \ \mathcal{G}(z) = (\omega\tau \, \mathcal{I} - \mathcal{P})^{-1} \tag{A18}$$

Here $\omega_0 \neq 0$ is the resonance frequency, τ the finite lifetime of the state and m the degeneracy, here also the Segrè characteristic. In principle there is no problem to take the limits $\tau \to \infty$ and $\omega_0 \to 0$. In order to derive an evolution law from Eqs. (A17) and (A18) we expand the propagator

$$e^{-i\mathcal{P}t/\tau} = e^{-i\omega_0 t} e^{-t/\tau} \sum_{k=0}^{m-1} \left(\frac{-it}{\tau} \right)^k \frac{1}{k!} \mathcal{J}^{(k)} \tag{A19}$$

and its Fourier–Laplace transform, or the associated resolvent

$$(\omega\tau \, \mathcal{I} - \mathcal{P})^{-1} = \sum_{k=1}^{m} ((\omega - \omega_0)\tau + i)^{-k} \, \mathcal{J}^{(k-1)} \tag{A20}$$

From Eqs. (A17–A20) one deduces that the evolution of a coherent-dissipative system generated by

$$\mathcal{T} = \mathcal{P}/\tau \tag{A21}$$

no longer follows the standard exponential decay law, cf. radioactive decay and the arguments that identify such behavior under some relevant time intervals [5]. Instead, the exponential above is multiplied by a polynomial leading to a modified principle (in the limit of an infinite dimensional Jordan block one might consider more general evolution laws). The Fourier–Laplace transform, (Eq. (A20)), contains, as a result multiple poles, which result in general spectral gain–loss functions that go beyond the simple Lorentzian (or distorted Lorentzian) shapes. It is straightforward to see the effect of the shift operator J. Borrowing some data from our m-dimensional problem defined in Section 2.2, we obtain

$$\mathcal{P}/\tau = (\mathcal{E}_o/\hbar - i/\tau)\mathcal{I} + (1/\tau)\mathcal{J} \tag{A22}$$

where $\tau = \hbar/k_B T$, is the average thermal lifetime, and $\mathcal{E}_o$ is the real part of the degenerate energy (set to zero previously for simplicity). To discuss the modified decay law we start by

$$N(t) = |\langle \varphi_0 | \varphi(t) \rangle|^2 = e^{-\frac{t}{\tau}} \tag{A23}$$

where

$$\varphi(t) = e^{-i\omega_0 t} e^{-t/\tau} \varphi_0 \tag{A24}$$

obtaining as usual

$$dN = -\frac{1}{\tau} N(t) dt \tag{A25}$$

In the degenerate case with $f_k(t) = e^{-i\omega_0 t} e^{-t/\tau} f_k$, one gets for the rth power of t (note that only f_1 is an eigenfunction, while the others complete the root manifold)

$$N(t) \propto |\langle f_1 | \mathcal{J}^{(r)} | f_{r+1} \rangle|^2 \left(\frac{t}{\tau}\right)^r \frac{1}{r!} e^{-\frac{t}{\tau}} = \left(\frac{t}{\tau}\right)^r \frac{1}{r!} e^{-\frac{t}{\tau}} \tag{A26}$$

where

$$\mathcal{J}^{(r)} = \sum_{k=1}^{m-r} |f_k\rangle \langle f_{k+r}|$$

For the highest power $m-1$ we obtain from Eq. (A26)

$$dN = t^{m-2} \left(m - 1 - \frac{t}{\tau}\right) N(t) dt \tag{A27}$$

and the new microscopic law of evolution

$$dN(t) > 0; t < (m-1)\tau \tag{A28}$$

This relation is consistent with the results of Section 2.3 and Appendix 1, i.e. identifying τ with τ_{lim} and $(m-1)\tau$ with τ_{rel}.

Hence we have proved that Jordan blocks appearing in the generator of the coherent-dissipative dynamical system yields a non-decaying evolution law that suggests microscopic self-organization. Furthermore this evolution is consistent with generic timescales obtained from the thermalized Bloch equation subject to a degenerate density operator exhibiting a canonical Jordan form with a Segrè characteristic, m, of the same order as the one generating the dynamics.

Acknowledgments The author thanks the organizers of the NATO Advanced Research Workshop for the invitation. He is further indebted to Prof. Nino Russo the NATO co-director, and Eugene Kryachko for proposing and enduring the present write-up. Financial support has been given, in addition to the NATO Science Program, from the Swedish Foundation for Strategic Research.

References

1. A. S. Davydov, *Quantum Mechanics* (Pergamon Press, Oxford, 1965, 2nd edition, 1976).
2. A. S. Davydov, *Solitons in Molecular Systems* (Kluwer, Dordrecht, 1985).
3. Stuart A. Kauffman, *Reinventing the Sacred* (Perseus Books Group/Basic Books, New York, 2008).
4. P. O. Löwdin, *Linear Algebra for Quantum Theory* (Wiley, New York, 1998).
5. E. J. Brändas, Relaxation Processes and Coherent Dissipative Structures. In: *Dynamics During Spectroscopic Transitions*, edited by E. Lippert and J. D. Macomber, Springer Verlag, Berlin, 148–193 (1995).
6. E. J. Brändas, Applications of CSM Theory. In: *Dynamics During Spectroscopic Transitions*, edited by E. Lippert and J. D. Macomber, Springer Verlag, Berlin, 194–241 (1995).
7. E. J. Brändas, Quantum Mechanics and the Special- and General Theory of Relativity, *Adv. Quant. Chem.* **54**, 115–132 (2008).
8. E. Schrödinger, Quantisierung als Eigenwertproblem, *Annalen der Physik* **79**, 361–376 (1926).
9. Harold V. McIntosh, Quantization as an Eigenvalue Problem. In *Group Theory and Its Applications*, Vol. 3, edited by Ernest M. Loebl, Academic, New York, 333–368 (1975).
10. H. Weyl, Über gewöhnliche Differentialgleichungen mit Singularitäten und die zuge- hörigen Entwicklungen willkürlicher Funktionen, *Math. Ann.* 68, 220–269 (1910).
11. E. C. Titchmarsh, *Eigenfunction Expansions Associated with Second-Order Differential Equations* (Oxford University Press, London, Vols. 1 and 2, 1946).
12. R. G. Newton, *Scattering Theory of Waves and Particles* (Springer Verlag, New York, 2nd edition, 1982).
13. M. Hehenberger, H. V. McIntosh and E. Brändas, Weyl's Theory Applied to the Stark Effect in the Hydrogen Atom, *Phys. Rev.* **A10**, 1494–1506 (1974).
14. E. Brändas and M. Hehenberger, Determination of Weyl's m-Coefficient for a Continuous Spectrum, *Lect Notes Math.* **415**, 316–322 (1974).
15. E. Brändas, M. Rittby and N. Elander, Titchmarsh-Weyl Theory And Its Relations to Scattering Theory: Spectral Densities and Cross Sections; Theory and Applications, *J. Math. Phys.* **26**, 2648–2658 (1985).
16. E. Engdahl, E. Brändas, M. Rittby and N. Elander, Resonances and Background: A Decomposition of Scattering Information, *Phys. Rev.* **A37**, 3777–3789 (1988).
17. E. Brändas and N. Elander, Editors: *Resonances—The Unifying Route Towards the Formulation of Dynamical Processes—Foundations and Applications in Nuclear, Atomic and Molecular Physics,* Springer Verlag, Berlin, *Lecture Notes in Physics*, Vol. **325**, 1–564 (1989).

18. E. Balslev and J. M. Combes, Spectral Properties of Many-Body Schrödinger Operators with Dilatation-Analytic Interactions, *Commun. Math. Phys.* **22**, 280–294 (1971).

19. P. O. Löwdin, Scaling Problem, Virial Theorem, and Connected Relations in Quantum Mechanics, *J. Mol. Spectrosc.* **3**, 46–66 (1959).

20. E. Brändas and P. Froelich, Continuum Orbitals, Complex Scaling, and the Extended Virial Theorem, *Phys. Rev.* **A16**, 2207–2210 (1977).

21. C. E. Reid and E. Brändas, On a Theorem for Complex Symmetric Matrices and its Relevance in the Study of Decay Phenomena, *Lect. Notes Phys.* Vol. **325**, 476–483 (1989).

22. E. Brändas, Resonances and Dilatation Analyticity in Liouville Space, *Adv. Chem. Phys.* **99**, 211–244 (1997).

23. F. R. Gantmacher, *The Theory of Matrices* (Chelsea, New York, Vols. I, II, 1959).

24. A. J. Coleman and V. I. Yukalov, *Reduced Density Matrices: Coulson's Challenge, Lecture Notes in Chemistry* (Springer-Verlag, Berlin, **72**, 2000).

25. C. N. Yang, Concept of Off-Diagonal Long-Range Order and the Quantum Phases of Liquid He and of Superconductors, *Rev. Mod. Phys.* **34**, 694–704 (1962).

26. C. H. Obcemea and E. Brändas, Analysis of Prigogine's Theory of Subdynamics, *Ann. Phys.* **151**, 383–430 (1983).

27. I. Prigogine, *From Being to Becoming*, (Freeman WH, San Fransisco, 1980).

28. E. B. Karlsson and E. Brändas, Modern Studies of Basic Quantum Concepts and Phenomena, *Phys. Scripta* **T76**, 7–15 (1998).

29. C. A. Chatzidimitriou-Dreismann and E. J. Brändas, Proton Delocalization and Thermally Activated Quantum Correlations in Water: Complex Scaling and New Experimental Results. *Ber. Bunsenges, Phys. Chem.* **95**, 263–272 (1991).

30. E. J. Brändas, Are Einstein's Laws of Relativity a Quantum Effect? In: *Frontiers in Quantum Systems in Chemistry and Physics*, edited by S. Wilson, P. J. Grout, J. Maruani, G. Delgado-Barrio and P. Piecuch, Springer Verlag, Vol. **18**, 238–256 (2008).

31. E. J. Brändas, Organization of Particle Beams. A Possible New Cooling Concept, *In Proceedings from the Workshop on Beam Cooling and Related Topics*, Oct 4–8, 1993, edited by J. Bosser, CERN 94–03, 106–117 (1994).

32. P. W. Anderson, The Resonating Valence Bond State in LaCuO and Superconductivity, *Science* **235** (4793), 1196–1198 (1987).

33. P. W. Anderson, *The Theory of Superconductivity in the High-T_C-Cuprate Super-Conductors*, (Princeton Series in Physics, edited by Sam B. Treiman, Princeton University Press, New Jersey, 1997).

34. J. Tahir-Kheli and W. A. Goddard III, Chiral Plaquette Polaron Theory of Cuprate Super-Conductivity, *Phys. Rev. B—Condens. Matter Mater. Phys.* **76** (1), Art. no. 014514 (2007).

35. L. J. Dunne, E. J. Brändas, J. N. Murrell and V. Coropceanu, Group Theoretical Identification of Active Localized Orbital Space in High-T_C-Cuprate Superconductors, *Solid State Commun.* **108**, 619–623 (1998).

36. L. J. Dunne and E. J. Brändas, Two-Fluid Model of Superconducting Condensates and Spin Gaps in $d_{x^2-y^2}$-Wave High T_c Cuprates from Repulsive Electronic Correlations, *Int. J. Quant. Chem.* **99**, 798–804 (2004).

37. E. Brändas and C. A. Chatzidimitriou-Dreismann, Coherence in Disordered Condensed Matter. V: Thermally Activated Quantum Correlations in High-T_C Superconductivity, *Ber. Bunsenges. Phys. Chem.* **95**, 462–466 (1991).

38. A. J. Coleman, Structure of Fermion Density Matrices, *Rev. Mod. Phys.* **35** (3), 668–686 (1963).

39. Y. J. Uemura, G. M. Luke, B. J. Sternlieb, J. H. Brewer, J. F. Carolan, W. N. Hardy, R. Kadono, J. R. Kempton, R. F. Kiefl, S. R. Kreitzman, P. Mulhern, T. M. Riseman, D. L. Williams, B. X. Yang, S. Uchida, H. Takagi, J. Gopalakrishnan, A. W. Sleight, M. A. Subramanian, C. L. Chien, M. Z. Cieplak, G. Xiao, V. Y. Lee, B. W. Statt, C. E. Stronach, W. J. Kossler and X. H. Yu, Universal Correlations Between T_C and n_s/m^* (Carrier Density over Effective Mass) in High-T_C Cuprate Superconductors, *Phys. Rev. Lett.* 2317–2321 (1989).

40. M. Svrček, P. Baňacký and A. Zajac, Nonadiabatic Theory of Electron-Vibrational Coupling: New Basis for Microscopic Interpretation of Superconductivity, *Int. J. Quant. Chem.* 393–414 (1992).

41. E. Karlsson, E. J. Brändas and C. A. Chatzidimitriou-Dreismann, The Interplay Between Universal Quantum Correlations and Specific Mechanisms, Involving Short-Lived Magnetic Clusters, in HTSC's, *Physica Scripta.* **44**, 77–87 (1991).

42. E. Brändas and C. H. Chatzidimitriou-Dreismann, On the Connection Between Certain Properties of the Second-Order Reduced Density Matrix and the Occurrence of Coherent-Dissipative Structures in Disordered Condensed Matter, *Int. J. Quant. Chem.* **40**, 649–673 (1991).

43. E. J. Brändas, Quantum Concepts and Complex Systems, *Int. J. Quant. Chem.* **98**, 78–86 (2004).

44. E. Brändas, P. Baňacký and E. Karlsson, Nonadiabaticity and Off-Diagonal Long-Range Order in High-Temperature Superconductors, preprint (1992). Unpublished.

45. D. Wechsler and J. Ladik, Correlation-Corrected Energy Bands of $YBa_2Cu_3O_7$: A Mutually Consistent Treatment, *Phys. Rev. B: Condens Matter* **55**, 8544–8550 (1997).

46. E. Brändas and C. A. Chatzidimitriou-Dreismann, Creation of Long Range Order in Amorphous Condensed Systems, *Lect. Notes Phy.* **325**, 486–540 (1989).

47. C. A. Chatzidimitriou-Dreismann and E. J. Brändas, Coherence in Disordered Condensed Matter. IV: Conductivities of Molten Alkali Chlorides—A Novel Relation, *Ber. Bunsenges. Phys. Chem.* **93**, 1065–1069 (1989).

48. C. A. Chatzidimitriou-Dreismann, Complex Scaling and Dynamical Processes in Amorphous Condensed Matter, *Adv. Chem. Phys.* **80**, 201–314 (1991).

49. R. Pfeifer and H. G. Hertz, Activation Energies of the Proton-Exchange Reactions in Water Measured with the ^{1}H-NMR Spin-Echo Technique, *Ber. Bunsenges. Phys. Chem.* **94**, 1349–1353 (1990).

50. H. Weingärtner and C. A. Chatzidimitriou-Dreismann, Anomalous H^+ and D^+ Conductance in H_2O–D_2O Mixtures, *Nature*, **346**, 548–550 (1990).

51. E. J. Brändas, Some Theoretical Problems in Chemistry and Physics, *Int. J. Quant. Chem.* **106**, 2836–2839 (2006).

52. I. P. Grant and H. M. Quiney, Application of Relativistic Theories and Quantum Electro-Dynamics to Chemical Problems, *Int. J. Quant. Chem.* **80**, 283–297 (2000).

53. G. D. Birkoff, *Relativity and Modern Physics* (Cambridge University Press, Cambridge, 1921).

54. E. J. Brändas, Dissipative Systems and Microscopic Selforganization, *Adv. Quant. Chem.* **41**, 121–138 (2002).

55. E. J. Brändas, Complex Symmetric Forms and the Emergence of Jordan Blocks in Analytically Extended Quantum Theory, *Int. J. Comp. Math.* **86:2**, 315–319 (2009).

56. A. Bohm, M. Gadella, M. Loeve, S. Saxon, P. Patuleanu and C. Puntmann, Gamow-Jordan Vectors and Nonreducible Density Operators from Higher Order S-Matrix Poles, *J. Math. Phys.* **38**, 6072–6100 (1997).

57. J. Kumicák and E. Brändas, Complex Scaling and Lyapunov Converters, *Int. J. Quant. Chem.* **46**, 391–399 (1993).

Directed Transport of the Davydov Solitons by Unbiased a.c. Forces

L.S. Brizhik, A.A. Eremko, B.M.A.G. Piette, and W.J. Zakrzewski

Abstract We show that in asymmetric molecular chains a periodic unbiased field causes a drift of the Davydov solitons. This directed current, known as ratchet phenomenon, has a threshold with respect to the intensity and the frequency of the field. In spatially symmetric chains a harmonic periodic electric field generates oscillations of solitons but does not result in their directed drift. Such a drift current can be induced in symmetric chains by a time periodic asymmetric external field. This complex dynamics of solitons is generated by the interplay between the Peierls–Nabarro barrier, external field and dissipative effects in the chain. The dependence of the amplitude of soliton oscillations and the velocity of the drift are shown to depend on the intensity of the field, its frequency and the coefficient of the energy dissipation.

Keywords Ratchet effect · Polaron · Solitons · Peierls–Nabarro potential · Field driven current

1 Introduction

The phenomenon of directed transport of charge carriers induced by alternating un-biased (zero mean) forces in various systems has been attracting a great deal of attention. This phenomenon is known also as the ratchet phenomenon [1]. Study of this effect is promising for technical applications in nanotechnologies, including molecular motors, and is important for understanding the functioning of biological motors. The conditions for the ratchet behaviour are now well understood and some interesting theoretical ratchet models have also been proposed. Moreover, a

L.S. Brizhik (✉) and A.A. Eremko
Bogolyubov Institute for Theoretical Physics, 14-b Metrologichna Str, 03680, Kyiv, Ukraine
e-mail: brizhik@bitp.kiev.ua

B.M.A.G. Piette and W.J. Zakrzewski
Department of Mathematical Sciences, University of Durham, Durham, DH13LE, UK

N. Russo et al. (eds.), *Self-Organization of Molecular Systems: From Molecules and Clusters to Nanotubes and Proteins*, NATO Science for Peace and Security Series A: Chemistry and Biology, © Springer Science+Business Media B.V. 2009

relatively large variety of experimental realizations is also available (see, e.g., review [1] and references therein). In particular, semiconducting heterostructures, such as diode (n, p)-junctions, semiconductor superlattices, Josephson junction arrays, SQUID ratchets, quantum dot arrays with broken spatial symmetry have been engineered and shown to possess ratchet properties.

Here we demonstrate that electron self-trapped states in quasi-one-dimensional molecular systems also exhibit the ratchet behaviour. Such self-trapped states, generally called polarons (in one-dimensional systems also known as the Davydov solitons) are localized states of electrons formed due to the electron-phonon interaction [2–5]. Their properties have been studied in great detail both theoretically and experimentally. The class of low-dimensional molecular systems in which polarons exist, includes quasi-1D organic and inorganic compounds (like conducting platinum chain compounds), conducting polymers (e.g., polyacetylene [3], polypyrrole [6], polythiophene [7]), biological macromolecules (α-helical proteins [2], DNA [8]) etc.

It is known that the necessary conditions for the ratchet effect under the action of stochastic or deterministic external forces, both in classical and quantum systems, involve the energy dissipation in the system and the breaking of spatial and/or temporal symmetries [1, 9]. The mechanism for the appearance of the directed motion caused by the zero-mean force, has been established for particles moving in a spatially periodic ratchet potential, and has been extended to soliton ratchets [10–13]. Some of the necessary requirements for the ratchet effect are naturally intrinsic to solitons in molecular systems. Energy dissipation in molecular systems is always present due to the interaction of the atoms with the many degrees of freedom of the surrounding medium which can be considered as a thermal bath. In molecular chains the spatial symmetry can be broken because of the complex structure of the elementary cell. The temporal symmetry can be broken by applying an asymmetric periodic force, e.g. by an appropriate biharmonic driver. This suggests that we should expect the ratchet phenomenon for the Davydov solitons to occur in complex molecular chains. Indeed, our numerical studies [14] have confirmed the existence of the drift of such solitons in an asymmetric diatomic chain under the action of unbiased periodic force. In molecular chains, due to their discreteness, solitons move in a Peierls–Nabarro potential which is periodic with a period equal to the lattice constant [15], and which plays the role of the ratchet potential. The presence of the Peierls–Nabarro potential plays an essential role in the ratchet behaviour of solitons in discrete systems [12, 13].

2 Hamiltonian and Equations

To demonstrate the existence of the ratchet effect for solitons we have performed numerical simulations of the polaron dynamics in asymmetric molecular chains in the presence of an external periodic unbiased electromagnetic field [14]. Thus we have considered a diatomic molecular chain that contains two different atoms or groups of atoms in a unit cell, periodically arranged along the chain axis at their equilibrium positions, $z_{n,1}^0 = na$, $z_{n,2}^0 = na + b$, where a is the lattice constant and b is the distance between the two atoms in a unit cell. The total Hamiltonian of a molecular chain in an external field is represented by the following sum of terms:

$$H = H_0 + H_{int}, \quad H_0 = H_e + H_{ph} + H_{e-ph}. \tag{1}$$

Here the Hamiltonian H_0 describes the electrons in a molecular chain interacting with the lattice vibrations. In the approximation of the nearest-neighbour hopping interaction, in the site representation, H_e is given by the expression

$$H_e = \sum_n \left[E_1 a_{n,1}^\dagger a_{n,1} + E_2 a_{n,2}^\dagger a_{n,2} - J_s (a_{n,1}^\dagger a_{n,2} + a_{n,2}^\dagger a_{n,1}) \right.$$
$$\left. - J_l (a_{n,1}^\dagger a_{n-1,2} + a_{n-1,2}^\dagger a_{n,1}) \right], \tag{2}$$

where $a_{n,j}^\dagger$ ($a_{n,j}$) are creation (annihilation) operators of an electron on the site (n, j); E_j is the on-site electron energy which takes into account the presence of the neighbouring atoms. Furthermore, J_s and J_l are the energies of the hopping interactions with the nearest neighbours from the same unit cell and from the neighbouring cell, respectively. In our work we have studied only one extra electron in the chain, and therefore, we omit the electron spin index. The Hamiltonian of the lattice vibrations, H_{ph}, in the harmonic approximation and in the approximation of the nearest-neighbour interaction, is given by

$$H_{ph} = \frac{1}{2} \sum_n \left[\frac{P_{n,1}^2}{M_1} + \frac{P_{n,2}^2}{M_2} + w_s (u_{n,1} - u_{n,2})^2 + w_l (u_{n,1} - u_{n-1,2})^2 \right], \tag{3}$$

where M_1 and M_2 are masses of atoms, and $u_{n,j}$ are the longitudinal displacements of these atoms from their equilibrium positions. Also, $z_{n,j} = z_{n,j}^0 + u_{n,j}$; $p_{n,j}$ are the momenta, canonically conjugate to $u_{n,j}$; w_s and w_l are the elasticity constants describing the strengths of the interactions between, respectively, the nearest-neighbour atoms belonging to one unit cell and to the neighbouring cells.

The electron–phonon interaction originates from the fact that the hopping interactions coefficients J_s and J_l, of the site energy E_j depend on the interatomic separation. Taking into account such a dependence of only the site energy, we obtain the electron–phonon interaction Hamiltonian, H_{int}, which, in the linear approximation with respect to the lattice displacements, takes the form:

$$H_{e-ph} = \sum_n \left[a_{n,1}^\dagger a_{n,1} \left[\chi_1 (u_{n,1} - u_{n-1,2}) - \chi_s (u_{n,1} - u_{n,2}) \right] \right.$$
$$\left. + a_{n,2}^\dagger a_{n,2} \left[\chi_1 (u_{n+1,1} - u_{n,2}) - \chi_s (u_{n,1} - u_{n,2}) \right] \right]. \tag{4}$$

Here χs and χl are the coefficients of the electron–phonon interaction between nearest neighbours. Finally, the Hamiltonian of the interaction with the external electric field $E(t)$ is given by:

$$H_{int} = -eE(t) \sum_n \left((na - z_0)a_{n,1}^\dagger a_{n,1} + (na + b - z_0)a_{n,2}^\dagger a_{n,2} \right). \qquad (5)$$

In the absence of an external field the Fröhlich Hamiltonian (Eq. (1)) describes the states of electrons interacting with lattice vibrations. Self-trapped states of electrons in such systems are usually described in the adiabatic approximation which is equivalent to the semiclassical consideration in which the vibrational subsystem is treated as a classical one. In this approximation the wave function of the system is represented in a multiplicative Born–Oppenheimer form

$$|\psi\rangle = U|\psi_e\rangle, \qquad (6)$$

where U is the unitary operator of the coherent atom displacements induced by the presence of quasiparticles, described by the vector state $|\psi_e\rangle$. In the case of one extra electron in the chain its state is described by

$$|\psi_e\rangle = \sum_{n,j} \Psi_{n,j} a_{n,j}^\dagger |0\rangle. \qquad (7)$$

Here $|0\rangle$ is the corresponding vacuum state and $\psi_{n,j}$ is the quasiparticle wave function in the site representation, which satisfies the normalization condition $\langle \psi_e | \psi_e \rangle = 1$.

Considering $\langle \psi | H | \psi \rangle$ as the Hamiltonian functional of the quasiparticle wavefunction and of the lattice variables, we obtain a system of nonlinear equations which describe the interaction between the quasiparticle and phonon subsystems. For numerical studies it is convenient to use dimensionless time measured in units of $\hbar/J$, energy measured in units of J, displacements measured in the length unit $l = \hbar\sqrt{2/JM}$, and to express all parameters in the dimensionless units using the following relations:

$$M_{1,2} = \frac{1}{2}M(1\pm m), w_{s,l} = \frac{1}{2}W(1\pm w), J_{s,l} = \frac{1}{2}J(1\pm d), \chi_{s,l} = \frac{1}{2}X(1\pm x), \quad (8)$$

or, respectively:

$$m = \frac{M_1 - M_2}{M}, w = \frac{w_s - w_l}{W}, d = \frac{J_s - J_l}{J}, x = \frac{\chi_s - \chi_l}{X}, X = \chi_s + \chi_l. \quad (9)$$

In these units the dynamic equations for the electron and lattice functions, $\psi_{n,j}$, and $u_{n,j}$, respectively, in the presence of an external electric field $E(t)$, take the form:

$$i\frac{d\Psi_{n,1}}{dt} = \left[-1 + \frac{D}{2} + (n - n_0)\,E(t)\right]\Psi_{n,1} + \frac{1}{2}(1 + d)\Psi_{n,2} + \frac{1}{2}(1-d)\Psi_{n-1,2}$$

$$+ G\left[(1+x)(u_{n,1}-u_{n,2})-(1-x)(u_{n,1}-u_{n-1,2})\right]\Psi_{n,1}$$

$$i\frac{d\Psi_{n,2}}{dt} = \left[-1 - \frac{D}{2} + (n-n_0+b)\,E(t)\right]\Psi_{n,2} + \frac{1}{2}(1+d)\Psi_{n,1} + \frac{1}{2}(1-d)\Psi_{n+1,1}$$

$$+ G\left[(1+x)(u_{n,1}-u_{n,2})-(1-x)(u_{n+1,1}-u_{n,2})\right]\Psi_{n,2},$$

$$\frac{d^2 u_{n,1}}{dt^2} = -\frac{C}{1-m}\left[(1+w)(u_{n,1}-u_{n,2})+(1-w)(u_{n,1}-u_{n-1,2})\right]$$

$$+ \frac{G}{1-m}\left[2x|\Psi_{n,1}|^2 - (1-x)|\Psi_{n-1,2}|^2 + (1+x)|\Psi_{n,2}|^2\right] - \eta\frac{du_{n,1}}{dt},$$

$$\frac{d^2 u_{n,2}}{dt^2} = \frac{C}{1+m}\left[(1+w)(u_{n,1}-u_{n,2})+(1-w)(u_{n+1,1}-u_{n,2})\right]$$

$$+ \frac{G}{1+m}\left[-2x|\Psi_{n,2}|^2 + (1-x)|\Psi_{n+1,1}|^2 - (1+x)|\Psi_{n,1}|^2\right] - \eta\frac{du_{n,2}}{dt}$$

$$(10)$$

Here the intensity of the electric field $E(t)$ is measured in units ea/J and we have defined:

$$G = \frac{Xl}{2J}, \quad C = \frac{\hbar^2 W}{MJ^2}, \quad D = \frac{E_2 - E_1}{J} \tag{11}$$

Moreover we have also introduced into the equations for the atom displacements a damping force, which is described by the terms proportional to η in Eq. (10). This force describes the interaction between the atoms and the thermal bath that generates the dissipation of the energy. In all our simulations we have taken $\eta = 0.2$, unless indicated differently.

3 Dynamics of Solitons in the a.c. Unbiased Field

To study the polaron dynamics, we have first derived the stationary solutions of Eq. (10) in the absence of an external field, i.e., by setting $E(t) = 0$. For our numerical simulations we have assumed free boundary conditions. For properly chosen values of the system parameters, when the polaron size is not too small, Eq. (10) can be studied analytically in the continuum (long-wave) approximation (see, e.g., [2]). In this case one can show that Eq. (10) can be reduced to the Schrödinger equation for the electron wave function in the self-consistent deformation potential. This potential is proportional to the electron probability at a given place and time so that the Schrödinger equation contains a cubic nonlinearity and is known as the nonlinear Schrödinger equation (NLSE). In the leading order approximation the NLSE has the soliton solution.

$$\psi_{n,j}(t) = \sqrt{\frac{(1-P)\kappa a}{4}}\,\frac{\exp(-iEt/\hbar)}{\cosh \kappa(z_{n,j} - R)}, \tag{12}$$

where $z_{n,j}$ are the atom positions along the chain and E and R are the eigenenergy and the c.m. coordinate of the soliton, respectively. The coefficient P in Eq. (12) is the weight coefficient determined by the contribution of the energy sublevels of the upper electron band to the formation of the soliton state and, for the properly chosen values of the parameters, is very small, $P \ll 1$. The localization parameter of the soliton, κ, in the notation (Eq. (9)) is given by the relation:

$$\kappa = (1 - p)\frac{4G^2(1 + x^2 - 2xw)\sqrt{1 + D^2/4}}{aC(1 + d^2)(1 - w)}. \tag{13}$$

We have taken the numerical values of the parameters (Eq. 11) so that the stationary state of the Hamiltonian H_0 (the stationary solution) is close to the analytical solution (Eq. 12) and is self-trapped within a few lattice sites. In particular, this is the case for $G = 0.4, C = 0.22, d = D = 0.1, x = 0.05, w = 0.15, m = 0.3$.

The numerically determined stationary solutions of Eq. (10) at $E(t) = 0$ were then used as the initial conditions (the initial excitations) for their dynamics in the presence of the field at $E(t) \neq 0$. We have studied numerically the time evolution of such an excitation by calculating its profile, half-width and the position of the quasiparticle centre of mass (c.m.) coordinate for various forms of the external periodic unbiased electric field. We have found that the field, itself, does not significantly affect the profile of the soliton, though it causes oscillations of the c.m. of the soliton and of its width. In Fig. 1 we present the values of the c.m. coordinate of a soliton as a function of time for an asymmetric chain ($d = 0.1, D = 0.1$) in

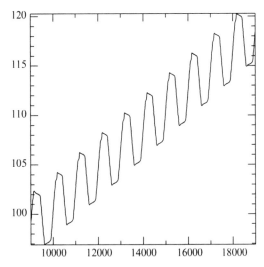

Fig. 1 Position of the c.m. of the soliton as function of time in an external harmonic field at $G = 0.4, C = 0.22, E_0 = 0.08, T = 1,000$ d.u. in an asymmetric chain with: $d = D = 0.1, x = 0.05, w = 0.15, m = 0.3$

the presence of an unbiased harmonic field $E(t) = E_0 \sin(2\pi t/T)$ at $E_0 = 0.08$, $T = 2000$ (note, all parameters are measured in dimensionless units, (d.u.)). This figure demonstrates very clearly that the unbiased harmonic field causes a drift of the soliton, i.e., it generates a uni-directed current in the chain. The fine structure of the drift (the details of the oscillations) depends not only on the parameters of the field and the dissipation coefficient, but also on the asymmetry properties of the system. Our results, shown in Fig. 1, clearly demonstrate the ratchet behaviour of polarons.

As one can see from Fig. 1, in the electric field periodic in time, the soliton trajectory is a sum of oscillations and a drift. The values of the drift velocity, the period and amplitude of oscillations as well as all the details of the drift profile are determined by the values of the chain and field parameters. This effect has a threshold with respect to the intensity of the field and its period, *i.e.*, the effect takes place provided that $E > E_{0,cr}, T > T_{cr}$. Thus, for instance, for $E_0 = 0.08$ and for the other values of the chain parameters as above we have $T_{cr} = 400$ d.u. Moreover, we have also found (see [14]) that the directed current of polarons under harmonic perturbations in molecular systems arises also when there is only an asymmetry in the electronic subsystem.

Note that chains with only one nonzero anisotropy parameter, d or D, possess a reflection symmetry. In such cases a harmonic electric field does cause soliton oscillations around its initial position but does not generate a soliton drift (this conclusion is confirmed by our numerical simulations). Instead, according to the symmetry consideration [9], one can expect the ratchet phenomenon to take place in a symmetric chain only if an external unbiased periodic field is asymmetric in time. As shown below, this is indeed the case,. In particular, Fig. 2 presents the trajectory with a unidirectional (on average) motion of a soliton in the unbiased biharmonic periodic

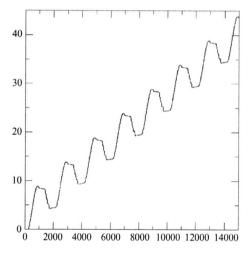

Fig. 2 Position of c.m. of the soliton as function of time in an external biharmonic field at $G = 0.4$, $E_0 = 0.087$, $\beta = 0.6$, $\varphi = \pi/2$, $T = 2,000$ in a symmetric chain at: $C = 0.22$, $d = 0.1, D = x = w = m = 0$

field, $E(t) = E_0 \sin(2\pi t/T) + \beta \sin(4\pi t/T - \varphi)(E_0 = 0.08, \beta = 0.6, \varphi = \pi/2)$. The figure shows that the unbiased biharmonic field causes a directed (on average) motion of solitons in the symmetrical chains.

It is worth to mention here that the case of $D = 0.1, d \neq 0$, applies to polyacetylene with alternating chemical bonds [3], while the case $d = D = 0$ corresponds to a simple chain with one atom per unit cell with the lattice constant $a/2$. From Fig. 2 we see that, even in such simple systems, the ratchet effect takes place!

Let us observe that this dynamics of solitons is induced by the discreteness of the molecular chain which leads to the Peirls–Nabarro potential, which is periodic with the same period as the lattice [15–18]. Notice that the Peierls–Nabarro potential is a characteristic feature of many discrete models and is responsible for the unidirectional motion in various discrete soliton ratchets described by, e.g., the discrete sine-Gordon [10] and the discrete nonlinear Schrödinger-type [13] equations.

The presence of the Peierls–Nabarro potential explains our numerical results and the existence of the threshold with respect to the intensity of the field and its period. For driving amplitudes below the threshold (depinning threshold) the soliton remains pinned to the lattice site and its c.m. oscillates around it. The drift is possible when the intensity of the field exceeds the depinning threshold and the amplitude of the soliton oscillations exceeds the lattice constant (the period of the Peierls–Nabarro barrier). The oscillation amplitude depends on the intensity of the field and on its period and is approximately proportional to the square of the period of the electric field, $A \propto E_0 T^2$. If the period of the external force is very small, this amplitude is less than the period of the ratchet potential and the soliton oscillates within the potential well of the Peierls–Nabarro barrier and, therefore, remains trapped on a lattice site.

To describe the soliton dynamics, both in the homogeneous and in the discrete cases, a collective coordinate approach is often used [11, 19]. In such an approach, a discrete soliton of the form (Eq. 12), is treated as a single particle and the c.m. coordinate of the soliton, $R(t)$, is its collective coordinate. In a molecular chain the dynamic equation for this c.m. coordinate of the soliton, $R(t)$, under the external force, $E(t)$, taking into account the lattice discreteness and the energy dissipation, takes the form:

$$M_s \ddot{R} = \gamma^* \dot{R} + f(R) + eE(t), \tag{14}$$

where $M_s = m^* + \Delta m$ is the effective mass of the soliton, 'dressed' with phonons, $\gamma^* \propto \eta$ [20], and $f(R) = -dU_{PN}/dR$ with U_{PN} being the periodic Peierls–Nabarro potential, $U_{PN}(R + a) = U_{PN}(R)$ (a is the lattice constant), [15]. Equation (14) is of a type well known to lead to the ratchet phenomenon [1] i.e. to a unidirectional (on average) motion of a particle whose trajectory is a limit cycle phase locked to the external periodic driver $E(t)$.

In a chain with one atom per unit cell, the Peierls–Nabarro potential is given by the expression $U_{PN}(R) = U_0 \cos(2\pi R/b)$, where b is the lattice constant of the chain, and the height of the barrier depends on the electron–phonon coupling of the system [15]. In a diatomic chain, this barrier is described by two periodic terms, and is given by:

$$U_{PN}(r) = U_1 \cos(2\pi r) + U_2 \cos(4\pi(r + \theta)), r = \frac{R}{a}, \qquad (15)$$

where U_i depend on the square of the electron–phonon coupling, similarly to the case of a uni-atomic chain.

It is known that the broken spatial symmetry in the ratchet potential, which in our case is the periodic potential of the Peierls–Nabarro barrier, and/or of the time correlations in the driving force are crucial and lead to the ratchet effect [1, 9]. Our numerical results, shown in Fig. 1, clearly demonstrate the ratchet behaviour of polarons in asymmetric diatomic molecular chains under an unbiased harmonic (temporarily symmetric) field $E(t) = E_0 \sin(2\pi t/T)$. This indicates that the Peierls–Nabarro potential (Eq. 15) in asymmetric chains is asymmetric. To show this we have studied the dynamics of the soliton governed by the discrete Eq. (11) in a constant field, $E(t) = E_0 = const$. We have found that $E_{th}(E) = E_{th}(-E)$ for symmetric chains, while $E_{th}(E) \neq E_{th}(-E)$ for asymmetric ones.

As is clear from Eq. (14), the soliton is pinned by the lattice and, in a static electric field, it can move if the external field exceeds the pinning threshold value. In a symmetric chain this threshold is symmetric for fields E and $-E$, but is asymmetric for asymmetric chains.

Results shown in Fig. 3, demonstrate that, indeed, as is well known from the analysis of Eq. (14) [9, 21], the uni-directional motion of the soliton corresponds to the limit cycle which is phase-locked to the frequency of the external driver: $R(t + T_c) = R(t) + ka$, $\dot{R}(t + T_c) = \dot{R}(t)$ (a is the period of the ratchet potential, T_c is the period of the cycle). On this orbit, the average soliton velocity is expressed as

$$\langle V \rangle = \frac{1}{T_c} \int_0^{T_c} \dot{R} dt = \frac{ka}{lT} \qquad (16)$$

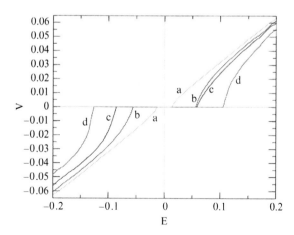

Fig. 3 Dependence of the average soliton velocity on the constant electric E field for $G = 0.4$, $C = 0.22$, $b = 0.5$ and: (a) $d = D = 0$; (b) $D = 0$, $d = 0.1$; (c) $D = 0.1$, $d = 0.1$; (d) $D = 0.2$, $d = 0.1$

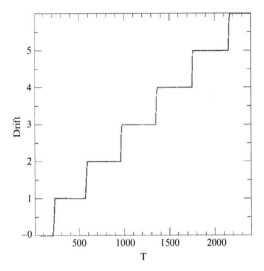

Fig. 4 Displacement (in the lattice units) of the c.m. of the soliton as function of the period of an external harmonic field at $G = 0.4, C = 0.22, d = D = 0.1, x = w = m = 0, E_0 = 0.88$

with k and l being integer numbers. In this resonance regime, the soliton passes k sites during l periods of the external drive so that, except for a shift in space, its profile is completely reproduced after this time interval.

The dependence of the soliton shift, $D = <V> T$, on the period of the applied field, obtained by numerical simulations of Eq. (10), is shown in Fig. 4. We note that the soliton displacement is not a smooth monotonic function of T but a piecewise function with plateaus. The plateau values of the soliton velocity satisfy Eq. (16) and correspond to dynamic regimes, which are limit cycles with rotation numbers $(\kappa = 1, 2, \ldots; l = 1)$ that are phase-locked to the driver.

The main characteristics of the soliton dynamics, obtained in numerical simulations of Eq. (10) and presented in Fig. 4, can be explained in terms of the solutions of the Eq. (14) for the collective coordinate. To demonstrate this we have studied Eq. (14) numerically and we present our results in Fig. 5.

For these studies we have chosen the ratchet potential in the form Eq. (15) and found that Eq. (14) describes qualitatively and often also quantitatively the dynamics of the soliton described by the discrete system of Eq. (10), as can be seen by comparing Fig. 5a and b with Figs. 2 and 4, respectively. The best agreement corresponds to the ratchet potential parameters $U_2/U_1 = 0.25, \varphi = -1/8$ at which the profile of the potential Eq. (15) has a well manifested asymmetry – see Fig. 6.

Generally speaking, the possibility of a soliton to drift in an unbiased field and the properties of this drift is a problem involving many parameters. To analyze the corresponding space diagram, we can only present some two-dimensional crossections of the corresponding parameter space. Some of such diagrams are given in [14], from which it follows that soliton oscillation amplitude as a function of the absorption, has plateaux due to the discreteness of the chain. The lengths of the

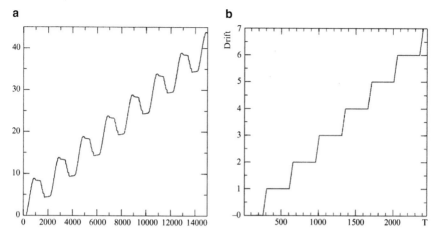

Fig. 5 Position of the c.m.c. as function of time in the field $E_e = 8.7$, $T = 2000$, $\gamma = 400$, (**a**) and the dependence of the velocity of the drift on the period of the field $E_e = 4.5$, $\gamma = 100$, (**b**) in the ratchet potential with the parameters $u_1 = 1.0$, $u_2 = 0.25$, $\theta = -1/8$

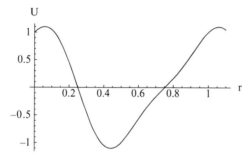

Fig. 6 Profile of the Peierls–Nabarro potential in an asymmetric chain at the parameters $u_1 = 1.0$, $u_2 = 0.25$, $\theta = -1/8$

plateaux depends on the parameters of the system. Beyond this plateaux, the amplitude decreases with the increase of the energy dissipation.

Soliton drift results from the superposition of various oscillating processes with different periods. As a result, soliton dynamics is a highly nonlinear processes, and the dependence of the soliton velocity on the damping coefficient and on the field intensity is nonmonotonous and the velocity can even change its sign!

In molecular chains, a soliton motion in the Peierls–Nabarro potential is accompanied not only by the oscillations in time of the soliton velocity but also by the oscillations of a soliton localization parameter κ [15] which determines also the soliton amplitude (Eq. (13)). Overcoming the Peierls–Nabarro barrier, the soliton gets broader and therefore, its amplitude decreases.

The dependence of the average soliton velocity on the damping coefficient and on the driver intensity, as has been reported in [14], strongly depends on the values of the chain parameters and of the parameters of the driving force $E(t)$. The important parameter here is the parameter of the asymmetry of the biharmonic field, φ.

4 Conclusions

In conclusion, our study has shown that the ratchet effect can take place in one-dimensional molecular systems which support the existence of large polarons (solitons). Such self-trapped electron states are formed at intermediate strengths of the electron interaction with lattice deformations. The coupling constant of this interaction has to be large enough as the Peierls–Nabarro barrier (Eq. (15)) is essential for the dynamics of the soliton; on the other hand, this coupling cannot be too large to prevent the formation of small polarons, whose transport properties are qualitatively different from those of the solitons. There is a large class of low-dimensional molecular systems, including biological macromolecules and some conducting polymers and inorganic compounds, which possess large polarons, as it has been demonstrated in many experiments.

It is also worth pointing out that, in a certain sense, in spite of their different physical nature, molecular ratchets, considered here, are similar to the ratchets of the discrete cavity optical solitons [13], based on the discrete nonlinear Schrödinger-type equation (stationary molecular solitons can be described by this equation as well).

We would like to add that, similarly to the deterministic fields considered here, symmetric white noise [22] can also cause the uni-directed current of solitons in low-dimensional molecular systems, though the dynamics of solitons in such cases is less symmetric and more complicated than for the harmonic fields (we plan to report on this in the near future). The role of the noise in inducing the ratchet effect is important for the understanding of the functioning of biological motors. Molecular solitons, which are the charge and energy carriers during the metabolic processes in biological systems according to the Davydov hypothesis [2], are formed in α-helical proteins. Such protein macromolecules possess a highly asymmetric structure and, so, their Peierls–Nabarro barrier is asymmetric too. Moreover, these macromolecules are immersed in the cellular cytoplasm and are subject to thermal fluctuations. Therefore, in these systems the presence of a symmetric stochastic noise can result in the formation of a directed current of solitons, i.e., such systems can be good candidates for molecular motors.

On the other hand, there is a class of low-dimensional compounds, such as polyacetylene (PA), polydiacetylene (PDA), polytiophene (PT), etc., which provide experimental evidence for the existence of large polarons and bipolarons [6, 7, 23]. Based on our results we expect that, in these compounds, the unbiased alternating electric field can induce a directed current. In the compounds with an asymmetric unit cell, such as polyphenylene–venillene, polythienylene–venillene, this directed current can be induced by a harmonic periodic field, while in compounds with a symmetric unit cell, such as PA, PDA, PT, polyphenylene, polypyrrole, polyaniline, polyfurane, polysilans etc., this effect can be observed in periodic biharmonic, or, in general, in asymmetric periodic in time fields.

Moreover, the ratchet effect can take place also in compounds with charge density waves (CDW) (see review [24]). Similarly to the case of polarons, such CDWs are also described within the adiabatic approximation for the many-electron wavefunctions and in the limit of low concentration of electrons their CDW wavefunction reduces to the wavefunction of separated bisolitons [25], as was shown in [26]. In fact, the pinning of the CDW by the lattice in these systems does exist and it has been proved that the sliding mode of the CDW is possible in constant fields only above some threshold [24]. Indeed, the d.c. signal produced by biharmonic microwaves in TTF-TCNQ (tetrathiofulvalene-tetracyanoquinodimethane) has been experimentally observed [27]. Some features of experimental results were explained in [28] where it was assumed that the charge transport in TTF-TCNQ occurs via a rigidly sliding CDW (a CDW was modelled as a Brownian motion of a classical particle in the periodic potential within the collective coordinate approach, Eq. (14). This agrees qualitatively with our results, though a further detailed study still has to be performed.

Acknowledgments The authors thank Y.O. Zolotaryuk for fruitful discussions. We acknowledge a Royal Society travel grant, and one of us, W.J.Z., acknowledges a NATO ARW "Molecular Self-Organization in Micro-, Nano, and Macro Dimensions: From Molecules to Water, to Nanoparticles, DNA and Proteins", dedicated to Alexander S. Davydov 95th birthday, travel grant.

References

1. P. Reinmann, Phys. Reports, **361**, 57 (2002)
2. A.S. Davydov, Solitons in Molecular Systems. Dordrecht, Reidel (1985)
3. W.P. Su, J.R. Schriefer, A.J. Heeger, Phys. Rev. Lett. **42**, 1698 (1979); Phys. Rev. B **22**, 2099 (1980)
4. A.J. Heeger, S. Kivelson, J.R. Schrieffer, W.-P. Su, Rev. Mod. Phys. **60**, 781 (1988)
5. M.J. Rice, S.R. Phillpot. Phys. Rev. Lett. **58**, 937 (1987); M.J. Rice, S.R. Phillpot, A.R. Bishop, D.K. Campbell. Phys. Rev. **B 34**, 4139 (1986)
6. J.L. Bredas, J.C. Scott, K. Yakushi, G.B. Strreet, Phys. Rev. B, **30**, 1023 (1984)
7. T.-C. Chung, J.H. Kaufman, A.J. Heeger, F. Wudle, Phys. Rev. **B 30**, 702 (1984)
8. E.M. Cowell, S.V. Rakhmanova, PNAS, **97**, 4556 (2000)
9. S. Denisov, S. Flach, A.A. Ovchinnikov, O. Yevtushenko, Y. Zolotaryuk, Phys. Rev. E **66**, 041104 (2002); S. Flach, O. Yevtushenko, Y. Zolotaryuk, Phys. Rev. Lett. **84**, 2358 (2000); S. Flach, Y. Zolotaryuk, A.E. Miroshnichenko, M.V. Fistul Phys. Rev. Lett. **88**, 184101 (2002); M. Salerno, Y. Zolotaryuk, Phys. Rev. E **65**, 056603 (2002)
10. A.V. Ustinov, C. Coqui, A. Kemp, Y. Zolotaryuk, M. Salerno, Phys. Rev. Lett. **99**, 087001 (2004); Y. Zolotaryuk, M. Salerno, Phys. Rev. E **73**, 066621 (2006)
11. L. Morales-Molina, N.R. Quintero, A. Sánchez, F.G. Mertens, Chaos **16**, 013117 (2006)
12. Ya. Zolotaryuk, M. Salerno, Phys. Rev. E **73**, 066621 (2006)
13. A.V. Gorbach, S. Denisov, S. Flach, Optics Lett. **31**, 1702 (2006)
14. L. Brizhik, A. Eremko, B. Piette, W. Zakrzewski. J. Phys.: Cond. Matt. 20 (2008) 255242
15. L. Brizhik, L. Cruzeiro-Hansson, A. Eremko, Yu. Olkhovska, Phys. Rev. **B 61**, 1129 (2002); Ukr. J. Phys. **43**, 667 (2003)
16. V.A. Kuprievich, Physica D **14**, 395 (1985)
17. A.A. Vakhnenko, Yu.B. Gaididey, Teor. Mat. Fiz. (in Russian) **68**, 350 (1986)
18. O.O. Vakhnenko, V.O. Vakhnenko, Phys. Lett. A **196**, 307 (1995)
19. O.M. Braun and Y.S. Kivshar, Phys. Rep. **306**, 2 (1998)
20. A.A. Davydov, A.A. Eremko, Teor. Mat. Fiz. **43**, 367 (1980) (in Russian)
21. G.M. Zaslavsky, Physics of Chaos in Hamiltonian Systems. Imperial College Press, London (1998)

22. J. Luczka, R. Bartussek, P. Hänggi, Europhys. Lett. **31**, 431 (1995); J. Luczka, T. Chernik, P. Hänggi, Phys. Rev. E **56**, 3968 (1997)
23. E.G. Wilson, J. Phys. C **16**, 6739 (1983)
24. G. Grüner, Rev. Mod. Phys. **60**, 1129 (1988)
25. L.S. Brizhik, A.S Davydov. Sov. J. Low Temp. Phys. **10**, 748 (1984); L.S. Brizhik, Sov. J. Low Temp. Phys. **12**, 437 (1986)
26. A.A. Eremko, Phys. Rev. B **46**, 3721 (1992); *ibid* **50**, 5160 (1994)
27. K. Seeger, W. Maurer, Sol. St. Comm. **27**, 603 (1978)
28. W. Wonneberger, Sol. St. Comm. **30**, 511 (1979)

Some Properties of Solitons*

L.S. Brizhik, A.A. Eremko, L.A. Ferreira, B.M.A.G. Piette,
and W.J. Zakrzewski

Abstract We present a general review of some aspects of the dynamics of topological solitons in 1 and 2 dimensions. We discuss some recent work on the scattering of solitons on potential obstructions and in the presence of some external fields.

Keywords Ratchet effect · Polaron · Solitons · Kinks · Breathers · Landau-Lifshitz equation · Peierls–Nabarro potential field

1 Introduction

Solitons arise in various areas of applied mathematics and in the mathematical description of some processes in physics [1] and in biology [2].

In many applications of mathematics to the description of physical processes one uses either point like objects or plane waves. Solitons are different. They describe objects that are localised in space (but not localised to a point). So their energy density is described by a function which is essentially nonzero in a finite region; i.e. is significantly nonzero only in a small region and goes to zero, exponentially or as an inverse power, as one moves away from this region.

L.S. Brizhik and A.A. Eremko
Bogolyubov Institute for Theoretical Physics, 03680 Kyiv, Ukraine
e-mail: brizhik@bitp.kiev.ua; eremko@bitp.kiev.ua

L.A. Ferreira
Instituto de Física de São Carlos, IFSC/USP, São Carlos, SP, Brazil
e-mail: laf@ifsc.usp.br

B.M.A.G. Piette and W.J. Zakrzewski (✉)
Department of Mathematical Sciences, University of Durham, Durham DH1 3LE, UK
e-mail: B.M.A.G.Piette@durham.ac.uk; W.J.Zakrzewski@durham.ac.uk

*Talk given by W.J. Zakrzewski at the NATO ARW 2008 (Davydov memorial meeting) in Kiev — June 2008

The stability of solitons is often guaranteed by topological considerations, most often associated with the topology of $S^N \rightarrow S^N$ maps.

The simplest examples of such maps (in 1 and 2 dimensions) involve, respectively, the sine-Gordon kinks and the solitons (baby skyrmions) of the S^2 sigma model. In the sine-Gordon case the Lagrangian density is given by

$$\mathcal{L} = \partial_\mu \phi \cdot \partial^\mu \phi - \lambda \sin(\phi). \tag{1}$$

Solutions of the sine-Gordon equations of motion are well known [3]. They involve kinks and antikinks, which are topological solitons; breathers, which can be thought of as interacting bound states of kinks and antikinks, further bound states of kinks and breathers and many other solutions, less interesting from our point of view. The Lagrangian of the sine-Gordon model is relativistically covariant; thus if we have its static solution we can always obtain a non-static solution by Lorentz-boosting the static one. Thus starting with a static kink we can obtain a moving kink. The same is true with nonstatic solutions, like stationary breathers. They can be Lorentz boosted to generate moving breathers. Thus this Lorentz covariance can be used to put all kinks and breathers in their centre of mass; i.e. when we talk about the scattering of two kinks we consider their motion relative to their centre of mass. This is important as we will mention later on.

In two spatial dimensions (i.e. for $N=2$) the solitons are based on sigma models. In the relativistic case the Lagrangian is given by:

$$\mathcal{L} = \partial_\mu \phi \cdot \partial^\mu \phi - \theta_S \left[(\partial_\mu \phi \cdot \partial^\mu \phi)^2 - (\partial_\mu \phi \cdot \partial_\nu \phi)(\partial^\mu \phi \cdot \partial^\nu \phi) \right] - V(\phi), \tag{2}$$

where for V we can take any 'simple' function of ϕ_3 which vanishes as $\phi_3 = 1$. In most of our discussions we have used

$$V(\phi) = \mu(1 - \phi_3^2). \tag{3}$$

Moreover, to have topology we require that the vector ϕ lies on the unit sphere S^2, i.e. that $\phi \cdot \phi = 1$. To have a finite potential energy the field at spatial infinity is required to go to $\phi_3 = \pm 1$, $\phi_1 = \phi_2 = 0$. In this work we choose "the vacuum" to be defined as $\phi_3 = +1$. The model with this choice of the potential, i.e with V given by Eq. (3) is called the "new baby skyrme model" [4].

The three terms in Eq. (2) are, from left to right, the pure S^2 sigma model, the Skyrme and the potential term. The last two terms are needed to stabilize the solitons. They have no influence on the topology — which is still based on the topology of $S^2 \rightarrow S^2$ maps as imposing the spatial infinity boundary condition has defined a one-point compactification of R_2, allowing us to consider ϕ on the extended plane $R_2 \bigcup \infty$ topologically equivalent to S^2. Incidentally, potential terms other than Eq. (3) have also been studied [4]. The results, in their cases, are slightly different but their generic features are the same. Hence here we discuss the results obtained for Eq. (3).

2 Dynamics

The dynamics of the sine-Gordon kinks is well known. We have analytical expressions for the two kinks moving towards each other and we can easily see what happens. In fact, the kinks reflect from each other, as can be seen by looking at the time dependence of their energy density. They come close to each other and then reflect. This can be further checked by performing the numerical simulation of the time evolution of two static kinks some distance appart. In this case the kinks gradually move away from each other thus showing that the forces between them are repulsive. If one sends them towards each other they gradually slow down and finally repel. Of course, as the motion is in one dimension not much can happen (they can repel or pass through each other, and in this case they choose the first option).

In two dimensions we have more possibilities, as the solitons can be sent towards each other at an impact parameter. But on top of that we often find applications when the dynamics is not based on the relativistic Lagrangian mentioned in the previous section but, instead, is determined by the Landau–Lifshitz equation.

In that case the equation of motion is given by

$$\frac{\partial \phi}{\partial t} = \phi \times \frac{\partial L}{\partial \phi} \tag{4}$$

where L now stands for only the spatial part of $\mathcal{L}$, i.e of Eq. (2).

The dynamics of both cases is very different. In the relativistic case, when the solitons are sent towards each other at zero impact parameter (head-on) we have the familiar 90° scattering. Thus, in this case the system evolves in such a way that after the scattering the two outgoing solitons are moving in the direction at 90° with respect to their motion before the scattering. When the solitons are sent at a very large impact parameter they miss each other completely; for intermediate values of the impact parameter their scattering interpolates between these two extremes.

The head-on 90° scattering of the solitons has been explained in many ways; the most compelling involves the indistinguishability of solitons [1]. As the system of two solitons is described by a function which is symmetric with respect of the interchange of their positions this is built into their phase space which, in terms of the relative position, is really described by R^2 mod a reflection with respect to the line joining their positions. Hence, effectively, the space is $\frac{R^2}{Z_2}$, where Z_2 describes this reflection, and so it is a cone. A straight line motion in this space, going through the vertex of the cone, is described by a 90° motion when viewed in R^2. At other values of the impact parameter the motion still involves a straight line but this time not going through the vertex etc.

In the case involving the Landau–Lifshitz equation the situation is completely different.

First of all the equations involve only first order time derivatives and so the motion takes place in a lower dimensional phase space.

This has been analysed in great detail by Papanicolaou and Tamaras [5] who showed that when one has a system of two solitons one can introduce $\mathbf{r} = (x_1, x_2)$ — a 2 dimensional vector describing their relative position and the relative momentum $\mathbf{p}$ and the momentum satisfies

$$p_i \sim \alpha \, \epsilon_{ij} x_j. \tag{5}$$

Here, α is nonzero if the system has a nonvanishing topological charge, which is the case when we have two solitons. The equation of motion is then of the form

$$\frac{d^2 x_i}{dt^2} \sim \alpha \epsilon_{ij} \frac{dx_j}{dt} \tag{6}$$

resulting in a motion along a circle.

Incidentally, in three spatial dimensions the dynamics of solitons is even more complicated (see, e.g. [1]) but, interestingly, many aspects of it can be related to the dynamics in two dimensions (through projections into various planes).

All this discussion concerned solitons moving in a free space, i.e. in a space with no potential obstructions. In the next section we discuss some aspect of the scattering of solitons when we do have a potential obstruction—either in the shape of a potential bump or a potential hole. Of course, such an obstruction is located in some region of space hence the system looses its translational invariance. So all our discussion will be performed in the frame in which the obstruction is at rest.

3 Potential Obstructions

There are various ways of introducing a potential obstruction; a hole or a barrier. However, given that the soliton field, strictly speaking, is never zero, even though it vanishes (exponentially, or as power) as we move away from its position, this potential has to be introduced in such a way that it does not change the "tail" of the soliton, i.e. it has to vanish when, in one dimension, $\phi \sim 0$ or π or, in two dimensions, when $\phi_3 \sim 1$.

3.1 Sigma Models

A possible way of introducing such an obstruction is to add an extra term to the Lagrangian. This extra term should vanish outside the obstruction, i.e. when $\phi_3 = 1$. Of course, there are many terms that we could use but given that our Lagrangian already contains a term with this property in [6] we exploited this fact and chose to add $\alpha(1 - \phi_3^2)$ in some region of x and y. In fact, we chose our obstruction to be located in some finite region of x, say at positive x and to be independent of y. So it is in the form of a trough in the "hole" case or a dam in the "barrier" case. In our work we chose the obstruction to be constant and located in a small range of

x, for $x \in (-L/2, L/2)$. This, effectively, corresponds to taking μ in the original Lagrangian to be given by μ_0 for x in the range of the obstruction and μ_1 elsewhere. Then the case of $\mu_0 > \mu_1$ corresponds to a barrier (dam), and for $\mu_0 < \mu_1$ we have a hole (trough).

Then sending the soliton from a point well away from this obstruction, i.e. initially placed at some sufficiently negative x, in the positive x direction, we can study the effects of the scattering on the obstruction.

We have performed many numerical simulations varying both the sign and value of $\mu_0 - \mu_1$ and the velocity of the incoming soliton. We have found that when we have a barrier the scattering is very elastic with the system essentially converting all its kinetic energy into the potential energy to 'climb the potential hill' and then releasing this energy back as either the kinetic energy of the reflected or the transmitted soliton. Hence the velocity of the outgoing soliton was always very close, in magnitude, to the original velocity.

For the hole the situation was different. Varying the incoming velocities we saw many trajectories. Sometimes the solitons were transmitted, sometimes they were trapped in the hole, sometimes they were reflected. In Fig. 1, we present plots of

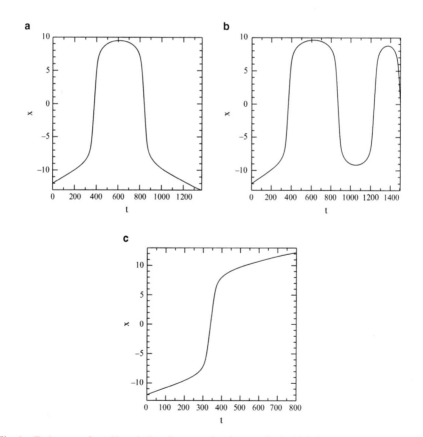

Fig. 1   Trajectory of a soliton during the scattering for a well of width $L = 10$ and $\mu_1 = 1$. and $\mu_0 = 0.8$. Here $\lambda = 0.5$ and (**a**) $v = 0.0102$, (**b**) $v = 0.0106$, (**c**) $v = 0.012$

positions in x as a function of time seen in three representative simulations (i.e. started with different values of initial velocities and corresponding to a different behaviour).

Thus we have seen that in addition to the transmission the soliton can get trapped in the hole. This trapping can lead to it being permanently stuck in the hole (with the soliton radiating its excess of energy) or after a bounce or two in the hole the soliton can be ejected either forwards or backwards so that the whole process looks like a transmission or a reflection! We have looked at this process in great detail [7]. In [7] it is argued that the interaction proceeds through the excitement of the vibrational modes of the soliton. Moreover, [7] presents a simple model of these modes and of their interaction with the soliton—we will comment on this later on. Before that, we look at other models and in particular at the sine-Gordon model in one dimension. This model has no genuine vibrational modes so we might have expected the behaviour of its solitons to be somewhat different. In the next section we show that this is, however, not the case.

3.2 Sine-Gordon Model

In the studies of the sine-Gordon case the obstruction potential has been introduced in two different ways — either by making λ in Eq. (1) position dependent [8] or by altering the basic metric [9]. Here we discuss the results reported in [8]; the results of [9] are qualitatively very similar.

Both papers have found that in the sine-Gordon model, like in the sigma model in two dimensions, the solitons can get trapped, be transmitted and bounce back. The process is inelastic and depends on the initial condition of the soliton. Of course, it also depends on the size and the depth of the hole. If the initial condition is taken in the form of a kink moving with a given velocity then there is a well defined critical velocity above which the kink get transmitted (with a certain loss of energy). Below this critical velocity it is trapped or reflected. The ranges of velocities, at which the kink is reflected are very narrow. As the hole becomes shallower the critical velocity decreases and as the hole becomes narrower the number of velocity windows at which we observe reflections gets larger. All the details are given in [8].

In Fig. 2 we present a plot of the outgoing velocity as a function of the incoming velocity in the case of a relatively narrow well (λ in the well is reduced from 1 down to 0.8). The hole is relatively narrow — i.e. a soliton fits in it about three times.

Looking in detail at the velocities at which the reflections take place we note that these velocities come in several small 'windows' which are very narrow. If we try to explain them by the excitation of the vibrational modes of the solitons we have a problem with the sine-Gordon model, in which a soliton has no such modes but the reflections do take place. Of course, although the model has no vibrational modes it has pseudovibrational ones which can get excited and can radiate (similar conclusions, in a different context, were reached by Romanczukiewicz [10]). An example

Fig. 2 Outgoing velocity of the kink as a function of its initial velocity

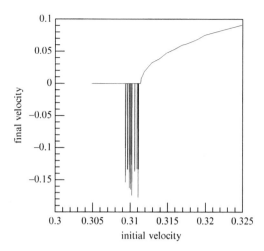

of such a mode is the variation of the slope of the kink. The usual kink solution is given by

$$\varphi(x) = 2 \arctan(\exp(\theta(x - x_0))), \tag{7}$$

where x_0 is the kink's position and θ is its slope. For Eq. (7) to be a solution of the equation of motion which follows from Eq. (1) we need to set $\theta = \lambda$. However, if we put θ different from λ we excite the mode which corresponds to the variation of θ. In fact, when the kink enters the hole, where λ is different, it automatically tries to adjust its slope and so it excites this mode. Of course, when this mode of the kink is excited — the kink begins to radiate and it is this interaction of the radiation with the kink itself which is reponsible for the final outcome of the scattering process.

4 Breathers

The sine-Gordon model, in addition to the kink, also possesses breathers as its solutions. These breathers are given by

$$\phi(x,t) = 2 \arctan\left(\frac{\sin(\omega t)}{\omega \cosh(\sqrt{1 - \omega^2}\, x)}\right). \tag{8}$$

Their energy is $16\sqrt{1 - \omega^2}$ and they can be thought of as bound states of a kink and an antikink with the binding increasing as $\omega \to 1$. Hence it is interesting to see what can happen as a breather is sent towards a hole; clearly, it can scatter by changing its ω or it can split leaving a kink (or an antikink) in the hole and allowing its partner to scatter forwards or backwards.

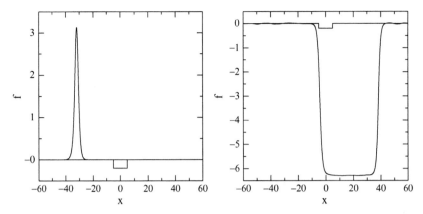

Fig. 3 Breather profile for $L = 10$, $\lambda = 0.8$, $v = 0.1$ and $\omega = 0.1$. (**a**) $t = 1$ before scattering with the well, (**b**) $t = 500$, the breather has split into a kink and a trapped anti-kink

We have performed several numerical experiments of such systems and they are described in great detail in [11]. Here we will mention only some of them, sending the interested reader, who would like to see more details, to [11]. As expected our results have shown that the breathers do get trapped, pass with, in general, a different ω (note that increasing ω releases some energy), or split with either a kink or an antikink being ejected from the hole. Sometimes the hole can also lead to the creation of a further breather.

In Fig. 3 we present a couple of pictures showing a breather (of frequency $\omega = 0.1$ moving with velocity $v = 0.1$) just before a scattering on a hole (of depth $\lambda = 0.8$ and size $L = 10$), and some time later, in the case when the interaction with the hole leads to the splitting of the breather.

5 Simple Models

Let us mention now two simple models which partially explain what we have seen in our numerical simulations.

5.1 Two Dimensional Case

First we consider the sigma model in two dimensions. In this case the soliton possesses many vibrational modes [12] and so our model treats the soliton as a system of four masses which are connected to each other by elastic springs [7]. To check its validity we took our system of four masses and sent it towards the potential hole. As the system falls into the hole the masses begin to oscillate. These oscillations are then found to model the soliton vibrations seen in full simulations. Some energy is transferred to these oscillations and if the energy of the centre of mass is too low the system becomes trapped in the hole. Sometimes, when the system reaches one of the edges of the hole it happens to be in a state that allows the energy of the oscillations to be transferred back to the system as a whole (the energy of its centre of mass) and the soliton can come out. Whether this happens or not depends on the flow of the energy between the vibrational modes and on the kinetic and potential energy of the soliton. As we showed in [7], the model reproduces quite well the main features of the scattering pattern seen in the full simulations.

5.2 Sine-Gordon Case

In the sine-Gordon case we do not have vibrational modes so our model is different. To construct it we looked at the old results on the scattering of kinks on point in-purities [13] (showing a similar trapping/transmission/reflection pattern) and their recent explanation by Goodman and collaborators [14]. In [14] the authors explained the observed results by invoking an interaction of the kink with the oscillation of the vacuum (around the impurity) which was described by a standing wave whose amplitude was a further degree of freedom (in addition to the position of the kink). This interaction proceeded through the excitation of the quasimode of the kink—namely its slope. The model of Goodman et al. reproduced all the features of the results of the original simulations reported in [13] and so the two models discussed by us in [8] are based on the adaptation of the ideas of Goodman et al. to our case. In both models we introduced degrees of freedom describing various standing waves in the hole (in one model the waves were restricted to the edges of the hole and in the other they described the global standing waves in the hole). The waves in both models were chosen somewhat arbitrarily as the idea was not to reproduce the pattern in any detail but just to check whether the mechanism of Goodman can be applied to our case too.

In fact both models worked surprisingly well. They reproduced the pattern quite well, although the critical value of velocity was a little too high. Given that these models involved only very few (3 or 4) degrees of freedom we were very encouraged by these results; they require further work to understand better which modes are important and which are less so.

6 Further Solutions (Wobbles etc.)

However, the kinks and breathers are not the only finite energy solutions of the sine-Gordon model. In fact, the model possesses also solutions which describe bound states of kinks and breathers. One of such solutions, the 'wobble' was recently discussed in detail by Kälberman [15].

The paper of Kälberman presents an analytic form of this solution, shows that it describes a static kink in interaction with a breather, and then discusses some of its properties.

Recently, three of us [16], have looked at these more general solutions of the sine-Gordon model describing kinks and breathers. The work in [16] was based on the Hirota method [17] of deriving such solutions. Using this method [16] has shown that

$$\phi = 2 \, \text{ArcTan} \, \frac{\left[2 \{\cotan\theta\} \cos \Gamma_I + e^{\tilde{\Gamma}_3} \{ e^{-\Gamma_R} + \rho^2 \, e^{\Gamma_R} \} \right]}{\left[\{ e^{-\Gamma_R} + e^{\Gamma_R} \} - 2 \{\cotan\theta\} \rho \, e^{\tilde{\Gamma}_3} \cos\{\Gamma_I + \varphi\} \right]}, \qquad (9)$$

where

$$\tilde{\Gamma}_3 = \gamma_K \{ x - v_K t \} + \eta_K, \qquad \gamma_K = \cosh \alpha_K, \qquad v_K = \tanh \alpha_K, \quad (10)$$

$$\rho = \frac{\cosh\{\alpha_B - \alpha_K\} - \cos \theta}{\cosh\{\alpha_B - \alpha_K\} + \cos \theta} \quad \text{and} \quad \varphi = 2 \, \text{ArcTan} \, \frac{\sin \theta}{\sinh\{\alpha_B - \alpha_K\}}. \quad (11)$$

and

$$\Gamma_R = \frac{1}{\sqrt{1 - v_B^2}} \cos \theta \, \{ x - v_B t \} + \eta_B,$$

$$\Gamma_I = \frac{1}{\sqrt{1 - v_B^2}} \sin \theta \{ t - v_B x \} + \xi_B$$

describes the most general kink/breather field configuration in which the kink and the breather have arbitrary velocities (and so move with respect to each other).

If one then takes

$$\eta_K = \eta_B = v_B = v_K = 0, \qquad \xi_B = \frac{\pi}{2}$$

and denotes

$$\omega = \sin \theta, \qquad -\frac{\pi}{2} \le \theta \le \frac{\pi}{2}$$

then

$$\rho = \frac{1 - \sqrt{1 - \omega^2}}{1 + \sqrt{1 - \omega^2}}, \qquad \varphi = \pm \pi. \qquad (12)$$

and then the expression above takes the form

$$\phi = 2 \, \text{ArcTan} \, \frac{\left[\frac{\sqrt{1-\omega^2}}{\omega} \sin\{\omega t\} + \frac{1}{2} e^x \{ e^{-\sqrt{1-\omega^2} x} + \rho^2 \, e^{\sqrt{1-\omega^2} x} \} \right]}{\left[\cosh\{ \sqrt{1-\omega^2} x \} + \frac{\sqrt{1-\omega^2}}{\omega} \rho e^x \sin\{\omega t\} \right]}, \qquad (13)$$

where ω is a frequency varying from -1 to 1. This agrees with the expression given by Kälberman which describes a stationary field configuration in which the kink and the breather sit on 'top of each other' and are not in relative motion.

As is clear from Eq. (13) the field configuration depends on one parameter (the frequency of the breather) and so, the stability of this field configuration was studied in [16]. This was done by calculating ϕ and its time derivative from Eq. (13) which were then used as the initial condition for the simulations. The results of these simulations were in complete agreement with the analytical expression thus showing that the solution was stable with respect to small perturbations (due to the discretisations).

The wobble was also found to be stable with respect to larger perturbations. To establish this the original slope of the kink was changed: (i.e. in the expression (13) the factor $exp(x)$ was replaced by $exp(\lambda x)$ where $\lambda \neq 1$) and several simulations with λ ranging from 1.05 to 1.3 were performed.

The 'new' initial conditions corresponded to an incorrect initial field configuration and so the system possessed some extra energy. The system then evolved towards a stable wobble emitting some radiation which was sent out towards the boundaries of the grid. For λ close to one—the perturbations were small—hence the system returned to its initial configuration (with $\lambda = 1.0$). For larger values of the perturbation the system was more perturbed and often not only kept on sending out its excess of energy but also, at regular intervals, altered its frequency of oscillation (increasing it) which allowed it to send out even more radiation. In Fig. 4 we present the plots of the time dependence of the total energy the potential energy as seen in the simulation in which λ was set at 1.15.

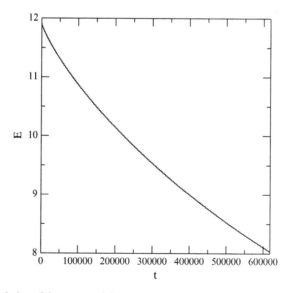

Fig. 4 Time evolution of the energy of the perturbed wobble as seen in a simulation started with $\lambda = 1.15$

One can also study what happens when one sends the initial ('wobble') configuration towards a potential hole. This was also studied in [16] and it was found that, for a single breather [16], the final outcome can be one of many possibilities. Namely, the hole can break the breather or it can separate the breather from the kink (as mentioned in [16] in one simulation the kink was trapped in the hole while the breather bounced off the kink and returned to where the wobble originally came from).

Even more interesting results were found when the initial field configurations involved a kink with more breathers. Such configurations are stable and [16] gives an explicit expression for a configuration corresponding to a kink and two breathers which is then shown to be stable.

7 The Energy

As we mentioned before the energies of the kink or the breather configurations have very simple forms. This is indeed also the case for the more general field configurations mentioned above. In fact, [18] shows that the energies of all field configurations that can be derived by the Hirota method are additive as they are determined entirely by the asymptotic values of the τ_i functions that arise in the construction of these field configurations. The τ_i functions arise when one sets $\phi = 2 \log \frac{\tau_1}{\tau_0}$ and then solves the relevant equations. An interested reader can find more detail in [18].

In the cases discussed above the Hamiltonian density is given by

$$H = (\partial_t \phi)^2 + (\partial_x \phi)^2 + [\sin (\phi)]^2 , \qquad (14)$$

and so the energy becomes

$$E = \int_{-\infty}^{\infty} dx \, H = 2\partial_x (\ln \tau_0 + \ln \tau_1) \, |_{x=-\infty}^{x=\infty} \qquad (15)$$

and so its value is determined entirely by the asymptotic values of τ_i functions.

This way, [18] has shown that the energies of solutions we have mentioned in this talk are:

1. For moving kink:

$$E_{1-soliton} = 4 \frac{1}{\sqrt{1 - v^2}}. \qquad (16)$$

2. For a moving breather (of frequency ω)

$$E_{breather} = 8 \frac{\sqrt{1 - \omega^2}}{\sqrt{1 - v^2}}. \qquad (17)$$

3. For the wobble

$$E_{wobble} = 4\frac{1}{\sqrt{1-v_K^2}} + 8\frac{\sqrt{1-\omega^2}}{\sqrt{1-v_B^2}}. \tag{18}$$

4. For a solution involving a kink and two breathers mentioned above (with their velocities set to zero)

$$E_{kink+2breathers} = 4 + 8\sqrt{1-\omega_1^2} + 8\sqrt{1-\omega_2^2}, \tag{19}$$

where $\omega_i = \sin\theta_i$, $i = 1, 2$.

8 Perturbed Field Configurations

The paper [16] discusses also more general field configurations. It suggests that a general field configuration corresponding to a perturbed kink evolves into a configuration of one kink and several breathers (moving with respect to each other). To see whether this is the case [16] looked at various perturbations of the kink, paying particular attention to configurations which were generated by adding to the kink an extra perturbation of the form

$$\delta\phi(t=0) = \frac{B}{\cosh(\mu x)}, \quad \delta\frac{\partial\phi}{\partial t}(t=0) = \frac{A}{\cosh(\nu x)}. \tag{20}$$

These simulations were performed for various values of A, B, μ and ν. In all cases the perturbation made the kink move and generated many moving breather-like configurations. To see what the system would finally settle at, the energy was absorbed at the boundaries of the grid. This had the effect of slowing down the kink and also of absorbing and/or altering some breather-like structures. The process was very slow and the results were somewhat inconclusive. It was clear that a general field configuration gradually split into moving kinks and breathers, and some radiation, which quickly moved out to the boundaries. However, the resultant field configuration was metastable; it still radiated, albeit very slowly, and gradually evolved towards a field configuration involving mainly a kink. Whether at the end of its evolution the system ended up with a kink or a kink with some breathers was hard to determine. The interested reader is encouraged to look at more details which are given in [16].

9 Ratchet Behaviour of Solitons

Another interesting property of solitons is their behaviour in an external harmonic or biharmonic field. This phenomenon, which goes under the name of ratchet behaviour, involves the appearance of a directed drift (motion) of solitons under the influence of an unbiased (zero mean) external force [19]. The first studies of the ratchet phenomenon for kinks of the sine-Gordon model were performed by Salerno and Zolotaryuk [20]. In their work they satisfied the necessary conditions for the ratchet effect; i.e. they added an external unbiased force which was periodic in time and also a further term which was responsible for energy dissipation. This has guaranteed the phenomenon and in [20] they described various aspects of this phenomenon, like dependence on the initial condition of the kink and of the field etc. Of course, these properties are important as the ratchet behaviour has found applications in many areas; from nanoscale technologies to the functioning of biological motors. Hence it is important to understand the generality of the phenomenon and of its applications. In fact, in a recent paper [21] we have looked at the ratchet behaviour in some low-dimensional molecular chains which support the existence of polarons. The class of such low-dimensional molecular systems is quite large, it includes biological macromolecules (α-helical proteins), quasi-1D organic and inorganic compounds (like conducting platinum chain compounds), conducting polymers (polyacetylene, polydiacetylene), etc.

We have also looked further at the ratchet behaviour in the sine-Gordon system - this time looking at breathers. This work is still in progress [22], so here we can present only very preliminary results.

9.1   Sine-Gordon Breathers

To do this we added to the sine-Gordon model two terms; one responsible for dissipation and one involving the external field. Hence the equation we used was.

$$\frac{\partial^2 \phi}{\partial^2 t} - \alpha \frac{\partial \phi}{\partial t} - \frac{\partial^2 \phi}{\partial^2 x} - \lambda \frac{\sin(2\phi)}{2} - E(t) = 0, \tag{21}$$

where α determines the strength of the absorption and $E(t)$ determines the external field.

First we looked at the behaviour of a breather in the presence of a bi-harmonic field. In this case we took the field $E(t)$ in the form

$$E(t) = A\sin(\omega t) + B\sin(2\omega t + \delta) \tag{22}$$

and performed various simulations starting with various breathers.

9.1.1   Biharmonic Field

First we considered the case of the bi-harmonic field. In this case we took both A and B nonzero. The results of the simulations depended a little on the initial conditions (i.e. how well the kinks, in the breather, were separated when the field was turned on).

Fig. 5 Trajectory of a kink, split of from the breather of frequency 0.01; electric field characterised by $A = B = 0.1, \delta = 0.1, \omega = 0.175$

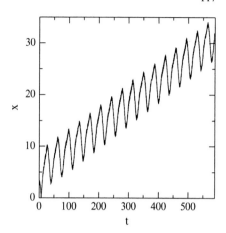

When they were reasonably split the kinks moved away from each other as if we had two independent topological structures (strictly speaking a kink and an antikink). In Fig. 5 we present a typical plot of the trajectory of the kink (as a function of t); the antikink follows a similar trajectory moving towards negative x.

We see that the kink, once it is away from the antikink, behaves as if they were completely independent.

9.1.2 Harmonic Field

Next we looked at the field involving one term for $E(t)$. This was implemented by putting $B = 0$. This time the results were different. To understand them we have to consider the discussion given in [20]. There it was shown that for a single kink of the sine-Gordon model we need the field to be biharmonic to see the ratchet behaviour. In our case the field is not of a simple kink but when the breather becomes sufficiently separated (i.e. its kinks are antikinks are sufficiently separated) then they behave effectively as free kinks and antikinks and then the discussion given in [20] applies. Thus, in a harmonic field, each kink will not exibit the ratchet behaviour. Thus, during the initial stage, due to the action of the field, the kink and antikink of the original breather get separated until, effectively, they behave as if they were independent of each other. At this stage the ratchet behaviour stops operating and they, independently of each other, oscillate in the external field. However, the very small (but still nonzero) atractive forces between the kink and antikink are still there so, having themselves separated, the kink and antikink may start moving towards each other. Once they are close again—they can annihilate (due to the action of the field), or … get separated again.

In the next few pictures (Figs. 6 and 7) we present our preliminary results (showing the time dependence of the trajectory of the field in $x \geq 0$ obtained in several simulations).

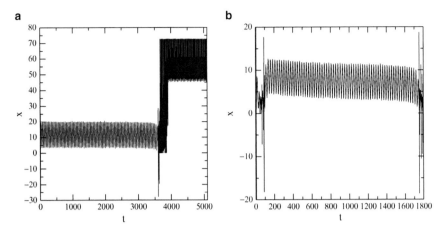

Fig. 6 Trajectory of a kink split of from the breather; electric field characterised by $A = 0.237$, $B = 0$. In (**a**) frequency of the breather is $\omega = 0.24$ and of the field $\omega 1 = 0.145$. In (**b**) they are 0.09 and 0.038, respectively

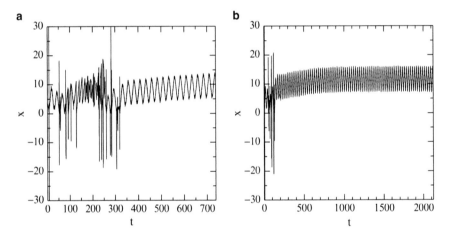

Fig. 7 Trajectory of a kink split of from the breather; electric field characterised by $A = 0.237$, $B = 0$. In (**a**) frequency of the breather is $\omega = 0.02$ and of the field $\omega 1 = 0.2495$. In (**b**) they are 0.09 and 0.2451 respectively

The first picture shows very clearly that after the initial separation the kink oscillates around $x \sim 10$ and then gradually starts moving towards $x \sim 0$ and then annihilates.

The following two pictures show very clearly that the separation of kinks can be delayed—but then once the field separates sufficently far from each other—they oscillate around their new positions like a kink in a harmonic field.

Finally, in the last three pictures we show two further aspects of the motion. In Fig. 8a we see that the motion of the kink is amazingly regular and after a long time

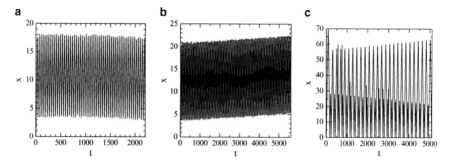

Fig. 8 Trajectory of a kink split of from the breather; electric field characterised by $A = 0.237$, $B = 0$. In (**a**) and (**b**) frequency of the breather is $\omega = 0.24$ and of the field $\omega 1 = 0.16$ and $\omega 1 = 0.14$, respectively. In (**c**) $\omega = 0.20$ and $\omega 1 = 0.032$

the kink is beginning to move towards the antikink. Fig. 8b shows that sometimes the split kinks move with a low velocity away from each other and Fig. 8c shows that sometimes the breather does not get split at all. Here we plot the $x > 0$ of one kink of the breather; as you can observe it comes back to $x = 0$ (where it is in a breather) and gradually moves out, i.e. the amplitude gradually increases. We have also seen the simulations in which the breather has never split (i.e. this growth was virtually nonexistent).

Clearly, these are only very preliminary results. This work is still in progress and we hope to produce more complete results sometime in the near future.

10   Final Comments and Conclusions

We started this talk with a very brief review of the topological solitons and of their dynamics. Then we reviewed the results of our studies of the scattering of topological solitons on a potential obstruction, of both a barrier- and a hole-type.

Our results have shown that when a soliton was sent towards the barrier its behaviour resembled that of a point particle. Thus at low energies the soliton was reflected by the barrier and at higher energy it was transmitted. The scattering process was very elastic. During the scattering the kinetic energy of the soliton was gradually converted into the energy needed to 'climb the barrier'. If the soliton had enough energy to get to the 'top' of the barrier then it was transmitted, otherwise it slid back regaining its kinetic energy.

In the hole case, the situation was slightly different. When the soliton entered the hole it gained extra energy. Some of this energy was converted into the kinetic energy of the soliton, some was radiated away. So when the soliton tried to 'get out' of the hole it had less kinetic energy than at its entry and, when this energy was too low, it remained trapped in the hole. During the scattering process, like in the case of a barrier, the soliton's size changed and so it started oscillating. Afterwards, even when the soliton left the hole, its size continued to oscillate. Hence, looking at this problem from the point of view of the scattering on a hole, it is clear some energy of the soliton was transferred to the oscillations resulting in the emission of radiation, i.e. in an inelastic scattering process.

We have also looked at the scattering of breathers on potential holes. As breathers can be thought of as bound states of kinks and antikinks they can split leaving a trapped kink or an antikink in the hole and allowing its partner to escape either forwards or backwards. In addition, as the energy of the breather depends on the frequency of its oscillation ('breathing') this frequency can change as well. All such phenomena were seen in our simulations and we hope to report on this more fully in the near future [11].

We have also looked at some field configurations, which are solutions of the equations of motion and which describe bound states of kinks and breathers. As the energy of each breather depends on its frequency (and vanishes in the limit of this frequency going to 1) the extra energy, due to these extra breathers, does not have to be very large. The solutions appear to be stable and this stability is guaranteed by the integrability of the model. We have tested this numerically and have found that small perturbations, due to the discretisations, do not alter this stability. To change it we need something more drastic—like the absorption or the space variation of the potential (i.e. the coefficient of the sin^2 term in the Lagrangian). But even then the effects are not very large—one sees splitting of breathers etc. but no 'global annihilation'.

Finally, we have also presented some preliminary results on the behaviour of breathers in the presence of an external electric field. When this electric field was biharmonic the breathers were split into their components (a kink and an antikink) and the movement of each of them exhibited a ratchet behaviour. When the field was harmonic the outcome was more complicated. Most of the time the breathers were split and then they moved to some distance from each other at which stage the forces between them were quite weak and effectively they were independent of each other. At that stage the harmonic electric field could not sustain the ratchet behaviour and so the kink and antikink oscillated around their positions. However gradually, the forces would pull them towards each other and they either annihilated or reformed as a breather which was then split again. Sometimes the electric field simply annihilated the breathers, at other times they took a while to get split. All this depended on the initial conditions of the breather and the electric field. We plan to investigate this further and present our results in one of our future publications.

Let us finish by adding that all our results (on the scattering of topological solitons on obstructions) generalise to other models; such as. e.g. the Landau–Lifschitz model with a position dependent potential or an external magnetic field, other models in $(1+1)$ dimensions, such as a $\lambda\phi^4$ model or even models describing ferro- and anti-ferro-magnets [23].

Clearly a lot of work still has to be done in this area and we can expect to get further interesting results.

Acknowledgements This talk was given at the Davydov memorial meeting in Kiev which was held in June 2008. One of us (WJZ) would like to thank the organisers for inviting him to this very enjoyable meeting.

References

1. N. Manton and P.M. Sutcliffe, *Topological Solitons*, Cambridge: Cambridge University Press, (2004)
2. N.S. Davydov, *Solitons in Molecular Systems*, Dordrecht: Reidel, (1985)
3. P.G. Drazin, *Solitons*, Cambridge University Press, Cambridge, UK (1983)
4. T. Weidig, *Nonlinearity* **12** (1999) 1489–1503 or
   P. Eslami, M. Sarbishaei and W.J. Zakrzewski, *Nonlinearity*, **13**, 1867–1881 (2000)
5. N. Papanicolaou and T.N. Tomaras, *Nucl. Phys.* **B 360**, 425 (1991)
6. J. Brand, B. Piette and W.J. Zakrzewski, *J. Phys.* **A 38**, 10403–10412 (2005)
7. B.M.A.G. Piette and W.J. Zakrzewski, hep-th/0610095, *J. Phys.* **A 40**, 329–346 (2007)
8. B.M.A.G. Piette and W.J. Zakrzewski, *J. Phys.* **A 40**, 5995–6010 (2007)
9. K. Javidan, *J. Phys.* **A 39**, 10565–10574 (2006)
10. T. Romanczukiewicz, *J. Phys.* **A 39**, 3479–3494 (2006)
11. B. Piette and W.J. Zakrzewski - preprint, arXiv:0710.4391 (2007)
12. B.M.A.G. Piette and R.S. Ward, *Physica* **D 201**, 45–55 (2005)
13. Z. Fei, Yu.S. Kivshar, and L. Vazquez, *Phys. Rev.* **A 45**, 6019–6030 (1992)
14. R.H. Goodman and R. Haberman, *Physica* **D 195**, 303–323 (2004)
15. G. Kälberman, *J. Phys.* **A 37**, 11607 (2004)
16. R. Hirota, "Direct Methods In Soliton Theory," In *Soliton*, eds. Bullough, R.K., Caudrey, P.J., 157–176 (1980)
17. L.A. Ferreira, B. Piette, and W.J. Zakrzewski preprint, arxiv:0708/1088 (hep-th), *Phys. Rev.* **E 77** 036613 (2008)
18. L.A. Ferreira and W.J. Zakrzewski arxiv:0707/1603, *JHEP* **09** 015 (2007)
19. P. Reinmann, Phys. Rev., **36**, 57 (2002)
20. M. Salerno and Y. Zolotaryuk, Phys. Rev., **65**, 056603 (2002); Phys. Rev. E, **73**, 066621 (2006)
21. L. Brizhik, A. Eremko, B.M.A.G. Piette, and W.J. Zakrzewski, *J. Phys. Cond. Matter* **20**, 255242 (2008)
22. L. Brizhik, A. Eremko, B.M.A.G. Piette, and W.J. Zakrzewski (in preparation)
23. J. Martin Speight, *Nonlinearity* **19**, 1565–1579 (2006)

Spectral Effects of Resonance Dynamic Intermolecular Interaction for Crystalline Carboxylic Acids at Temperature Phase Transitions

V.D. Danchuk, A.P. Kravchuk, and G.O. Puchkovska

Abstract In the present paper we report on temperature dependent FTIR spectra studies of Davydov splitting value for the in-phase CH_2 rocking vibrations of methylene chains in crystalline n-carboxylic acids $CH_3(CH_2)_{n-2}COOH$ with odd ($n = 15, 17, 19$) and even ($n = 10, 14, 16, 22$) numbers n of carbon atoms in the temperature region from 100 K to the crystal melting point.

The analysis of obtained temperature dependencies allows to determine the following regularities. For all acids in the region of low temperatures the Davydov splitting value practically does not depend on temperature. When temperature increases for acids with odd number $n = 10, 14, 16$ of carbon atoms the splitting value decreases to zero without sharp changes. For acids with odd number and even number $n = 22$ of carbon atoms, the slow decreasing of the splitting value and then sharp decreasing to zero in the nearest region to the crystal melting point takes place.

A statistic-dynamic model is proposed which provides an adequate description of the observed effects. In the frameworks of this model two different mechanisms are responsible for the temperature changes of the vibrational modes splitting value. In addition to the thermal expansion of crystals at heating, the damping of vibrational excitons on orientational defects of different nature takes place. Genesis of such defects is related to the excitation of conformational, librational and rotational degrees of freedom of H-bonded molecular dimers at different temperatures.

Theoretical analysis of resonance dynamical intermolecular interaction effect on the intramolecular vibrations spectra of the crystals was performed in the terms of stochastic equations with account of such mechanisms. The explicit expression

V.D. Danchuk (✉) and A.P. Kravchuk
National Transport University, 1 Suvorov str., 01010 Kyiv, Ukraine
e-mail: vdanchuk@ukr.net; apk3@ukr.net

G.O. Puchkovska
Institute of Physics of National Academy of Sciences of Ukraine, 46 Nauki Prospect, 03022 Kyiv, Ukraine
e-mail: puchkov@iop.kiev.ua

N. Russo et al. (eds.), *Self-Organization of Molecular Systems: From Molecules and Clusters to Nanotubes and Proteins*, NATO Science for Peace and Security Series A: Chemistry and Biology, © Springer Science+Business Media B.V. 2009

for the theoretical dependence of Davydov splitting value on temperature was obtained. Computer simulation of such dependence was performed for crystalline normal chain carboxylic acids. Good agreement between the experimental data and computer simulation results takes place.

Keywords Carboxylic acid crystals · Davydov splitting · Damping of vibrational excitons · Conformational transitions · Librations · Rotations · IR spectra

1 Introduction

It is known that in long-chain aliphatic crystals in the temperature range below the melting point the molecules can perform the orientational disordered motions of different types. It is clear that any distortions in the periodic location of atoms in the plane perpendicular to the long axis of the molecules should result in some peculiarities of intermolecular interactions. In particular, one could expect changes in the resonance dynamic intermolecular interaction (Davydov splitting of vibrational excitons) in the orientational disordered phases.

Earlier, in studying the effects of the intermolecular interaction and the problems of molecular dynamics in crystals of homologous series of odd-numbered normal paraffins, α-olefins, n-carboxylic acids, cholesteryl n-alkanoates, and aromatic compounds by the method of infrared spectroscopy, in particular, the temperature dependence of spectral peak positions for Davydov components of the intramolecular vibrations was observed [1–4]. These spectral components approach each other in the region of the order—orientational disorder phase transition. For nowadays, there is no commonly accepted explanation of the observed temperature dependence.

In this work we report on temperature dependent FTIR spectra studies of Davydov splitting value for CH_2 rocking vibrations of crystalline long-chain n-carboxylic acids $CH_3(CH_2)_{n-2}COOH$ with odd (n = 15, 17, 19) and even (n = 10, 14, 16, 22) numbers of carbon atoms, n, in the temperature region from 100 K to the crystal melting temperature.

Also we propose mechanisms, which adequately describe the observed effects for carboxylic acid crystals as examples. These mechanisms are connected with the damping of vibrational excitons in crystals due to their interaction with orientational lattice defects, which appear due to the excitation of conformational and librational–rotational degrees of freedom of organic molecules in the region of the phase transition.

2 Resonance Dynamical Intermolecular Interaction in Carboxylic Acid Crystals (Theory)

2.1 Low-Temperature (Order) Phase

Let's consider a low-temperature monoclinic modification of the crystalline n-carboxylic acids, whose alignment of methylene chains is described by the orthorhombic O-subcell with two $(CH_2)_2$ chain fragments [5] (Fig. 1).

In general, the eigenfrequencies for transversal vibrations, which correspond to the components of Davydov multiplets, can be found with the use of the standard procedure, which involves the diagonalization of the matrix [6]

$$\left\{ \hat{\omega}_0^2 - \hat{D} \right\},$$

(1)

when passing to the coordinates of symmetry, which are transformed according to irreducible representations of corresponding factor-group. Here $\hat{\omega}_0^2$ is a diagonal (in the coordinates of symmetry) matrix of squares of frequencies for intramolecular vibrations in the crystal in the absence of the resonance dynamic intermolecular interaction; $\hat{D}$ is the corresponding matrix of the resonance dynamic intermolecular interaction.

For a crystal with two molecules per unit cell, one can present the frequencies of Davydov multiplets components for jth normal fundamental vibrational mode $\omega^{(j)}$ (wave vector $\vec{k} = 0$) in the convenient form [7].

$$\omega_1^{(j)} \approx \omega_0^{(j)} + \Delta\omega^{(j)},$$
$$\omega_2^{(j)} \approx \omega_0^{(j)} - \Delta\omega^{(j)}.$$

(2)

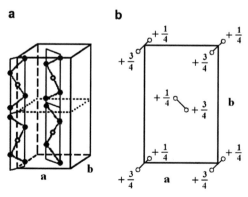

Fig. 1 Low-temperature structure of carboxylic acid crystals: (**a**) arrangement of methylene chains in the orthorhombic O-subcells (**b**) projection of carbon skeleton planes on the plane of crystalline axes (a, b) for orthorhombic O-subcell

Here $\omega_0^{(j)}$ is the frequency of jth intramolecular vibrations, not perturbed by the resonance dynamic intermolecular interaction;

$$\Delta\omega^{(j)} \approx \frac{\widehat{D}}{2\,\omega_0}$$

is a frequency shift, caused by the resonance dynamic intermolecular interaction. The expression (Eq. (2)) is written, when neglecting the values of second order of smallness.

And, at some approximations, the vibrational Davydov splitting value, $\Delta\nu$, from Eq. (2) is defined as

$$\Delta\nu = \omega_1^{(j)} - \omega_2^{(j)} = \frac{D_{(1,n)}^{(j)}}{\omega_0^{(j)}}. \tag{3}$$

2.2 High-Temperature (Orientational Disorder) Phase

Here we describe the contribution of the intermolecular interaction effects and molecular dynamics for carboxylic acid crystals in the orientationally disordered phase on the formation of the spectral bands of intramolecular vibrations in the framework of stochastic equations [8–10].

For this purpose, we consider $(N + 1)$ reactively coupled identical oscillators, which form an actual system (cluster) with $(N + 1)$ states of the selected optical (IR) active mode in the bath, which includes also the molecules, which undergo reorientation. At some approximations, the time correlation function of this selected mode is described by the equation

$$\frac{d}{dt}\widehat{G}(t) = \widehat{G}(t)\left[i\,\widehat{\Omega} + \widehat{\Gamma} + \gamma\widehat{1}\right], \quad \widehat{G}(0) = 1. \tag{4}$$

The time correlation function, $\widehat{G}(t)$, includes oscillations ($\widehat{\Omega}$ is the matrix of oscillations); reactive relaxations due to the interaction between the oscillators of the actual system ($\widehat{\Gamma}$ is a matrix of reactive relaxations); pure relaxations, γ, due to the anharmonic interaction of the selected vibrational mode with vibrations of the bath and due to the reorientation of molecules.

The corresponding normalized spectral function, $\widehat{J}(\omega)$, can be determined from $\widehat{G}(t)$ with the use of Fourier transformation:

$$\widehat{J}(\omega) = \frac{1}{\pi}\int_0^\infty \widehat{G}(t)\,e^{-i\omega t}dt. \tag{5}$$

To determine the matrix of frequency oscillations, $\widehat{\Omega}$, which is a part of the expression (Eq. (4)), we write the stochastic Hamiltonian of the system in the form

$$\widehat{H}(t) = \widehat{H}^{(0)}(t) + V^{(1)}(t). \tag{6}$$

Here $\widehat{H}^{(0)}(t) = \widehat{P}(t) + V^{(0)}(t)$ is the Hamiltonian of the system, which describes a weak influence of the thermostat; $V^{(0)}(t)$ is the stochastic potential energy of interactions between the molecules of the actual system and the bath; $V^{(1)}(t)$ is the stochastic potential energy of interactions between the molecules of the actual system. For convenience we assume that it is precisely the resonance dynamic intermolecular interaction that gives the main contribution to the perturbation of frequencies, which are elements of the oscillations matrix, $\widehat{\Omega}$, and determines the magnitudes and the structure (symmetry) of the actual system.

Then, taking into account the processes of the discrete frequency modulation, we write the elements of the oscillations matrix, $\widehat{\Omega}$, for ith component of Davydov multiplet of jth normal vibration in the form

$$\left(\Omega_i^{(j)}\right) = \delta_{l,n}\left(\omega_0^{(j)} + n\,\Delta\omega_i^{(j)}\right), \; l, n = 0, 1, 2, \ldots, N. \tag{7}$$

The elements of the reactive relaxations matrix, $\widehat{\Gamma}$, which are determined by the rate of molecules transition to their excited (conformational, librational, rotational) states can be written in the following form:

$$(\Gamma)_{l,n} = R\left[(N-n)\,\delta_{l+1,n} + k\,n\delta_{l-1,n} - (N-n+k\,n)\,\delta_{l,n}\right]. \tag{8}$$

Here the frequency modulation of the intramolecular vibrational mode due to the resonance dynamic intermolecular interaction is switched on with the rate R and switched off with the rate $k\,R$. These rates can be presented in the form of the adapted Arrhenius law:

$$R = \sum_{m=1}^{p} A_m \exp\left(-\frac{E_{Am}}{k_B T}\right), \; k = \sum_{m=1}^{p} \exp\left(\frac{\Delta S_m T - \Delta E_{Am}}{k_B T}\right), \tag{9}$$

where E_{Am} is the activation energy; k_B is Boltzmann constant; T is absolute temperature; ΔS_m, ΔE_{Am} are, respectively, variations of entropy and activation energy, required for the turning out the discrete modulation processes in the excited state, m, which corresponds each conformational, librational or rotational transition of a molecule.

Having taken into account the binomial distribution of states for the actual system and having performed the averaging over the ensemble, we obtain the corresponding spectral function

$$\hat{J}(\omega) = \frac{1}{\pi}\left\langle\left[i\left(\omega - \left\langle\Omega_i^{(j)}\right\rangle\right) - \gamma - \Gamma\right]^{-1}\right\rangle, \tag{10}$$

where $\langle.\rangle$ denotes averaging. Then the average value of the vibration frequency for the ith component of Davydov multiplet of jth normal vibration is determined as follows

$$\left\langle\Omega_i^{(j)}\right\rangle = \omega_0^{(j)} + \Delta\omega_i^{(j)}\frac{N}{1 + \sum_{m=1}^{p}\exp\left(\frac{\Delta S_m T - \Delta E_{Am}}{k_B T}\right)} \tag{11}$$

The analysis of the expressions (Eq. (11)) shows that, when the temperature increases, the values of Davydov components frequencies, $\left\langle\Omega_i^{(j)}\right\rangle$, tend to reach the value of the frequency of the vibrational mode nonperturbed by the intermolecular resonance, ω_0.

In this case, the role of entropy variations, ΔS_m, as a measure of system disordering becomes dominant in the comparison with the intermolecular interaction energy.

3 Results and Discussion

It is known [2], that most n-carboxylic acids in the condensed state form hydrogen bonded central symmetric cyclic dimers (Fig. 2), which are layered in the crystal, ended by methyl groups on both sides. The quantum-chemical calculations show, that the rotational barrier for C–C$_\alpha$ carbon bond, namely, the nearest to a dimer ring, is essentially less than that for other C–C carbon bonds of molecules [11]. This fact provides the possibility for an existence of different conformers due to the molecules deformation nearby the dimer ring. So for some temperatures the conformation transitions can lead to the molecular oriental disordering in sites of the crystalline lattice.

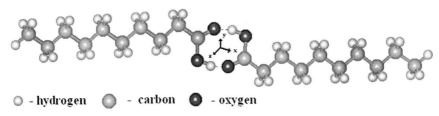

○ - hydrogen   ◯ - carbon   ● - oxygen

Fig. 2 Structure and natural coordinates of dimer $CH_3(CH_2)_4COOH$

The procedure and equipment for measuring IR-spectra of long-chain molecules in the temperature range from 100 to 350 K were described in [1, 2]. IR-spectra of n-carboxylic acids have been measured with the use of UR-20 spectrophotometer in the spectral range of 400–4,000 cm^{-1} (the spectral split program is 2). The spectral width of a split in the range of 700 cm^{-1} is 1.5–2 cm^{-1}.

IR-spectral parameters of Davydov's components, namely, peak positions, spectral widths, intensities, have been measured in the temperature range of 103–373 K. The nitrogen cryostat has been used for the spectral measurements at low temperatures. In order to get the intermediate temperatures a special vacuum cell was constructed. Its internal part was cooled with liquid or gas-like nitrogen with the control of its supply or heated up to the desired temperature. The temperature was measured by the copper–constantan thermocouple, which was placed directly on the cell with the sample. Samples of investigated acids were obtained by sealing of their melts between KBr plates, which were glued at the side ends. In the case of kC14, kC16 and kC18 samples it was possible to obtain single-crystal films and to measure their IR-spectra in the polarized light at the normal light incident on the (001) plane at $\mathbf{E} \| \mathbf{a}$ and $\mathbf{E} \| \mathbf{b}$ (Fig. 3). IR-spectra have been recorded at each 4–5 K, and near the phase transition the temperature step has been decreased to 1–2 K. The accuracy of temperature control was ± 1 K. The accuracy for the determination of Davydov's doublet component spectral positions was not worse than 0.2–0.3 cm^{-1}.

According to [12, 13], all investigated acids crystallize from the melt in the monoclinic modification. Parameters of O-subcell are practically the same for the crystals of the majority of long-chain aliphatic compounds.

It is experimentally observed that the most sensitive IR-bands to the conformational and orientational transitions in the IR-spectra of n-carboxylic acid crystals $CH_3(CH_2)_{n-2}COOH$ are the bands at 720 cm^{-1}, which correspond to the in-phase rocking vibrations of methylene chains of these molecules (Fig. 3).

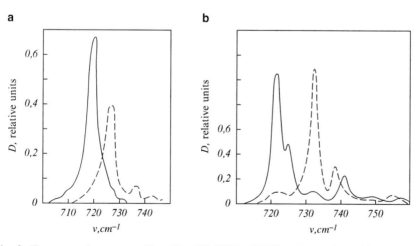

Fig. 3 IR-spectra of monocrystalline film $CH_3(CH_2)_{14}COOH$ at T $=$ 300 K (**a**) and T $=$ 100 K (**b**) in polarized light: dash curve - $\mathbf{E} \| \mathbf{a}$, solid curve - $\mathbf{E} \| \mathbf{b}$

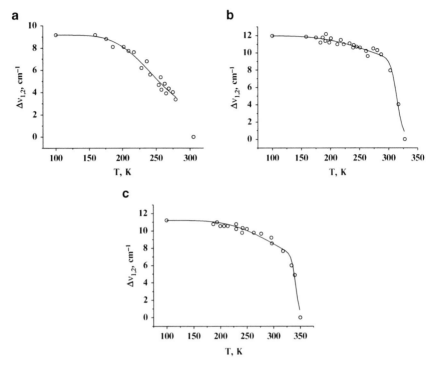

Fig. 4 Computer simulation for temperature dependence of Davydov splitting value $\Delta\nu_{1,2}$ for CH_2 in-phase rocking vibrations of the molecules in n-carboxylic acids. (**a**) n = 10, (**b**) n = 15, (**c**) n = 22 (dots—experiment [2], curve—theory)

In Fig. 4 (dots) we present the experimental data of the temperature changes in Davydov splitting value, $\Delta\nu_{1,2}$, for these vibrations of different carboxylic acid crystals, when the samples are heated from 100 K to their melting temperatures.

The analysis of obtained temperature dependencies for n-carboxylic acid crystals has allowed to draw the following regularities. For all acids in the region of low temperatures Davydov splitting value practically does not depend on temperature. When temperature increases, for acids with odd (n = 15, 17, 19) and even (n = 22) numbers of carbon atoms, the slow decreasing of the splitting value takes place. In the nearest region ($\Delta T = 20 \div 30K$) to the crystal melting point we observe the sharp decreasing of this value to zero. For acids with even numbers n = 10, 14, 16 of carbon atoms the splitting value decreases to zero without sharp changes.

In our case the theoretical temperature dependence of the corresponding Davydov splitting value has the following form

$$\Delta\Omega_{1,2}^{(j)}(T) = \frac{\Delta\omega_{1,2}^{(j)}}{1 + \exp\left\{\dfrac{\Delta S_1 T - \Delta E_{A1}}{k_B T}\right\} + \exp\left\{\dfrac{\Delta S_2 T - \Delta E_{A2}}{k_B T}\right\}}, \quad (12)$$

where $\Delta\omega_{1,2}^{(j)}$ is the value of Davydov splitting of intramolecular vibrations for the crystals in the ordered low-temperature monoclinic phase; ΔS_1, ΔE_1 are, respectively, variations of entropy and activation energy at conformation transitions of molecular dimers; ΔS_2, ΔE_2—corresponding energetic parameters at orientational transitions of molecular dimers as a whole in sites of the crystal lattice.

The fitting of the theoretical dependency (Eq. (12)) to experimental data was realized by the variation of ΔS_1, ΔE_1, ΔS_2, ΔE_2 values with the help of Levenberg–Marquardt method in order to obtain the minimum of the average square deviation of the theoretical dependency from experimental data.

In Fig. 4 (lines) we present the computer simulation results of observed Davydov splitting value temperature dependencies by Eq. (12). For n-carboxylic acid crystals with even number of carbon atoms ($n = 10, 14, 16$) the best agreement of the theoretical dependency with the experiment was obtained in the two parameter approach for Eq. (12), taking into account only ΔS_1 and ΔE_1. For carboxylic acid crystals with numbers of carbon atoms $n = 15, 19, 22$ a satisfactory theoretical description of the experiment was obtained in the four parametric approach for Eq. (12), taking into consideration not only $\Delta S_1, \Delta E_1$, but also $\Delta S_2, \Delta E_2$.

Thereby, the analysis of computer simulation results shows, that for light carboxylic acids with even number of carbon atoms ($n = 10, 14, 16$) the decreasing of the resonance dynamic intermolecular interaction efficiency at heating up to the melting point is caused mainly by the conformational dynamics of molecular dimers (see Fig. 4). Here in the temperature region of the "crystal-liquid" phase transition the positional melting of the crystalline lattice is probably realized.

In the case of carboxylic acids with $n = 22$ and with odd numbers $n = 15, 17, 19$, the damping of vibrational excitons is caused not only by conformational transitions, but also in the temperature region near the melting point by the excitation of librational and rotational motions of molecular dimers as a whole. The absence of the molecular rotation in sites of the crystalline lattice for the light carboxylic acids with even numbers of carbon atoms ($n = 10, 14, 16$) up to the melting point is, probably, caused by the denser molecule packing in these lattices. As the calculations of intermolecular interactions, carried out in the frames of the atom—atom assumption [12], showed, in the case of acids with an even number of carbon atoms, n, the interaction between end methyl groups is on 60% larger, than that of the acids with an odd n value. This effect results in the larger hindering of methylene chain rotations in the first case. For the acids with the even n such rotation is significantly hampered, since not only volumetric dimer rings, but also the stronger interactions between the layers of molecules hinder it. The increase of the carbon atoms number, n, up to 22 leads to the decrease of the specific contribution from the interactions of methyl groups, and thus, to the smaller difference in the behavior (dynamics) of even and odd acids, which becomes apparent in the shape of the temperature dependence. The similar temperature dependence of Davydov splitting value is also observed for the acid with $n = 24$ in [14].

Herewith the estimated values for corresponding energetic parameters of conformational and orientational dynamics in carboxylic acid crystals were obtained (Table 1).

Table 1 The theoretical estimations of energetic parameters for conformational and orientational dynamics of molecular dimers in n-carboxylic acid crystals

Number of carbon atoms, n	ΔE_1, kJ/mol	ΔS_1, kJ/mol K	ΔE_2, kJ/mol	ΔS_2, kJ/mol K
10	14.5 ± 1.3	0.06 ± 0.005	–	–
14	32.0 ± 3.3	0.11 ± 0.01	–	–
15	9.0 ± 1.5	0.019 ± 0.006	156.9 ± 4.4	0.50 ± 0.01
16	11.6 ± 0.9	0.040 ± 0.004	–	–
19	10.1 ± 0.9	0.031 ± 0.003	185.3 ± 5.0	0.57 ± 0.01
22	12.4 ± 1.6	0.032 ± 0.006	271.9 ± 3.2	0.80 ± 0.01

Earlier, in [15] estimated values of activation energy and entropy variations for exciting the molecule rotations at the transition into the rotator phase for crystalline n-paraffin $C_{19}H_{40}$ were obtained. They are $\Delta E_A = (167.4 \pm 2.9)$ kJ/mol, $\Delta S = (0.5651 \pm 0.0084)$ kJ/(mol $\cdot$ K). This variation of activation energy, ΔE_A, for $n - C_{19}H_{40}$ on order of magnitude agrees with the corresponding value, which is obtained by the proton spin relaxation spectroscopy method ($\Delta E_A = 96$ kJ/mol) [16].

As seen from Table 1, the estimated values for energetic parameters of n-carboxylic acids obtained in present work agree with those ones for n-paraffins.

The good agreement of computer modelling results with experimental data, as well as the correlation between obtained in the present work the estimated values for energetic parameters of carboxylic acids and corresponding ones for normal paraffins [15, 16], probably, indicates the adequacy of the proposed here models for the description of resonance dynamic intermolecular interactions in orientationally condensed media.

4 Conclusions

1. In FTIR-spectra for the region of the "order-orientation disorder" phase transition the temperature dependence of Davydov splitting value for the in-phase CH_2 rocking vibrations of crystalline long-chain n-carboxylic acids $CH_3(CH_2)_{n-2}COOH (n = 10, 14, 15, 16, 17, 19, 22)$ has been found out.
2. A statistic and dynamic model is proposed, which provides an adequate description of the observed effect. In the framework of this model, the damping of vibrational excitons on orientational defects of different nature takes place. Genesis of such defects is caused by the excitation of conformational, librational and rotational freedom degrees of H-bonded molecular dimers at the heating of the crystal.
3. The theoretical analysis of the effects of resonance dynamical intermolecular interactions on the spectra of intramolecular vibrations of the crystals was performed in the terms of stochastic equations, taking into account the oriental disordering of the crystalline lattice. Computer simulations of such dependence were performed for pure carboxylic acids. Good agreement between the experimental and computer simulation results was obtained.

References

1. S.P. Makarenko, G.A. Puchkovskaya, Ukr. Fiz. Zhurnal 19 (1974) 421, in Russian.
2. S.P. Makarenko, G.A. Puchkovskaya, Ukr. Fiz. Zhurnal 20 (1975) 474, in Russian.
3. G.A. Puchkovskaya, A.A. Yakubov, J. Mol. Struc. 218 (1990) 141.
4. A.V. Sechkarev, V.V. Ovcharenko, Optica i spektroskopiya 43 (1977) 500, in Russian.
5. V. Vand, Acta Cryst. 4 (1951) 104.
6. H. Poulet, J.-P. Mathieu. Spectres de Vibration et Symetrie des Cristaux, Paris-Londres-New York, 1970, p. 437.
7. P. Dawson, J. Phys. Chem. Sol. 36 (1975) 1401.
8. R.Kubo. Relaxation and Resonance in Magnetic Systems, Plenum, New York, 1962, pp. 23–68.
9. W.G. Rotshild, J. Chem. Phys. 65 (1976) 455.
10. E.W. Knapp, S.F. Fischer, J. Chem. Phys. 74 (1981) 89.
11. N.L. Allinger, S.H.M. Chang, Tetrahedron 33 (1977) 1561.
12. E. von Sydov, Arkiv Kemi 9 (1955) 231.
13. E. Stenhagen, E. von Sydov, Arkiv Kemi 6 (1952) 309.
14. G. Zerbi, G. Conti, G. Minoni, S. Pison, A. Bigotto, J. Phys. Chem. 91 (1987) 2386.
15. G.O. Puchkovska, V.D. Danchuk, A.P. Kravchuk, Jan I. Kukielski, J. Mol. Struct. 704 (2004) 119.
16. M. Stohrer, F. Noask, J. Chem. Phys. 67 (1977) 3729.

Linear Augmented Cylindrical Wave Method for Electronic Structure of Isolated, Embedded, and Double-Walled Nanotubes

Pavel D'yachkov and Dmitry Makaev

Abstract The results of study of the band structure of single-walled nanotubes, both isolated and embedded into a crystal matrix, and double-walled nanotubes are surveyed. The mathematical apparatus of the linear augmented cylindrical wave (LACW) method is described, and its application to prediction of the electronic properties, semiconducting and metallic, of nanotubes is considered. The method uses the local density functional approximation and the muffin-tin (MT) approximation for the electron potential and is implemented as a quantum-mechanical program package.

Keywords Nanotubes · Double-walled · Embedded · Electronic structure

1 Introduction

An understanding of the structure of nanomaterials is very important for science and technological applications. Carbon nanotubes are nanomaterials with unique physical properties and possible technological applications. It was experimentally confirmed in 1991 that carbon can exist in the form of cylindrical nanofilaments obtained by rolling up of a graphene sheet. Such structures were identified in solid products of arc-discharge evaporation of graphite [1].

The small size of carbon nanotubes and the quantum character of their electrical properties determined the possibilities of using nanotubes as elements of new generation integrated circuits with characteristic size on the order of tens of nanometers. It is expected that nanotubes will facilitate a technological revolution in integrated circuit elements from a micrometer to a nanometer scale [3–8].

In the study of the electronic properties of nanotubes, theory has always been ahead of experiment. Quantum-chemical methods [9–12] predicted (and it was

P. D'yachkov (✉) and D. Makaev
Kurnakov Institute of General and Inorganic Chemistry, Russian Academy of Sciences,
Leninskii pr. 31, Moscow 119991 Russian Federation
e-mail: p_dyachkov@rambler.ru; dima@lester.ru

N. Russo et al. (eds.), *Self-Organization of Molecular Systems: From Molecules
and Clusters to Nanotubes and Proteins*, NATO Science for Peace and Security
Series A: Chemistry and Biology, © Springer Science+Business Media B.V. 2009

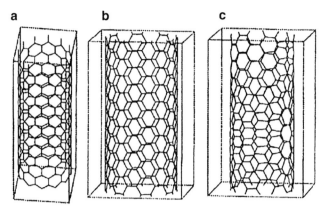

Fig. 1 (**a**) Zigzag, (**b**) armchair, and (**c**) chiral nanotubes

then experimentally established) that the character of conductivity—metallic or semiconducting—is determined by the geometry of the graphene framework. Another very important class is double-walled nanotubes composed of two concentric cylindrical graphene layers. From the standpoint of nanoelectronics, double-walled nanotubes are molecular analogues of coaxial cables.

A perfect single-walled carbon nanotube is a cylinder obtained by rolling up a graphite sheet (graphene). There are two types of nonchiral single-walled nanotubes: armchair of (n,n) configuration and zigzag of $(n,0)$ configuration; in addition, there is a family of different chiral tubulenes (Fig. 1). In armchair tubes, carbon hexagons are oriented with respect to the nanotube axis so that two C–C bonds of each hexagon are perpendicular to the tube axis; in zigzag tubes, these bonds are parallel to the axis. In chiral nanotubes, the orientation of hexagons with respect to the tube axis is intermediate. Nanotubes are capable of conducting electrical current without scattering of charge carriers [13–17]. There are two types of electron transport through matter: ordinary transport, in which the energy of charge carriers is close to the energy of electrons at the bottom of the conduction band and their velocity is proportional to the applied voltage, and time-of-flight transport, where the energy of electrons is not lost due to electron–phonon scattering. The limiting case of such transport is ballistic conductance, where the measured resistance of a nanotube is equal to the resistance of ohmic contacts and heat is released only through contacts but not through the conductor itself. The nanotube thereby acts as a waveguide: each subband of the conduction band makes a contribution to the conductance equal to the conductance quantum $G_0 = 2e^2/h$. In carbon nanotubes, the Kondo effect [15] and superconductivity were observed. A junction of two nanotubes, semiconducting and metallic, conducts current in one direction [18].

In summary, carbon nanotubes are an important class of nanomaterials. This underlies the importance of the development of accurate methods of calculation of the electronic structure of nanotubes. The linear augmented plane wave (LAPW) method is one of the most accurate quantum-chemical methods of calculation of the band structure of crystal. It was developed for crystals with three-dimensional translational symmetry (3D-LAPW) [19, 20]. Later, the basic ideas of this method were used for systems with another geometry. In the variant of the LAPW method for quasi-one-dimensional systems (1D-LAPW), it is assumed that an infinite system extended along one direction is confined in a potential box in the form of a parallelepiped [21]. In the directions normal to the translation axis, the movement of an electron is restricted. A generalization of the LAPW method for surface electronic states and layered crystals was developed in [22]. In this case, the movement of an electron in the interatomic region is described by plane waves in the directions of translations and by a standing wave in the third direction. The augmented wave method was formulated for cubic and spherical clusters [23, 24]. For calculation of the electronic band structure of nanotubes, the linear augmented cylindrical wave (LACW) method was developed in 1999–2007 at the Kurnakov Institute of General and Inorganic Chemistry, RAS [25–31]. In the present paper, we summarize these results.

2 Single-Walled Nanotubes

2.1 Computational Method

2.1.1 One-Electron Approximation

In the LACW method, a one-electron model is used, implying that a many-electron wave function is described by the determinant of one-electron functions:

$$
\Psi_k(q_1, q_2, \ldots, q_N) = \frac{1}{\sqrt{N_e!}} \det
\begin{Vmatrix}
\varphi_1(q_1) & \varphi_2(q_1) & \cdots & \varphi_{N_e}(q_1) \\
\varphi_1(q_2) & \varphi_2(q_2) & \cdots & \varphi_{N_e}(q_2) \\
\cdots & \cdots & \cdots & \cdots \\
\varphi_1(q_{N_e}) & \varphi_2(q_{N_e}) & \cdots & \varphi_{N_e}(q_{N_e})
\end{Vmatrix} . \tag{1}
$$

Then, the problem of calculation of electron levels is reduced to solution of the one-electron Schrödinger equation

$$
H\varphi_\mu = E_\mu \varphi_\mu, \mu = 1, 2 \ldots N_e, \tag{2}
$$

where H includes the operators of the kinetic and potential energy of N_e electrons.

2.1.2 Electron Potential

In calculation of a many-electron system, the key problem is the choice of electron potential. In the LACW method, this potential is constructed with the use of the muffin-tin (MT) approximation and the local density functional approximation for exchange interaction. The MT approximation implies that the crystal space is

Fig. 2 Nanotube (5, 5) in
a potential well confined by
cylindrical potential barriers

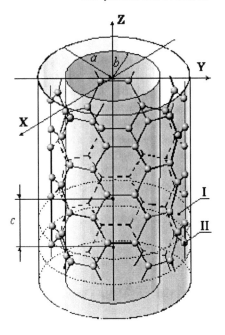

divided into regions of two types: atomic regions and interatomic regions. Each
atom of a polyatomic system is surrounded by a sphere (MT sphere). In the MT
spheres, the potential is taken as spherically symmetric. In the interatomic region,
the electron potential is taken to be constant. This potential is chosen as the energy
reference. The radii of MT spheres are selected so that the spheres of neighboring
atoms are in contact. Such a choice is physically rather evident: information on the
chemical nature of atoms constituting a polyatomic system is contained only inside
the MT spheres. In a nanomaterial, the movement of electrons is restricted by its
dimensions and geometry. In a nanotube, electron motion is confined to an approximately cylindrical layer with a thickness on the order of the doubled van der Waals
radius of the atom. Correspondingly, in the LACW method, the motion of electrons
in the space between MT spheres is limited by two cylindrical barriers impenetrable for electrons: the external barrier Ω_a of radius a and the internal barrier Ω_b of
radius b which are chosen so that the region confined by these barriers accommodates a significant portion of the electron density of the system under consideration
(Fig. 2). Such a potential is referred to as a cylindrical MT potential.

2.1.3 Coulomb and Exchange Interaction

To calculate the potential in the MT spheres, the electron density distribution $\rho(r)$ of
a system is constructed as a superposition of electron densities of its atoms. Inside
the MT spheres, its spherically symmetric part $\rho(r)$ is taken. The electrostatic potential $V_e(r)$ created by the distribution $\rho(r)$ is determined from the solution of the

Poisson equation. The Coulomb potential $V_C(r) = V_e(r) + V_n(r)$ in the MT spheres is obtained by adding the electrostatic potential $V_n(r)$ created by the positive charges of atomic nuclei. The electron density distribution $\rho(r)$ is also used for calculation of the exchange interaction in the local density functional approximation.

2.1.4 Solution of the Schrödinger Equation for the Interspherical Region

Let us derive the basic equations of LACW method, beginning with the solution of the Schrödinger equation for the interspherical and MT regions. In the interspherical region, the basis functions are the solutions of the Schrödinger equation for free movement of electrons inside an infinite tube with outer and inner radii a and b, respectively. When expressed in Rydberg's (Planck constant $\hbar = 1$, electron mass m $= 1/2$, electron charge $e = \sqrt{2}$) and cylindrical coordinates {Z, Φ, R}, this equation takes the form

$$\left\{ -\left[\frac{1}{R}\frac{\partial}{\partial R}\left(R\frac{\partial}{\partial R} \right) + \frac{1}{R^2}\frac{\partial^2}{\partial \Phi^2} + \frac{\partial^2}{\partial Z^2} \right] + U(R) \right\}$$
$$\times \Psi(Z, \Phi, R) = E\Psi(Z, \Phi, R) \tag{3}$$

The potential U(R), determining the region in which electrons of an isolated nanotube are allowed to move, takes the form

$$U(R) = \begin{cases} 0, & b \le R \le a \\ \infty, & R < b, R > a \end{cases}. \tag{4}$$

The solution of Eq. (3), taking into account Eq. (4), has the form $\Psi(Z, \Phi, R) = \Psi_P(Z)\Psi_M(\Phi)\Psi_{MN}(R)$. Here,

$$\Psi_P(Z) = \frac{1}{\sqrt{c}}exp[i(k + k_P)Z], \quad k_P = (2\pi/c)P, \quad P = 0, \pm 1, \pm 2, \ldots \tag{5}$$

is the wave function that describes the free movement of an electron along the translational symmetry axis Z with the period c. The wave vector k belongs to the one-dimensional Brillouin zone: $-\pi/c \le k \le \pi/c$. The function

$$\Psi_M(\Phi) = \frac{1}{\sqrt{2\pi}}e^{iM\Phi}, \quad M = 0, \pm 1, \pm 2, \ldots \tag{6}$$

describes the rotation of the electron about the symmetry axis of the system, and the function $\Psi_{MN}(R)$, determining the radial movement of the electron, is the solution of the equation

$$\left[-\frac{1}{R}\frac{d}{dR}R\frac{d}{dR} + \frac{M^2}{R^2} \right]\Psi_{MN}(R) + U(R)\Psi_{MN}(R) = E_{|M|,N}\Psi_{MN}(R). \tag{7}$$

Here, N is the radial quantum number and $E_{|M|,N}$ is the energy spectrum; the energy

$$E = (k + k_P)^2 + E_{|M|,N},\tag{8}$$

corresponds to the wave function $\Psi(Z, \Phi, R)$.

At $b \le R \le a$, Eq. (6) is written as

$$\left[\frac{d^2}{dR^2} + \frac{1}{R}\frac{d}{dR} + \kappa^2_{|M|,N} - \frac{M^2}{R^2}\right]\Psi_{MN}(R) = 0,\tag{9}$$

where $\kappa_{|M|,N} = \{E_{|M|,N}\}^{1/2}$. After substituting $\kappa R = x$ and $\Psi(R) = y(x)$ into Eq. (9), it reduces to the Bessel equation. Its solutions are referred to as cylindrical functions of the Mth order [32, 33]. Any solution of the Bessel equation can be represented as a linear combination of cylindrical Bessel functions of the first J_M and second Y_M kinds:

$$\Psi_{MN}(R) = C^J_{MN} J_M\left(\kappa_{|M|,N} R\right) + C^Y_{MN} Y_M\left(\kappa_{|M|,N} R\right).\tag{10}$$

In Eq. (10), C^J_{MN} and C^Y_{MN} are constants chosen so as to ensure the normalization of the wave function $\Psi_{MN}(R)$,

$$\int_b^a |\Psi_{MN}(R)|^2 R dR = 1,\tag{11}$$

and its vanishing at the interior and exterior potential barriers,

$$\begin{aligned} C^J_{MN} J_M\left(\kappa_{|M|,N}a\right) + C^Y_{MN} Y_M\left(\kappa_{|M|,N}a\right) = 0,\\ C^J_{MN} J_M\left(\kappa_{|M|,N}b\right) + C^Y_{MN} Y_M\left(\kappa_{|M|,N}b\right) = 0.\end{aligned}\tag{12}$$

From the set of Eqs. (12), the relationship between C^J_{MN} and C^Y_{MN}:

$$C^Y_{MN} = -C^J_{MN}\frac{J_M\left(\kappa_{|M|,N}a\right)}{Y_M\left(\kappa_{|M|,N}a\right)},\tag{13}$$

and the equation for $\kappa_{|M|,N}$:

$$J_M\left(\kappa_{|M|,N}a\right) Y_M\left(\kappa_{|M|,N}b\right) = J_M\left(\kappa_{|M|,N}b\right) Y_M\left(\kappa_{|M|,N}a\right),\tag{14}$$

are derived.

For integral (Eq. (11)), the expression:

$$\begin{aligned} &\int_a^b \Psi^*_{MN}(R)\,\Psi_{MN}(R)\,R dR\\ &= \frac{a^2}{2}\{\Psi'^*{}_{MN}(a)\,\Psi'_{MN}(a)\} - \frac{b^2}{2}\{\Psi'^*{}_{MN}(b)\,\Psi'_{MN}(b)\} = 1\end{aligned}\tag{15}$$

or

$$\frac{a^2}{2} \left[C^J_{MN} J'_M \left(\kappa_{|M|,N} a \right) + C^Y_{MN} Y'_M \left(\kappa_{|M|,N} a \right) \right]^2$$

$$-\frac{b^2}{2} \left[C^J_{MN} J'_M \left(\kappa_{|M|,N} b \right) + C^Y_{MN} Y'_M \left(\kappa_{|M|,N} b \right) \right]^2 = 1 \tag{16}$$

is valid. From this equation after the substitution (Eq. (13)), we find the coefficients C^J_{MN} and C^Y_{MN}. Thus, the basis function Ψ (k, P, M, N) in the Ω_{II} region in the general cylindrical coordinate system takes the form

$$\Psi_{II} (k, P, M, N) = \frac{1}{\sqrt{2\pi c}} \exp\{i (K_P Z + M\Phi)\}$$

$$\times \left[C^J_{MN} J_M \left(\kappa_{|M|,N} R \right) + C^Y_{MN} Y_M \left(\kappa_{|M|,N} R \right) \right] \tag{17}$$

Here, $K_P = k + k_P$.

2.1.5 Solution of the Schrödinger Equation for the MT Spheres

Inside the MT sphere α in the local spherical coordinate system $\{r, \theta, \varphi\}$, the basis function is expanded in spherical harmonics $Y_{lm}(\theta, \varphi)$ [19, 20]:

$$\Psi_{l\alpha} (r, \theta, \varphi \,|\, k, P, M, N) =$$

$$\sum_{l=0}^{\infty} \sum_{m=-l}^{l} [A_{lm\alpha} u_{l\alpha} (r, E_{l\alpha}) + B_{lm\alpha} \dot{u}_{l\alpha} (r, E_{l\alpha})] \, Y_{lm} (\theta, \varphi) . \tag{18}$$

In Eq. (18), $u_{l\alpha}$ is the solution of the radial Schrödinger equation in the MT sphere α for the energy $E_{l\alpha}$:

$$H_{MT\alpha} u_{l\alpha}(r) = E_{l\alpha} u_{l\alpha}(r). \tag{19}$$

Inside the MT sphere of radius r_α, the $u_{l\alpha}$ function is taken to be normalized:

$$\int_0^{r_\alpha} [u_{l\alpha} (r)]^2 \, r^2 dr = 1. \tag{20}$$

The $\dot{u}_{l\alpha} = [\partial u_{l\alpha}/\partial E]_{E_{l\alpha}}$ function is found from the equation $h \, \dot{u}_{l\alpha}(r) = u_{l\alpha}(r) + E_l \dot{u}_{l\alpha}(r)$. The $u_{l\alpha}(r)$ and $\dot{u}_{l\alpha}(r)$ functions are orthogonal:

$$\int_0^{r_\alpha} \dot{u}_{l\alpha} (r) \, u_{l\alpha} (r) \, r^2 dr = 0. \tag{21}$$

This can be easily verified by differentiating the left and right-hand sides of Eq. (20). The integral of the squared $\ddot{u}_{l\alpha}(r)$ function is designated as $N_{l\alpha}$:

$$N_{l\alpha} = \int_{0}^{r_{\alpha}} |\dot{u}_{l\alpha}(r)|^2 r^2 dr. \tag{22}$$

2.1.6   Basis Functions and Overlap and Hamiltonian Integrals

The desired solutions of the Schrödinger equation must be everywhere continuous and differentiable; therefore, to construct basis functions, the solutions of the wave equation for the interspherical region and MT spheres should be sewn together. This can be achieved by selecting the coefficients $A_{lm\alpha}$ and $B_{lm\alpha}$ in Eq. (18).

A major mathematical difficulty here is that function (Eq. (17)) is expressed in a general cylindrical coordinate system and function (Eq. (18)), in a local spherical system. However, using the theorem of addition for cylindrical functions, we can express Ψ_{II} through the cylindrical coordinates Z_{α}, Φ_{α} and R_{α} of the sphere α and the local spherical system r, θ, φ:

$$
\begin{aligned}
\Psi_{II\alpha}(r, \theta, \varphi \,|k, P, M, N) &= \frac{1}{\sqrt{2\pi c}} \exp\{i\,(K_P Z_{\alpha} + M\Phi_{\alpha})\} \\
&\times \exp\{iK_P\, r\cos\theta\}\,(-1)^M \\
&\times \sum_{m=-\infty}^{+\infty} \left[C_{MN}^J J_{m-M}\left(\kappa_{|M|,N}\,R_{\alpha}\right) + C_{MN}^Y Y_{m-M}\left(\kappa_{|M|,N}\,R_{\alpha}\right) \right] \\
&\times J_m\left(\kappa_{|M|,N}\,r\sin\theta\right) e^{im\varphi}
\end{aligned} \tag{23}
$$

Equating functions (Eqs. (18) and (23)) and their derivatives at the boundary of the MT spheres, we obtain

$$A_{lm\alpha} = D_{MNP,lm\alpha}\, r_{\alpha}^2 \left\{ I_2\, \dot{u}_{l\alpha}(r_{\alpha}, E_{l\alpha}) - I_1\, \dot{u}'_{l\alpha}(r_{\alpha}, E_{l\alpha}) \right\}, \tag{24}$$

$$B_{lm\alpha} = D_{MNP,lm\alpha}\, r_{\alpha}^2 \left\{ I_1\, u'_{l\alpha}(r_{\alpha}, E_{l\alpha}) - I_2\, u_{l\alpha}(r_{\alpha}, E_{l\alpha}) \right\}. \tag{25}$$

Here, the following designations are used:

$$
\begin{aligned}
I_1(MNP; lm\alpha) &= \int_{0}^{\pi} \exp\{iK_P r_{\alpha} \cos\theta\} \\
&\times J_m\left(\kappa_{|M|,N} r_{\alpha} \sin\theta\right) P_l^{|m|}(\cos\theta) \sin\theta d\theta,
\end{aligned} \tag{26}
$$

$$D_{MNP,lm\alpha} = \frac{1}{\sqrt{2}c} (-1)^{\frac{m+|m|}{2}+l} i^l \left[(2l+1) \frac{(l-|m|)!}{(l+|m|)!} \right]^{1/2}$$

$$\times \exp \{ i \left(K_P Z_\alpha + M\Phi_\alpha \right) \} (-1)^M \tag{27}$$

$$\times \left[C_{MN}^J J_{m-M} \left(\kappa_{|M|,N} R_\alpha \right) + C_{MN}^Y Y_{m-M} \left(\kappa_{|M|,N} R_\alpha \right) \right]$$

$$I_2(MNP;lm\alpha) = 2 \int_0^{\pi/2} \exp i (K_P r_\alpha \cos\theta) i K_P \cos\theta$$

$$\times J_m(\kappa_{|M|,N} r_\alpha \sin\theta) + (1/2)\kappa_{|M|,N} \sin\theta \tag{28}$$

$$\times [J_{m-1}(\kappa_{|M|,N} r_\alpha \sin\theta) - J_{m+1}(\kappa_{|M|,N} r_\alpha \sin\theta)] P_l^{|m|}$$

$$\times (\cos\theta) \sin\theta d\theta.$$

The integrals I_1 and I_2 are calculated by numerical integration methods. The overlap and Hamiltonian integrals are calculated by the formula:

$$\langle P_2 M_2 N_2 | P_1 M_1 N_1 \rangle = \delta_{P_2 P_1} \delta_{M_2 M_1} \delta_{N_2 N_1} - \frac{1}{c} (-1)^{M_1 + M_2}$$

$$\times \sum_\alpha \exp \{ i [(K_{P_1} - K_{P_2}) Z_\alpha + (M_1 - M_2) \Phi_\alpha] \}$$

$$\tag{29}$$

$$\times \sum_{m=-\infty}^{+\infty} \left[C_{M_2 N_2}^J J_{m-M_2} \left(\kappa_{|M_2|,N_2} R_\alpha \right) + C_{M_2 N_2}^Y Y_{m-M_2} \left(\kappa_{|M_2|,N_2} R_\alpha \right) \right]$$

$$\times \left[C_{M_1 N_1}^J J_{m-M_1} \left(\kappa_{|M_1|,N_1} R_\alpha \right) + C_{M_1 N_1}^Y Y_{m-M_1} \left(\kappa_{|M_1|,N_1} R_\alpha \right) \right]$$

$$\times \left\{ I_3 \left(P_1 - P_2, N_2, N_1, M_2, M_1; m, r_\alpha \right) - r_\alpha^4 \sum_{l=|m|}^{\infty} \frac{(2l+1)(l-|m|)!}{(l+|m|)!} c_{lm\alpha} \right\},$$

where

$$I_3 = 2 \int_0^{\pi/2} \int_0^{r_\alpha} \cos \left[r \left(K_{P_1} - K_{P_2} \right) \cos\theta \right] J_m \left(\kappa_{|M_2|,N_2} r \sin\theta \right)$$

$$\times J_m \left(\kappa_{|M_1|,N_1} r \sin\theta \right) r^2 \sin\theta \, d\theta dr, \tag{30}$$

$$c_{lm\alpha} = \{ I_2^* (M_2 N_2 P_2; r_\alpha, l, |m|) \, \dot{u}_{l\alpha}(r_\alpha, E_{l\alpha})$$

$$- I_1^* (M_2 N_2 P_2; r_\alpha, l, |m|) \, \dot{u}'_{l\alpha}(r_\alpha, E_{l\alpha}) \}$$

$$\times \{ I_2 (M_1 N_1 P_1; r_\alpha, l, |m|) \, \dot{u}_{l\alpha}(r_\alpha, E_{l\alpha})$$

$$- I_1 (M_1 N_1 P_1; r_\alpha, l, |m|) \, \dot{u}'_{l\alpha}(r_\alpha, E_{l\alpha}) \}$$

$$+ N_{lm\alpha} \{ I_1^* (M_2 N_2 P_2; r_\alpha, l, |m|) \, u'_{l\alpha}(r_\alpha, E_{l\alpha}) \tag{31}$$

$$- I_2^* (M_2 N_2 P_2; r_\alpha, l, |m|) \, u_{l\alpha}(r_\alpha, E_{l\alpha}) \}$$

$$\times \{ I_1 (M_1 N_1 P_1; r_\alpha, l, |m|) \, u'_{l\alpha}(r_\alpha, E_{l\alpha})$$

$$- I_2 (M_1 N_1 P_1; r_\alpha, l, |m|) \, u_{l\alpha}(r_\alpha, E_{l\alpha}) \},$$

$$\left\langle P_2 M_2 N_2 \left| \widehat{H} \right| P_1 M_1 N_1 \right\rangle$$

$$= \left(K_{P_1} K_{P_2} + \kappa_{|M_1|,N_1} \kappa_{|M_2|,N_2} \right) \delta_{P_2 P_1} \delta_{N_2 N_1} \delta_{M_2 M_1} - \frac{1}{c} (-1)^{M_1 + M_2}$$

$$\times \sum_\alpha \exp \left\{ i \left[\left(K_{P_1} - K_{P_2} \right) Z_\alpha + (M_1 - M_2) \Phi_\alpha \right] \right\}$$

$$\times \sum_{m=-\infty}^{+\infty} \left[C_{M_2 N_2}^J J_{m-M_2} \left(\kappa_{|M_2|,N_2} R_\alpha \right) + C_{M_2 N_2}^Y Y_{m-M_2} \left(\kappa_{|M_2|,N_2} R_\alpha \right) \right]$$

$$\times \left[C_{M_1 N_1}^J J_{m-M_1} \left(\kappa_{|M_1|,N_1} R_\alpha \right) + C_{M_1 N_1}^Y Y_{m-M_1} \left(\kappa_{|M_1|,N_1} R_\alpha \right) \right]$$

$$\times \left\{ K_{P_1} K_{P_2} I_3 + \kappa_{|M_2|,N_2} \kappa_{|M_1|,N_1} I'_3 + m^2 I_{10} - \right.$$

$$\left. - r_\alpha^4 \sum_{l=|m|}^\infty \frac{(2l+1)(l-|m|)!}{(l+|m|)!} \left[E_{l\alpha} c_{lm\alpha} + \gamma_{lm\alpha} \right] \right\}. \tag{32}$$

Here,

$$I'_3 = 2 \int_0^{\pi/2} \int_0^{r_\alpha} \cos \left[r \left(K_{P_1} - K_{P_2} \right) \cos \theta \right] J'_m \left(\kappa_{|M_2|,N_2} r \sin \theta \right)$$

$$\times J'_m \left(\kappa_{|M_1|,N_1} r \sin \theta \right) r^2 \sin \theta \, d\theta \, dr, \tag{33}$$

$$I_4 = 2 \int_0^{r_\alpha} \int_0^{\pi/2} \cos \left\{ r \left(K_{P_1} - K_{P_2} \right) \cos \theta \right\} J_m (\kappa_{|M_2|,N_2} r \sin \theta)$$

$$\times J_m (\kappa_{|M_1|,N_1} r \sin \theta) (\sin \theta)^{-1} \, dr \, d\theta, \tag{34}$$

$$\gamma_{lm\alpha} = \left\{ I_2^* \left(r_\alpha, K_{P_2}, M_2, N_2, l, |m| \right) I_1 \left(r_\alpha, K_{P_1}, M_1, N_1, l, |m| \right) \right.$$

$$+ I_1^* \left(r_\alpha, K_{P_2}, M_2, N_2, l, |m| \right) I_2 \left(r_\alpha, K_{P_1}, M_1, N_1, l, |m| \right) \right\}$$

$$\times \dot{u}_{l\alpha} (r_\alpha, E_{l\alpha}) u'_{l\alpha} (r_\alpha, E_{l\alpha}) - I_2^* \left(r_\alpha, K_{P_2}, M_2, N_2, l, |m| \right)$$

$$\times I_2 \left(r_\alpha, K_{P_1}, M_1, N_1, l, |m| \right) \dot{u}_{l\alpha} (r_\alpha, E_{l\alpha}) u_{l\alpha} (r_\alpha, E_{l\alpha})$$

$$- I_1^* \left(r_\alpha, K_{P_2}, M_2, N_2, l, |m| \right) I_1 \left(r_\alpha, K_{P_1}, M_1, N_1, l, |m| \right) \ddot{u}'_{l\alpha} (r_\alpha, E_{l\alpha})$$

$$\times u'_{l\alpha} (r_\alpha, E_{l\alpha}), \tag{35}$$

Finally, from the equation

$$det \, \|\langle P_2 M_2 N_2 \,|H|\, P_1 M_1 N_1 \rangle - E(k) \, \langle P_2 M_2 N_2 \,|\, P_1 M_1 N_1 \rangle\| = 0, \qquad (36)$$

the electron dispersion law $E(k)$ of an isolated nanotube is determined.

2.2 Results

The use of the cylindrical MT approximation makes the calculation depending on the radii a and b of the outer and inner potential barriers. They were varied so as to reproduce the full width of the valence band of a (5, 5) carbon nanotube (22 eV). In the basis set including all LACWs with an energy up to 50 eV ($E_{cut} = 50\,eV$), coincidence was achieved at $b = R_\alpha - 1.21\text{Å}$ and $a = R_\alpha + 1.21\text{Å}$. At $E_{cut} = 100\,eV$, coincidence was achieved at $b = R_\alpha - 1.38\text{Å}$ and $a = R_\alpha + 1.38\text{Å}$. In calculations of the other carbon nanotubes, the radii a and b were not optimized; rather, they were calculated by these formulas depending on the selected E_{cut} value, 50 or 100 eV.

2.2.1 Metallic Armchair (n, n) Nanotubes

We calculated (n, n) systems with n from 3 to 12. Figures 3 and 4 show the band structure and density of states in the vicinity of the Fermi level for the (12,12) carbon nanotube [34]. The computation results show that carbon nanotubes with n from 4 to 12 have a metal-type band structure with the Fermi level located at the intersection of two π-bands at the point $k = (2/3)(\pi/c)$. The density of states near the Fermi level between the first singularities of the valence band and conduction band is constant. In the center of the Brillouin zone, the upper occupied σ level $\Gamma_v(\sigma)$ is located above the upper occupied π level $\Gamma_v(\pi)$ in all (n, n) carbon nanotubes. In π-electron models, S_{v1} and S_{c1} are the boundary singularities of the valence and conduction bands, respectively, and the minimal gap is $E_{11} = E_{11}(\pi\pi^*) = E(S_{v1}) - E(S_{c1})$. However, as can be seen in Figs. 3 and 4, in the center Γ and at the boundary K of the Brillouin zone, the lower $\Gamma_{c1}(\pi)$ and $K_{c1}(\pi)$ states are located below the S_{c1} singularity and form a shoulder under the S_{c1} peak of the density of states. The $S_{v1} - S_{c1}$ gap still corresponds to the direct transition with the minimal energy. The gap $E_{11}(\sigma\pi^*) = E[\Gamma_{c1}(\pi)] - E[\Gamma_v(\sigma)]$ corresponds to the second direct transition (Fig. 3). For (n, n) carbon nanotubes with $n = 8$–10, the $E_{11}(\pi\pi^*)$ and $E_{11}(\sigma\pi^*)$ energies are almost the same, and for tubes of smaller diameter, $E_{11}(\sigma\pi^*)$ is 0.2–0.5 eV smaller than $E_{11}(\pi\pi^*)$. The plots of the direct transition energies $E_{11}(\pi\pi^*)$ and $E_{11}(\sigma\pi^*)$ versus the diameter d of nanotubes are shown in Fig. 5. There are significant deviations from the relationship $E_{11} \sim d^{-1}$. The situation is complicated by close $E_{11}(\pi\pi^*)$ and $E_{11}(\sigma\pi^*)$ values and the intersection of these characteristics in the range 0.7 nm$^{-1} < d^{-1} < 1.0$ nm^{-1} ($n = 8$–10).

Fig. 3 Band structure of the
(12,12) nanotube

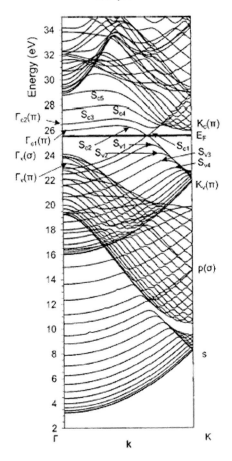

Fig. 4 Density of states
(DOS) of the (12,12) nan-
otube near the Fermi level

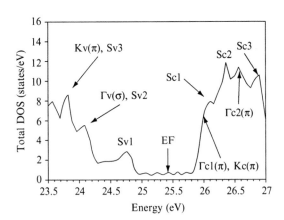

Fig. 5 Correlation between the diameter d and direct transition energies. Series 1: $E_{11}(\pi\pi^*) = E(S_{v1}) - E(S_{c1})$; series 2: $E_{11}(\sigma\pi^*) = E[\Gamma_{c1}(\pi)] - E[\Gamma_v(\sigma)]$; series 3: experimental data

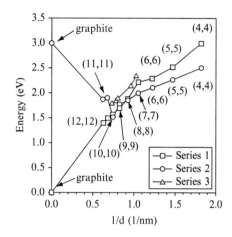

The electronic structure of the (3,3) nanotube is sharply different [34]. Here, the S_{c1} singularity coincides with the Fermi level, which leads to a sharp increase in the density of states at this level and correlates with the experimentally observed superconductivity of the (3,3) nanotube.

2.2.2 Semiconducting Zigzag (n, 0) Nanotubes

The $(n, 0)$ carbon nanotubes with $n = 5$–26 not a multiple of three are semiconducting (otherwise, the nanotubes are semimetals) [35]. As an example, Figs. 6 and 7 show the band structure and density of states of $(13, 0)$ carbon nanotube. The boundary singularities of the valence band (S_{v1}) and conduction band (S_{c1}) correspond to the direct $\pi\pi^*$ transition at the point Γ, and $E_{11} = E(S_{c1}) - E(S_{v1})$. The dependence of E_{11} on d^{-1} is oscillating (Fig. 8): the $E_{11}(d^{-1})$ function alternates between two curves corresponding to $(n, 0)$ nanotubes for which the remainder upon division of n by 3 is equal to 1 or 2 (n mod $3 = 1$ and n mod $3 = 2$, respectively). The curve mod $3 = 1$ is located above the curve mod $3 = 2$. The maximal values $E_{11} = 0.90\,\text{eV}$ and $E_{11} = 0.56\,\text{eV}$ for mod $3 = 1$ and mod $3 = 2$ correspond to the tubules with $d = 1.25$ ($n = 16$) and 11Å ($n = 14$), respectively. A further decrease in tube diameter leads to a sharp decrease in the gap E_{11} (Table 1). For carbon nanotubes with $n \leq 8(d \leq 6.3$ Å$)$, the gap is closed. Figure 8 shows that there is not a one-to-one correspondence between E_{11} and d. For example, the E_{11} gap of about 0.3 eV corresponds to four zigzag tubes with $d = 7.8, 8.6, 20.4,$ and 40.8 Å. The same is observed for the second direct gap $E_{22} = E(S_{c2}) - E(S_{v2})$. The amplitudes of oscillations of the $E_{22}(d^{-1})$ function are even greater (approximately by a factor of 3) than in the case of $E_{11}(d^{-1})$ (Fig. 9).

Fig. 6 Band structure of the (13,0) nanotube

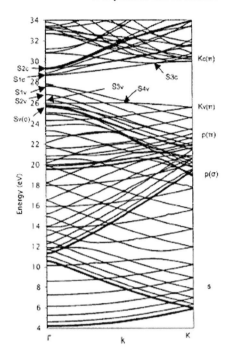

Fig. 7 Density of states of the (13,0) nanotube near the Fermi level

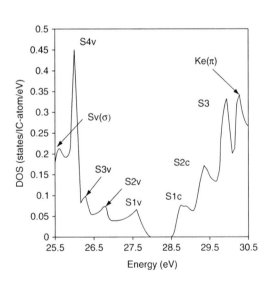

Fig. 8 Minimal optical gap vs. the diameter of semiconducting (n,0) nanotubes

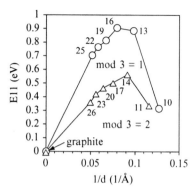

Table 1 Minimal energy gap of single-walled nanotubes

SWNT	E_{11}, eV
(10,0)	0.31
(11,0)	0.32
(13,0)	0.83
(14,0)	0.56
(16,0)	0.89
(17,0)	0.50
(19,0)	0.80
(20,0)	0.46
(22,0)	0.75
(23,0)	0.35
(25,0)	0.70
(26,0)	0.41
(28,0)	0.66
(29,0)	0.38
(31,0)	0.62

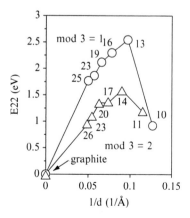

Fig. 9 Second direct optical gap E_{22} vs. the diameter d of semiconducting (n,0) nanotubes

3 Embedded Single-Walled Carbon Nanotubes

Interest has arisen in the design of hybrid electronic devices in which a carbon nanotube is embedded into a common bulk semiconductor. Such devices can be exemplified by electronic elements consisting of a single-walled carbon nanotube embedded into an epitaxially grown semiconductor heterostructure [37]. Let us consider how the interaction with a surrounding crystal can change the band structure of the embedded carbon nanotube [30, 36].

3.1 Computational Method

For an isolated nanotube, there are two vacuum regions Ω_v, on the outside and inside of the nanotube. The nanotube and the vacuum regions are separated by impenetrable (infinite) cylindrical potential barriers. For an embedded nanotube, it is surrounded on the outside with the region of a single-crystal matrix Ω_m (Fig. 10). The barrier V_m between the nanotube and the matrix is penetrable (finite), so that tunneling of electrons from the nanotube into the matrix is possible. Let us find the solutions of the Schrödinger equation for the orbitals and electronic energies of the nanotube in the matrix. The matrix is assumed to be a homogeneous medium with a constant potential V_m, which corresponds to the model of a single-walled carbon

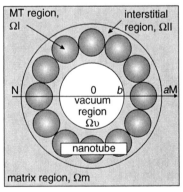

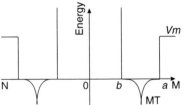

Fig. 10 (*Top*) Nanotube embedded into a matrix and (*bottom*) the cross section of the electron potential along the N0M line

nanotube in contact with an electron gas. Let us consider the case where the barrier V_m is located noticeably above the Fermi level, so that the matrix has a relatively weak effect on the states of the valence and conduction bands of the nanotube.

In the interspherical region of the nanotube and in the matrix region, the LACWs are the solutions of Schrödinger equation (Eq. (3)) with the potential

$$U(R) = \begin{cases} 0, & b \leq R \leq a \\ \infty, & R < b \\ V_m, & R > b \end{cases}. \tag{37}$$

Due to the cylindrical symmetry of the potential $U(R)$, the solutions of Eq. (3) are presented in the form $\Psi(Z, \Phi, R) = \Psi_P(Z)\Psi_M(\Phi)\Psi_{MN}(R)$. The $\Psi_{MN}(R)$ function describes the radial movement of an electron in the interspherical region Ω_{II} of the nanotube and in the matrix region Ω_m.

In the nanotube region, $U(R) = 0$ and Eq. (3) takes the form of Bessel equation (Eq. (9)). Its solutions, as in the cased of an isolated nanotube, are represented by Eq. (10).

In the matrix region $(U(R) = V_m)$, the $\Psi_{|M|N}(R)$ functions must obey the equation:

$$\left[\frac{d^2}{dR^2} + \frac{1}{R}\frac{d}{dR} - \left(V_m - \kappa^2_{|M|,N}\right) - \frac{M^2}{R^2}\right]\Psi_{|M|,N}(R) = 0. \tag{38}$$

Equation (38) at $V_m > \kappa^2_{MN}$ is a modified Bessel equation. Its solutions when R tends to infinity are modified Bessel functions of the first kind K_M:

$$\Psi_{MN}(R) = C^K_{MN} K_M\left(\kappa^K_{|M|,N} R\right), \tag{39}$$

where $\kappa^K_{|M|,N} = \left(V_m - \kappa^2_{|M|,N}\right)^{1/2}$. The function $\Psi_{MN}(R)$ should vanish at $R = b$, be continuous and differentiable at $R = a$, and normalized. As a result, we obtain the set of equations for the coefficients C^J_{MN}, C^Y_{MN} and C^K_{MN} and for the energy $\kappa^2_{|M|N}$:

$$\begin{cases} C^J_{MN} J_M\left(\kappa_{|M|,N} b\right) + C^Y_{MN} Y_M\left(\kappa_{|M|,N} b\right) = 0 \\ C^J_{MN} J_M\left(\kappa_{|M|,N} a\right) + C^Y_{MN} Y_M\left(\kappa_{|M|,N} a\right) = C^K_{MN} K_M\left(\kappa^K_{|M|,N} a\right) \\ \kappa_{|M|,N}\left[C^J_{MN} J'_M\left(\kappa_{|M|,N} a\right) + C^Y_{MN} Y'_M\left(\kappa_{|M|,N} a\right)\right] = \kappa^K_{|M|,N} C^K_{MN} K'_M\left(\kappa^K_{|M|,N} a\right). \\ \int\limits_b^\infty \left|\Psi_{|M|,N}(R)\right|^2 R dR = 1, \end{cases} \tag{40}$$

Thus, the form of the basis function in the interspherical region of the nanotube and in the matrix is finally determined.

As in the case of an isolated nanotube, the LACW inside the MT spheres is expanded in spherical harmonics $Y_{l,m}$ (Eq. (18)). Coefficients A_{lm} and B_{lm} are selected so that both $\Psi(k,P,M,N)$ and its derivative have no discontinuities at the boundaries of the MT spheres. However, the analytical form of the cylindrical wave near the MT spheres of the nanotube remains unaltered in going from the isolated nanotube to the nanotube embedded into the matrix. Therefore, the analytical expressions for the coefficients $A_{lm\alpha}$ and $B_{lm\alpha}$ (Eqs. (24) and (25), respectively) remain valid. Thus, at $b \leq R \leq a$, the LACWs Ψ_{MNP} have the same analytical form as for a separate nanotube, whereas, in the matrix region,

$$\Psi_{PMN}(R) = \frac{C_{MN}^K}{\sqrt{2\pi c}} e^{i(k+k_P)Z} e^{iM\Phi} K_M\left(\kappa_{|M|,N}^K R\right), \qquad (41)$$

the analytical expressions for the overlap and Hamiltonian integrals (Eqs. (29) and (32), respectively) remain valid, but the $\kappa_{|M|,N}$, C_{MN}^J and C_{MN}^Y values should be calculated by Eqs. (40).

3.2 Calculation Results

We studied the effect of the crystalline matrix on the electronic states of metallic (n,n) nanotubes with $4 \leq n \leq 12$ and semiconducting $(n,0)$ nanotubes with $10 \leq n \leq 25$ (n is not a multiple of three). Representative results are shown in Figs. 11 and 12. To characterize the barrier V_m, we used the dimensionless parameter $\varepsilon_m = V_m/\Delta$, where Δ is the position of the Fermi level in an isolated nanotube relative to the constant interspherical potential. It can be seen that the delocalization of electrons of a metallic (5,5) nanotube into the matrix region leads to a strong disturbance of the band structure. The most important matrix effect is the shift of the σ states located at the point Γ toward higher energies. As a result, the top of the valence σ band Γ_{v1} is shifted into the conduction band and σ electrons start participating in charge transfer. The point of intersection of boundary π bands is shifted toward the edge of the Brillouin zone, and the full width of the valence band is reduced. The metallic character of the band structure of an armchair nanotube persists. In the pristine nanotube, the Fermi level is located at a minimum, and tunneling of electrons into the matrix region leads to an increase in the density of states at the Fermi level.

For semiconducting nanotubes, the minimal gap E_{11} in the center of the Brillouin zone is sensitive to the matrix effect. With a decrease in the barrier V_m, the initial gap E_{11} of a (13,0) nanotube first slightly increases and then sharply decreases and collapses. The afore described matrix effect is common to nanotubes of all diameters.

Metallization of nanotubes under the action of a matrix as predicted by the model is consistent with the electrical properties of hybrid elements consisting of single-walled nanotubes in semiconducting layers [37]. In all 20 experimentally studied elements, the conductivity at room temperature was independent of the gate voltage;

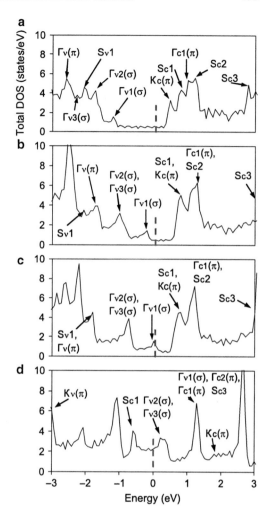

Fig. 11 Density of states of the single-walled (5,5) nanotube embedded into a crystal matrix for different barriers (6, 4, and 2, respectively)

i.e., all nanotubes in crystals turned out to be metallic. (According to statistics, one third of the tubes, i.e., about seven nanotubes, should be metallic, whereas the rest of them, i.e., about 13 nanotubes, should be semiconducting.)

4 Double-Walled Nanotubes

Double-walled nanotubes are the simplest case of multiwalled nanotubes. They consist of two concentric cylindrical graphene layers with a strong covalent bond between C atoms in each layer and a weak van der Waals interaction between

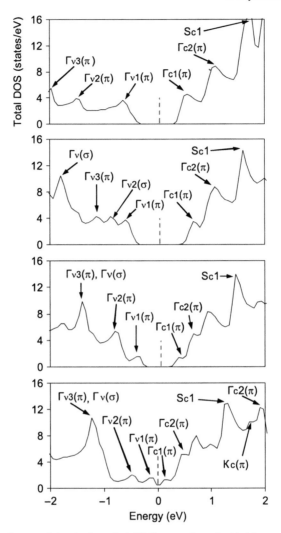

Fig. 12 Density of states of the single-walled (13,0) nanotube embedded into a crystal matrix for different barriers (6, 4, and 2, respectively)

the layers. From the standpoint of nanoelectronics, double-walled nanotubes are of interest since they are molecular analogues of coaxial cables. The interlayer interaction in a double-walled nanotube has an effect on both the optical and electrical properties of a nanocable [29, 38].

4.1 Computational Method

Let us assume that the atoms of a double-walled nanotube are confined between two infinite cylindrical barriers impenetrable to electrons Ω_{b1} and Ω_{a2}, beyond which

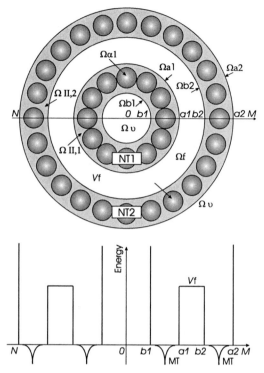

Fig. 13 (*Top*) Cross section of a double-walled nanotube and (*bottom*) the electron potential along the line N0M. Here, HT$_1$ and HT$_2$ are the inner and outer nanotubes, respectively; $\Omega_{II,1}$ and $\Omega_{II,2}$ are the interspherical regions of these tubes; Ω_{a1} and Ω_{a2} are the muffin-tin (MT) regions of tubes 1 and 2, respectively; Ω_{b1} and Ω_{a2} are the internal and external cylindrical impenetrable barriers between the double-walled nanotube and the vacuum regions Ω_v; V_f is the potential energy in the Ω_f region confined by the cylindrical potential barriers Ω_{a1} and Ω_{b2}

the vacuum region Ω_v is located (Fig. 13). The cylindrical potential barriers Ω_{a1} and Ω_{b2} on the outer side of the internal tube and on the inner side of the external tube are penetrable and, hence, tunneling exchange of electrons between the layers of a double-walled tube is possible. The radii of these barriers a_1, b_1 and a_2, b_2 are selected as in the case of single-walled nanotubes. The potential V_f of the interlayer region Ω_f in this model is the only parameter. We select it taking into account the graphite band structure. In graphite, the interlayer interaction splits and shifts the bands by about 1–2 eV. Calculations of three-dimensional graphite in the MT approximation and by the full potential method show that disturbances of the valence π and σ bands are 4–2 and 0.5–1.0 eV, respectively [39]. In the double-walled nanotube with the same interlayer distance, the band splittings and shifts will be about two times smaller since each graphene layer in graphite interacts with two nearest layers, whereas there is only one interlayer interaction in the double-walled nanotube. In addition, the effects of interlayer interaction are expected to be even smaller as compared to graphite due to the hybridization of the π and σ

states caused by the curvature of the tubes. Therefore, the potential V_f, the same for all double-walled nanotubes, was selected so that the level splitting and shifts in the (5,5)@(10,10) nanotube were, on average, 0.5 eV; the interlayer distance in this double-walled tube, 3.4 Å, is the same as in graphite.

In the interspherical region of the nanotube and in the classically impenetrable region between these spheres, the wave functions are the solutions of the Schrödinger equation for the free electron motion, which has its previous form (Eq. (3)); however, the potential $U(R)$ in this case has a more complicated form:

$$U(R) = \begin{cases} 0, & b_1 \leq R \leq a_1, b_2 \leq R \leq a_2 \\ \infty, & R < b_1, R > a_2 \\ V_f, & a_1 \leq R \leq b_2 \end{cases} . \tag{42}$$

Due to the symmetry of the potential $U(R)$, the solution of Eq. (3) is again presented in the form $\Psi(Z, \Phi, R) = \Psi_P(Z)\Psi_M(\Phi)\Psi_{MN}(R)$. The $\Psi_{MN}(R)$ function describes the radial movement of an electron in the interspherical regions $\Omega_{II,j}$ of two nanotubes ($j=1,2$) and in the classically forbidden region Ω_f. In the region $\Omega_{II,j}$, $U(R) = 0$, Eq. (7) is the Bessel equation and its solutions are linear combinations of cylindrical Bessel functions of the first J_M and second Y_M kinds:

$$\Psi^j_{II,|M|,N}(R) = C^{J,j}_{M,N} J_M\left(\kappa_{|M|,N} R\right) + C^{Y,j}_{M,N} Y_M\left(\kappa_{|M|,N} R\right), j = 1, 2. \tag{43}$$

In the region Ω_f, $U(R) = V_f$, and the $\Psi_{f,|M|,N}(R)$ functions must obey the equation

$$\left[\frac{d^2}{dR^2} + \frac{1}{R}\frac{d}{dR} - \left(V_f - \kappa^2_{|M|,N}\right) - \frac{M^2}{R^2}\right]\Psi_{f,|M|,N}(R) = 0. \tag{44}$$

Let us consider the electronic levels of a double-walled nanotube located below the potential V_f of the classical forbidden region. Equation (44) at $V_f > \kappa^2_{|M|,N}$ is a modified Bessel equation [32, 33]. Its solution is a linear combination of modified Bessel functions of the first K_M and second I_M kinds:

$$\Psi_{f,|M|,N}(R) = C^K_{M,N} K_M\left(\kappa^f_{|M|,N} R\right) + C^I_{M,N} I_M\left(\kappa^f_{|M|,N} R\right), \tag{45}$$

where $\kappa^f_{|M|,N} = \left(V_f - \kappa^2_{|M|,N}\right)^{1/2}$. The function $\Psi_{II,f|M|N}(R)$ should vanish at the inner and outer barriers of a double-walled tube:

$$C^{J,1}_{M,N} J_M\left(\kappa_{|M|,N} b_1\right) + C^{Y,1}_{M,N} Y_M\left(\kappa_{|M|,N} b_1\right) = 0, \tag{46}$$

$$C^{J,2}_{M,N} J_M\left(\kappa_{|M|,N} a_2\right) + C^{Y,2}_{M,N} Y_M\left(\kappa_{|M|,N} a_2\right) = 0, \tag{47}$$

be continuous and differentiable at the boundaries between the tubes at $R = a_1$ and $R = b_2$:

$$C_{M.N}^{J,1} J_M \left(\kappa_{|M|,N} a_1 \right) + C_{M.N}^{Y,1} Y_M \left(\kappa_{|M|,N} a_1 \right)$$
$$= C_{M.N}^{K} K_M \left(\kappa_{|M|,N}^{f} a_1 \right) + C_{M.N}^{I} I_M \left(\kappa_{|M|,N}^{f} a_1 \right), \tag{48}$$

$$C_{M.N}^{J,2} J_M \left(\kappa_{|M|,N} b_2 \right) + C_{M.N}^{Y,2} Y_M \left(\kappa_{|M|,N} b_2 \right)$$
$$= C_{M.N}^{K} K_M \left(\kappa_{|M|,N}^{f} b_2 \right) + C_{M.N}^{I} I_M \left(\kappa_{|M|,N}^{f} b_2 \right), \tag{49}$$

$$\kappa_{|M|,N} \left[C_{M.N}^{J,1} J_M' \left(\kappa_{|M|,N} a_1 \right) + C_{M.N}^{Y,1} Y_M' \left(\kappa_{|M|,N} a_1 \right) \right]$$
$$= \kappa_{|M|,N}^{f} \left[C_{M.N}^{K} K_M' \left(\kappa_{|M|,N}^{f} a_1 \right) + C_{M.N}^{I} I_M' \left(\kappa_{|M|,N}^{f} a_1 \right) \right], \tag{50}$$

$$\kappa_{|M|,N} \left[C_{M.N}^{J,2} J_M' \left(\kappa_{|M|,N} b_2 \right) + C_{M.N}^{Y,2} Y_M' \left(\kappa_{|M|,N} b_2 \right) \right]$$
$$= \kappa_{|M|,N}^{f} \left[C_{M.N}^{K} K_M' \left(\kappa_{|M|,N}^{f} b_2 \right) + C_{M.N}^{I} I_M' \left(\kappa_{|M|,N}^{f} b_2 \right) \right], \tag{51}$$

and normalized:

$$\int_{b_1}^{a_2} |\Psi_{IIf,|M|N}(R)|^2 R dR = 1. \tag{52}$$

From Eqs. (46)–(52), we can calculate and $C_{M.N}^{J,j}$ and $C_{M.N}^{Y,j}$ $(j = 1, 2)$, $C_{M.N}^{K}$, $C_{M.N}^{I}$, and $\kappa_{|M|,N}$.

Thus, in the regions $\Omega_{II,1}$, $\Omega_{II,2}$, and Ω_f, the form of the basis function $\Psi_{IIf,PMN}$ is finally determined. Inside the MT spheres α of the jth tube, the LACW Ψ_{PMN} of the double-walled tube is expanded in spherical harmonics Y_{lm} (18).

Both the cylindrical wave $\Psi_{II,PMN}$ and the spherically symmetric part $\Psi_{I,j\alpha,PMN}$ of the augmented cylindrical wave in the MT regions for the double-walled nanotubes have the same form as for the constituent tubes. Therefore, the expressions for the overlap and Hamiltonian integrals obtained for the single-walled tube (Eqs. (31) and (32), respectively) can be rewritten for the double-walled tube:

$$\langle \Psi_{P_2 M_2 N_2} | \Psi_{P_1 M_1 N_1} \rangle = \delta_{P_2 M_2 N_2, P_1 M_1 N_1} - \frac{1}{c} (-1)^{M_2 + M_1}$$
$$\times \sum_{j=1,2} \sum_{\alpha} \exp \left\{ i \left[(k_{P_1} - k_{P_2}) Z_{j\alpha} + (M_1 - M_2) \Phi_{j\alpha} \right] \right\}$$
$$\times \sum_{m=-\infty}^{\infty} \left[C_{M_2,N_2}^{J,j} J_{m-M_2} \left(\kappa_{|M_2|,N_2} R_{j\alpha} \right) + C_{M_2,N_2}^{Y,j} Y_{m-M_2} \left(\kappa_{|M_2|,N_2} R_{j\alpha} \right) \right]$$
$$\times \left[C_{M_1,N_1}^{J,j} J_{m-M_1} \left(\kappa_{|M_1|,N_1} R_{j\alpha} \right) + C_{M_1,N_1}^{Y,j} Y_{m-M_1} \left(\kappa_{|M_1|,N_1} R_{j\alpha} \right) \right]$$
$$\times \left\{ I_{3,j\alpha}^{P_2 M_2 N_2, P_1 M_1 N_1} - r_{j\alpha}^4 \sum_{l=|m|}^{\infty} \frac{(2l+1) [(l-|m|)!]}{2 [(l+|m|)!]} S_{lm,j\alpha}^{P_2 M_2 N_2, P_1 M_1 N_1} (r_{j\alpha}) \right\}, \tag{53}$$

$$\left\langle \Psi_{P_2 M_2 N_2} | \hat{H} | \Psi_{P_1 M_1 N_1} \right\rangle = \left(K_{P_2} K_{P_1} + \kappa_{|M_2|,N_2} \kappa_{|M_1|,N_1} \right) \delta_{P_2 M_2 N_2, P_1 M_1 N_1}$$

$$- \frac{1}{c} (-1)^{M_2 + M_1} \sum_{j=1,2} \sum_{\alpha} \exp \left\{ i \left[(k_{P_1} - k_{P_2}) Z_{j\alpha} + (M_1 - M_2) \Phi_{j\alpha} \right] \right\}$$

$$\times \sum_{m=-\infty}^{\infty} \left[C_{M_2,N_2}^{J,j} J_{m-M_2} \left(\kappa_{|M_2|,N_2} R_{j\alpha} \right) + C_{M_2,N_2}^{Y,j} Y_{m-M_2} \left(\kappa_{|M_2|,N_2} R_{j\alpha} \right) \right]$$

$$\times \left[C_{M_1,N_1}^{J,j} J_{m-M_1} \left(\kappa_{|M_1|,N_1} R_{j\alpha} \right) + C_{M_1,N_1}^{Y,j} Y_{m-M_1} \left(\kappa_{|M_1|,N_1} R_{j\alpha} \right) \right]$$

$$\times \left\{ K_{P_2} K_{P_1} I_{3,j\alpha}^{P_2 M_2 N_2, P_1 M_1 N_1} + \kappa_{|M_2|,N_2} \kappa_{|M_1|,N_1} I_{3,j\alpha}^{\prime P_2 M_2 N_2, P_1 M_1 N_1} \right.$$

$$+ m_4^2 I_{4,j\alpha}^{P_2 M_2 N_2, P_1 M_1 N_1} - r_{j\alpha}^4 \sum_{l=|m|}^{\infty} \frac{(2l+1) \left[(l-|m|)! \right]}{2 \left[(l+|m|)! \right]}$$

$$\left. \times \left(E_{l,j\alpha} S_{lm,j\alpha}^{P_2 M_2 N_2, P_1 M_1 N_1} (r_{j\alpha}) + \gamma_{lm,j\alpha}^{P_2 M_2 N_2, P_1 M_1 N_1} (r_{j\alpha}) \right) \right\}. \tag{54}$$

The electronic structure of the double-walled tube can be described as consisting of the band structure of its constituent single-walled nanotubes. Indeed, for the eigenvalues $\Psi_{nk}(r)$, we can calculate the probabilities $w_{j,nk}$ and $w_{f,nk}$ of electrons occurring in the jth region of the tube and in the classically impenetrable region:

$$w_{j,nk} = \sum_{PMN} |c_{nk,PMN}|^2 \frac{\left(C_{M,N}^{J,j} \right)^2 + \left(C_{M,N}^{Y,j} \right)^2}{\sum_{i=1,2} \left[\left(C_{M,N}^{J,i} \right)^2 + \left(C_{M,N}^{Y,i} \right)^2 \right] + \left(C_{M,N}^{I} \right)^2 + \left(C_{M,N}^{K} \right)^2},$$

$$j = 1, 2 \tag{55}$$

$$w_{f,nk} = \sum_{PMN} |c_{nk,PMN}|^2 \frac{\left(C_{M,N}^{I} \right)^2 + \left(C_{M,N}^{K} \right)^2}{\sum_{i=1,2} \left[\left(C_{M,N}^{J,i} \right)^2 + \left(C_{M,N}^{Y,i} \right)^2 \right] + \left(C_{M,N}^{I} \right)^2 + \left(C_{M,N}^{K} \right)^2}. \tag{56}$$

At a high barrier V_f, the probabilities $w_{j,nk}$ are close to zero or unity for different dispersion curves and are almost independent of the wave vector k. Hence, each dispersion curve $E_{j,n}(k)$ of the double-walled tube can be characterized by the number j of the tube on which the electrons of a given band are mainly localized. The band structures of double-walled nanotubes can be represented by two structures corresponding to the state of the inner and outer tubes. The full band structure of a double-walled nanotube is a superposition of the band structures of the core and shell tubes.

4.2 Computational Results

4.2.1 Semiconducting Double-Walled Nanotubes

We calculated semiconducting $(n,0)@(n',0)$ nanotubes with $10 \leq n \leq 23$ and $19 \leq n' \leq 32$, where n and n' are not multiples of 3. Table 2 presents the minimal gaps E_{11} in double-walled nanotubes and the shifts ΔE_{11} of these gaps caused by the interlayer interaction. The densities of states near the Fermi level of the single-walled $(13,0)$ and $(22,0)$ nanotubes can be compared with analogous data for the core $(13,0)$ and shell $(22,0)$ tubes in the double-walled system (Fig. 14).In the $(13,0)@(22,0)$ double-walled nanotube, both the inner and outer tubes belong to the series of n mod $3 = 1$; the minimal optical gap (0.83 eV) of the smaller single-walled $(13,0)$ nanotube is wider than the gap (0.76 eV) of the larger $(22,0)$ tube, which is consistent with the simple approximate equation $E_{11} \sim d^{-1}$. Our calculations show that the minimal optical gap E_{11} of the $(13,0)$ nanotube increases by 0.19 eV, while that of the $(22,0)$ nanotube decreases by 0.19 eV after the formation of the double-walled system. The density of states curves near the Fermi level show an analogous decrease in the energy shift of the second gap E_{22} by 0.3 and 0.4 eV for the $(13,0)$ and $(22,0)$ nanotubes, respectively. The interlayer interaction leads to an even stronger disturbance of the band structure of the inner nanotube as compared

Table 2 Minimal energy gap E_{11} of the core and shell nanotubes in the double-walled nanotube and gap shifts ΔE_{11}, after its formation

Double-walled nanotube	E_{11}, eV		ΔE_{11}, eV	
	Inner	Outer	Inner	Outer
(10,0)@(19,0)	0.64	0.65	0.32	−0.15
(10,0)@(20,0)	0.63	0.53	0.32	0.07
(10,0)@(19,0)	0.64	0.65	0.32	−0.15
(10,0)@(20,0)	0.63	0.53	0.32	0.07
(11,0)@(19,0)	0.71	0.65	0.39	−0.16
(11,0)@(20,0)	0.71	0.53	0.39	0.07
(13,0)@(22,0)	1.02	0.55	0.19	−0.19
(13,0)@(23,0)	1.02	0.50	0.19	0.15
(14,0)@(22,0)	0.70	0.56	0.14	−0.19
(14,0)@(23,0)	0.70	0.50	0.14	0.15
(16,0)@(25,0)	0.94	0.52	0.04	−0.18
(16,0)@(26,0)	0.93	0.48	0.04	0.07
(17,0)@(25,0)	0.45	0.52	−0.05	−0.18
(17,0)@(26,0)	0.45	0.48	−0.05	0.07
(19,0)@(28,0)	0.76	0.46	−0.05	−0.20
(19,0)@(29,0)	0.76	0.46	−0.05	0.07
(20,0)@(28,0)	0.42	0.46	−0.05	−0.20
(20,0)@(29,0)	0.42	0.46	−0.05	0.07
(22,0)@(31,0)	0.75	0.40	0.00	−0.22
(23,0)@(31,0)	0.40	0.40	0.06	−0.22

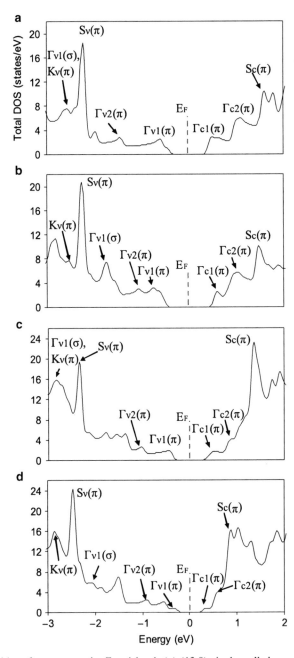

Fig. 14 Densities of states near the Fermi level: (**a**) (13,0) single-walled nanotube; (**b**) (13,0) core nanotube nested into the (22,0) nanotube; (**c**) (22,0) single-walled nanotube; and (**d**) the outer (22,0) nanotube with the nested (13,0) tube

to the outer nanotube. The reason for this is that the extra space located between the barriers Ω_{b2} and Ω_{a2} accessible for the electrons of the inner $(13,0)$ nanotube is about twice as large as the new accessible region between the Ω_{b1} and Ω_{a1} barriers for the outer $(22,0)$ nanotube. Upon the formation of the $(13,0)@(22,0)$ double-walled nanotube, the valence band width $E_v = E_F - E(\Gamma_{2vs})$ of the $(13,0)$ nanotube decreases by 1.40 eV and that of the $(22,0)$ nanotube, only by 0.04 eV.

In the double-walled $(14,0)@(22,0)$ nanotube, the inner tube belongs to the series of $n \bmod 3 = 2$. Here, the gap of the inner $(14,0)$ nanotube is narrower than the gap of the outer $(22,0)$ single-walled nanotube with $n \bmod 3 = 1$. Due to the interlayer interaction, the gap of the inner tube increase by 0.14 eV and that of the outer tube decrease by 0.19 eV. For the core and shell nanotubes, the gap shifts ΔE_{11} caused by the interlayer interaction are oppositely directed in the double-walled $(13,0)@(22,0)$ and $(14,0)@(22,0)$ nanotubes: The ΔE_{11} values are positive and negative for the inner and outer nanotubes, respectively.

In the double-walled $(13,0)@(23,0)$ nanotube with the wide-gap inner tube and the narrow-gap outer tube, the gap shifts ΔE_{11} are 0.19 and 0.15 eV, respectively, i.e., almost equal and positive. The same is true for the $(14,0)@(23,0)$ nanotube, in which both tubes belong to the series of $n \bmod 3 = 2$. Here, ΔE_{11} is 0.14 and 0.15 eV for the core and shell nanotubes, respectively.

Table 2 shows that, whatever the type of the inner tube, the energy gap E_{11} of the outer tube decreases by 0.15–0.22 eV if $n \bmod 3 = 2$. On the other hand, for the outer tube with $n \bmod 3 = 1$, the gap shift ΔE_{11} is always negative: $-0.15 \le \Delta E_{11} \le -0.05$ eV. In both cases, the ΔE_{11} shifts do not decrease, but slightly oscillate in going to tube of larger diameter. For the inner tubes, the ΔE_{11} shift directly depends on d. For the series with $n \bmod 3 = 2$ and $n \bmod 3 + 1$ with $10 \le n \le 16$, the ΔE_{11} shift is positive, and the maximal ΔE_{11} value is equal to 0.39 and 0.32 eV, respectively. In going to inner nanotubes of larger diameter, the ΔE_{11} shift sharply decreases and then varies from -0.05 to 0.06 eV.

4.2.2 Metallic Double-Walled Nanotubes

Let us consider the metallic double-walled $(5,5)@(10,10)$ nanotube. Figure 15 shows the influence of interlayer interaction on the density of states. The interlayer interaction does not disturb the metal character of the band structure of $(5,5)$ and $(10,10)$ nanotubes. The Fermi level is located between the π bands near $k=(2/3)(\pi/c)$ both in each single-walled nanotube and in the double-walled tube. The formation of the double-walled nanotube leads to an increase in the valence band width by 1.3 eV for the $(5,5)$ nanotube and only 0.15 eV for the $(10,10)$ nanotube. The high-energy shift of the σ states from the occupied π states is the most significant effect of the interlayer interaction in double-walled armchair nanotubes. In the center of the Brillouin zone, the highest-lying occupied σ states $\Gamma_{v1}(\sigma)$ are above the highest-lying occupied π states $\Gamma_{v1}(\pi)$ in all single walled armchair nanotubes.

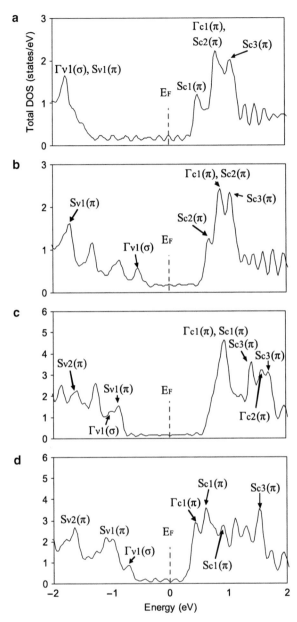

Fig. 15 Densities of states near the Fermi level: (**a**) (5,5) single-walled nanotube; (**b**) (5,5) core nanotube nested into the (10,10) nanotube; (**c**) (10,10) single-walled nanotube; and (**d**) the outer (10,10) nanotube with the nested (5,5) tube

5 Single-Walled Chiral Nanotubes

For even small-diameter chiral nanotubes, the number of atoms in the translational unit cell can be very large. For example, the translational cell of the achiral (10,10) nanotube comprises 40 atoms, whereas the translational cell of the chiral (10,9) tube of somewhat smaller diameter comprises 1,084 atoms. The basis set required for convergence rapidly increases with an increase in the number of atoms in the unit cell, which renders impracticable calculations of chiral tubes. These facts indicate that all rather than only translational symmetry properties of nanotubes should be considered in development of the theory of their electronic structure [31, 39].

5.1 Structure of Nanotubes

The atomic structure of any single-walled carbon nanotube is determined by two indices (n_1, n_2) and the bond length between carbon atoms $d_{C-C} = 1.42$ Å and can be generated in the following way [12]. Let us consider a cylinder of radius R_{NT} infinite along the Z axis:

$$R_{NT} = \frac{d_{C-C}\sqrt{3}}{2\pi} \left(n_1^2 + n_2^2 + n_1 n_2\right)^{1/2} \tag{57}$$

Let us select a point at this surface that determines the cylindrical coordinates Z, Φ, R of the first atom of the nanotube. The coordinates of the second atom of the nanotube are obtained by rotation of the coordinates of the first point through the angle Φ_{T_2} around the Z axis with the translation δ_{T_2} along this axis:

$$\Phi_{T_2} = \pi \frac{n_1 + n_2}{n_1^2 + n_2^2 + n_1 n_2}, \quad \delta_{T_2} = \frac{d_{C-C}}{2} \frac{n_1 - n_2}{\left(n_1^2 + n_2^2 + n_1 n_2\right)^{1/2}}. \tag{58}$$

The (n_1, n_2) nanotube has an axis of symmetry C_n, where n is the greatest common factor of the n_1 and n_2 indices; therefore, the coordinates of another $2(n-1)$ atoms are found from the rotational symmetry of the nanotube by means of $n-1$ rotations of the coordinates of the first two points around the Z axis through the angle $\omega_n = 2\pi/n$. These $2n$ atoms determine the repeating motif of the nanotube structure. The arrangement of the other atoms of the nanotube is determined by means of screw translations $S(\omega, h)$ of the coordinates of the $2n$ atoms. The screw axis of the (n_1, n_2) nanotube is characterized by an angle $0 < \omega < 2\pi$ and translation h

$$\omega = 2\pi \frac{n_1 p_1 + n_2 p_2 + (n_2 p_1 + n_1 p_2)/2}{n_1^2 + n_2^2 + n_1 n_2},$$

$$h = \frac{3d_{C-C}}{2} \frac{n}{\left(n_1^2 + n_2^2 + n_1 n_2\right)^{1/2}} = \frac{3\sqrt{3}d_{C-C}^2}{4\pi} \frac{n}{R_{NT}}. \tag{59}$$

where $p_1 \geq p_2$ are integers obeying the equation $p_2 n_1 - p_1 n_2 = n$.

Thus, when rotational and screw symmetries are considered, the actual rather than translational unit cell of any nanotube contains only two atoms. If these symmetry properties are taken into account when writing basis wave functions, the electronic structures of any nanotube can be calculated.

5.2 Computational Method

Let us discuss the symmetry properties of the wave function of the nanotube. Due to the existence of the n-fold symmetry axis, we can determine the discrete wave vector k_Φ corresponding to the irreducible representations of a cyclic group. The wave function Ψ should satisfy the equation

$$\Psi(Z, \Phi + t\omega_n, R) = e^{ik_\Phi t\omega_n} \Psi(Z, \Phi, R) \qquad (60)$$

where t is an integer. Substituting $t = n$ and using the cyclic conditions $\Psi(Z, \Phi + 2\pi, R) = \Psi(Z, \Phi, R)$, we find that the k_Φ values should be integers and can be written as $k_\Phi = L + nM$, where $M = 0, \pm 1, \ldots$ and $L = 0, 1, \ldots, n - 1$.

An ideal nanotube infinite along the Z axis is also invariant with respect to operations of screw translation $S(h, \omega)$, which is the displacement h along this axis with rotation about it through the angle ω. The screw translations form an Abelian group isomorphic to the group of primitive translations $T(h)$; therefore, for the wave function, the Bloch theorem should be valid:

$$\Psi(Z + th, \Phi + t\omega, R) = e^{iK_P th} \Psi(Z, \Phi, R), \qquad (61)$$

where

$$K_P = k + k_P, k_P = \frac{2\pi}{h} P, P = 0, \pm 1, \ldots \qquad (62)$$

and k is the wave vector in the first Brillouin zone.

Again, the potential is assumed to be spherically symmetric in the vicinity of atoms and constant in the interatomic space. It is also assumed that the atoms in the nanotube are located between two cylindrical barriers impenetrable for electrons. In the interspherical region, the basis wave function can be represented as a symmetrized cylindrical wave being the solution of the Schrödinger equation for the free electron motion in a cylindrical potential well that meets symmetry properties (Eqs. (60)–(62)):

$$\Psi_{II,PMN}(Z, \Phi, R | k, L)$$
$$= \frac{1}{\sqrt{2\pi h/n}} \exp i \left\{ \left[k + k_P - (L + nM) \frac{\omega}{h} \right] Z + (L + nM) \Phi \right\}$$
$$\times \left[C_{M,N}^{J,L} J_{L+nM} \left(\kappa_{|L+nM|,N} R \right) + C_{M,N}^{Y,L} Y_{L+nM} \left(\kappa_{|L+nM|,N} R \right) \right]. \qquad (63)$$

The basis function inside the MT spheres can be represented by a linear combination of spherical harmonics (Eq. (18)). Then, with the use of the explicit analytical form of basis functions, we can calculate the overlap and Hamiltonian matrix elements:

$$
\langle \Psi_{P_2 M_2 N_2} | \Psi_{P_1 M_1 N_1} \rangle |_{k,L} = \delta_{P_2 M_2 N_2, P_1 M_1 N_1} - \frac{n}{h} (-1)^{n(M_2+M_1)}
$$

$$
\times \sum_\alpha \exp\left\{ i \left[\left(k_{P_1} - k_{P_2} - n(M_1 - M_2) \frac{\omega}{h} \right) Z_\alpha + n(M_1 - M_2) \Phi_\alpha \right] \right\}
$$

$$
\times \sum_{m=-\infty}^{\infty} \left[C_{M_2 N_2}^{J,L} J_{m-(L+nM_2)} \left(\kappa_{|L+nM_2|, N_2} R_\alpha \right) \right.
$$

$$
\left. + C_{M_2 N_2}^{Y,L} Y_{m-(L+nM_2)} \left(\kappa_{|L+nM_2|, N_2} R_\alpha \right) \right] \tag{64}
$$

$$
\times \left[C_{M_1 N_1}^{J,L} J_{m-(L+nM_1)} \left(\kappa_{|L+nM_1|, N_1} R_\alpha \right) \right.
$$

$$
\left. + C_{M_1 N_1}^{Y,L} Y_{m-(L+nM_1)} \left(\kappa_{|L+nM_1|, N_1} R_\alpha \right) \right]
$$

$$
\times \left\{ I_{3,m\alpha}^{P_2 M_2 N_2, P_1 M_1 N_1} (r_\alpha) \right.
$$

$$
\left. - r_\alpha^4 \sum_{l=|m|}^{\infty} \frac{(2l+1)[(l-|m|)!]}{2[(l+|m|)!]} S_{lm,\alpha}^{P_2 M_2 N_2, P_1 M_1 N_1} (r_\alpha) \right\},
$$

$$
\left\langle \Psi_{P_2 M_2 N_2} | \widehat{H} | \Psi_{P_1 M_1 N_1} \right\rangle |_{k,L}
$$

$$
= \left[k + k_{P_2} - (L + nM_2) \frac{\omega}{h} + \kappa_{|L+nM_2|, N_2} \right]
$$

$$
\times \left[k + k_{P_1} - (L + nM_1) \frac{\omega}{h} + \kappa_{|L+nM_1|, N_1} \right]
$$

$$
\times \delta_{P_2 M_2 N_2, P_1 M_1 N_1} - \frac{n}{h} (-1)^{n(M_2+M_1)}
$$

$$
\times \sum_\alpha \exp\left\{ i \left[\left(k_{P_1} - k_{P_2} - n(M_1 - M_2) \frac{\omega}{h} \right) Z_\alpha + n(M_1 - M_2) \Phi_\alpha \right] \right\}
$$

$$
\times \sum_{m=-\infty}^{\infty} \left[C_{M_2,N_2}^{J,L} J_{m-M_2} \left(\kappa_{|L+nM_2|, N_2} R_\alpha \right) \right.
$$

$$
\left. + C_{M_2,N_2}^{Y,L} Y_{m-M_2} \left(\kappa_{|L+nM_2|, N_2} R_\alpha \right) \right]
$$

$$
\times \left[C_{M_1,N_1}^{J,L} J_{m-M_1} \left(\kappa_{|L+nM_1|, N_1} R_\alpha \right) + C_{M_1,N_1}^{Y,L} Y_{m-M_1} \left(\kappa_{|L+nM_1|, N_1} R_\alpha \right) \right]
$$

$$\times \left\{ \begin{array}{l} \left[k + k_{P_2} - (L + nM_2)\frac{\omega}{\hbar} \right] \left[k + k_{P_1} - (L + nM_1)\frac{\omega}{\hbar} \right] I_{3,\alpha}^{P_2 M_2 N_2, P_1 M_1 N_1} \\[2mm] + \kappa_{|L+nM_2|,N_2} \kappa_{|L+nM_1|,N_1} I_{3,\alpha}^{'P_2 M_2 N_2, P_1 M_1 N_1} + m^2 I_{4,\alpha}^{P_2 M_2 N_2, P_1 M_1 N_1} \\[2mm] -r_\alpha^4 \sum_{l=|m|}^{\infty} \frac{(2l+1)\,[(l-|m|)!]}{2\,[(l+|m|)!]} \\[2mm] \times \left(E_{l,\alpha} S_{lm,\alpha}^{P_2 M_2 N_2, P_1 M_1 N_1}(r_\alpha) + \gamma_{lm,\alpha}^{P_2 M_2 N_2, P_1 M_1 N_1}(r_\alpha) \right) \end{array} \right\}.$$

$$(65)$$

5.3 Results

As typical computation results, the band structure of the (15, 5) nanotube and the total and partial densities of states are shown in Figs. 16–18. This system has a fivefold symmetry axis and a screw axis. The translational cell contains 260 atoms; however, due to taking into account the screw symmetry of the nanotube, only 150 basis functions are required for convergence of electronic levels with an accuracy of 0.01 eV. As the cell contains only two atoms and, thus, only eight valence electrons, the band structure becomes very simple. The valence band of the nanotube contains only four dispersion curves that can be attributed to the completely filled s, $p_{1\sigma}$, $p_{2\sigma}$, and p_π atomic states, and the low-energy region of the conduction band contains only one p_π^* band. (Figs. 2 and 3).

Due to the introduction of the new quantum number L, the use of the symmetry properties makes it possible to give a more detailed classification of electronic eigenstates in nanotubes. The dispersion curves of this system are characterized by the wave vector k, which enumerates the irreducible representations of the group of screw translations, and $L = 0, \ldots, 4$, corresponding to the rotational symmetry of the system. It can be seen that this nanotube is a semiconductor in which the minimal

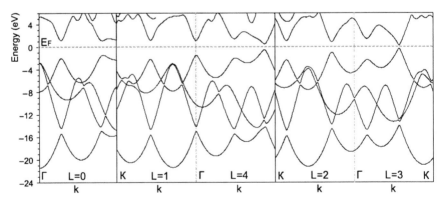

Fig. 16 Band structure of the (15,5) nanotube calculated using the basis set containing 150 functions

Fig. 17 Total density of states of the (15,5) nanotube

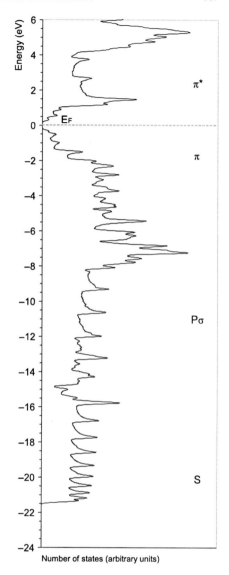

optical gap $E_{11} = 0.58$ eV corresponds to the direct transition at k $\approx 0.57(\pi/h)$ and $L = 3$, the second gap $E_{22} = 1.01$ eV corresponds to the direct transition at k $\approx 0.86(\pi/h)$ and $L = 4$, and the third gap $E_{33} = 1.82$ eV corresponds to the direct transition at k $\approx 0.28(\pi/h)$ and $L = 2$. Other example of calculations of carbon nanotubes, including the $(100, 99)$ system containing more that 100 thousand atoms in the unit cell were described in [31, 32].

Fig. 18 Partial densities of states of the (15,5) nanotube

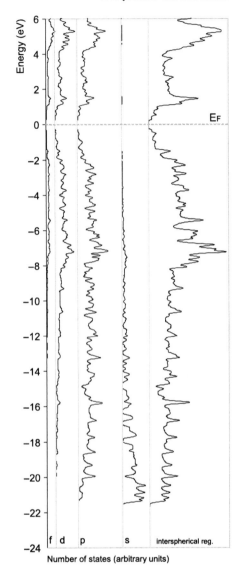

Acknowledgment This study was supported in part by the Russian Foundation for Basic Research (Grants no. 08-03-00262, 08-08-90411).

References

1. S. Iijima, *Nature* **354**, 56 (1991).
2. T. W. Ebeson, *Phys. Today* **49**, 26 (1996).
3. P. G. Collins, A. Zettl, H. Bando, A. Thess, R. E. Smalley, *Science* **278**, 100–102 (1997).

4. C. Dekker, *Phys. Today*, **52**(5), 22–28 (1999).
5. P. M. Ajayan, *Chem. Rev.* **99**, 1787–1799 (1999).
6. S. J. Tans, A.R.M. Verschueren, C. Dekker, *Nature* **393**, 49–51 (1998).
7. S. Frank, P. Poncharal, Z. L. Wang, W. A. de Heer, *Science* **280**, 1744–1746 (1998).
8. S. J. Tans, M. H. Devoret, H. Dai, A. Thess, R. E. Smalley, L. J. Geerligs, *Nature* **386**, 474–477 (1997).
9. J. W. Mintmire, B. I. Dunlap, C. T. White, *Phys. Rev. Lett.* **68**(5), 631–634 (1992).
10. R. Saito, M. Fujita, G. Dresselhaus, M. S. Dresselhaus, *Phys. Rev. B* **46**(3), 1804–1811 (1992).
11. R. Saito, M. Fujita, G. Dresselhaus, M. S. Dresselhaus, *Appl. Phys. Lett.* **60**, 2204 (1992).
12. C. T. White, D. H. Robertson, J. W. Mintmire, *Phys. Rev. B* **47**, 5485 (1993).
13. M. Bockrath et al., *Science* **275**, 1922–1925 (1997).
14. M. Bockrath et al., *Nature* **397**, 598–601 (1999).
15. J. Nygard, D. H. Cobden, P. E. Lindelof, *Nature* **408**, 342–345 (2000).
16. A. Y. Kasumov et al., *Science* **284**, 1508–1511 (1999).
17. A. F. Morpurgo, J. Kong, C. Marcus, H. Dai, *Science* **286**, 263–265 (1999).
18. Z. Yao, H.W.Ch. Postma, L. Balents, C. Dekker, *Nature* **402**, 273–276 (1999).
19. D. D. Koelling, G. O. Arbman, *J. Phys. F: Metal Phys.* **5**(11), 2041 (1975).
20. V. V. Nemoshkalenko, V. N. Antonov, (Naukova Dumka, Kiev, 1985) [in Russian].
21. P. N. D'yachkov, A. V. Nikolaev, *Dokl. Chem.* **344**(4–6), 235 (1995).
22. H. Krakauer, M. Posternak, A. J. Freeman, *Phys. Rev. B* **19**(4), 1706 (1979).
23. P. N. D'yachkov, B. S. Kuznetsov, *Dokl. Phys. Chem.* **395**(1), 57 (2004).
24. P. N. D'yachkov, in *Encyclopedia of Nanoscience and Nanotechnology*, edited by H. S. Nalwa. American Scientific Publishers, **1**,192–212, (2003).
25. P. N. D'yachkov, O. M. Kepp, *Dokl. Chem.* **365**(1–3), 62 (1999).
26. P. N. D'yachkov, O. M. Kirin, *Dokl. Phys. Chem.* **369**(4–6), 326 (1999).
27. P. N. D'yachkov, O. M. Kepp, in *Science and Application of Nanotubes*, edited by D. Tomanek and R. J. Enbody (Kluwer/Plenum. New York), 77 (2000).
28. P. N. D'yachkov, D. V. Kirin, *Ital. Phys. Soc., Conf. Proc.* **74**, 203 (2000).
29. P. N. D'yachkov, D. V. Makaev, *Phys. Rev. B* **74**, 155442 (2006).
30. P. N. D'yachkov, D. V. Makaev, *Phys. Rev. B* **71**, 081101(R) (2005).
31. P. N. D'yachkov, D. V. Makaev, *Phys. Rev. B* **76**, 057743 (2007).
32. G. N. Watson, *A Treatise on the Theory of Bessel Functions* (1945).
33. G. A. Korn, T. M. Korn, Mathematical handbook for scientists and engineers (McGraw-Hill, New York, 1961).
34. P. N. D'yachkov, H. Hermann, D. V. Kirin, *Appl. Phys. Lett.* **81**, 5228 (2002).
35. P. N. D'yachkov, H. Hermann. *J. Appl. Phys.* **95**, 399 (2004).
36. P. N. D'yachkov, D. V. Makaev, *Dokl. Phys. Chem.* **402**(2), 109 (2005).
37. A. Jensen, J. R. Hauptmann, J. Nygard, J. Sadowski, P. E. Lindelof., *Nano Lett.* **4**, 349 (2004).
38. D. V. Makaev, P.N. D'yachkov, JETF Lett. **84**(5–6), 397 (2006).
39. D. V. Makaev, P. N. D'yachkov, *Dokl. Phys. Chem.* **419**(1), 47–53 (2008).
40. R. C. Tatar, S. Rabii, *Phys. Rev. B* **25**, 4126 (1982).

Anharmonicity and Soliton-Mediated Transport: Thermal Solitons, Solectrons and Electric Transport in Nonlinear Conducting Lattices

W. Ebeling, M.G. Velarde, A.P. Chetverikov, and D. Hennig

Abstract We report here results about the excitation and survival of solitons in one-dimensional (1d) lattices with Morse interactions in a temperature range from low to physiological or room temperature (ca. 300 K). We also study their influence on added free electrons moving in the lattice. The lattice units (considered as "atoms" or "screened ion cores") are treated by classical (Newton–)Langevin equations. Then representing the densities of the core (valence) electrons of lattice units by Gaussian distributions we visualize lattice compressions as enhanced density regions. The local potentials created by the solitonic excitations are estimated as well as the classical and quantum–mechanical occupations. Further we consider the formation of solectrons, i.e. dynamic electron–soliton bound states. Finally, we add Coulomb repulsion and study its influence on solectrons. A discussion is also given about soliton-mediated electron pairing.

Keywords Morse interaction · Polaron · Soliton · Solectron · Electron pairing

1 Introduction

Excitation energy transfer processes in biological systems are problems of basic and long-standing interest [1–3], and especially the functional primary processes in photosynthetic reaction centers, drug metabolism, cell respiration, enzyme activities

W. Ebeling
Institut für Physik, Humboldt-Universität Berlin, Newtonstrasse 15, Berlin-12489, Germany
e-mail: ebeling@physik.hu-berlin.de

M.G. Velarde (✉)
Instituto Pluridisciplinar, Paseo Juan XIII, n. 1, Madrid-28040, Spain
e-mail: mgvelarde@pluri.ucm.es

A.P. Chetverikov
Dept. of Physics, Saratov State University, Astrakhanskaya 83, Saratov-410012, Russia
e-mail: chetverikovap@info.sgu.ru

D. Hennig
Institut für Physik, Humboldt-Universität Berlin, Newtonstrasse 15, Berlin-12489, Germany
e-mail: hennigd@physik.hu-berlin.de

N. Russo et al. (eds.), *Self-Organization of Molecular Systems: From Molecules and Clusters to Nanotubes and Proteins*, NATO Science for Peace and Security Series A: Chemistry and Biology, © Springer Science+Business Media B.V. 2009

and gene regulation have been studied intensively. In this context understanding the mechanism of electron transfer (ET) in biomolecules has attracted considerable attention during the last years [4–23]. The exploitation of the ET processes to construct technological devices has already been proposed and for such an achievement a theoretical understanding of the transfer mechanism in nature is needed and/or it has to be invented.

Inspired by the success of biomolecule modifications along with the determination of their three-dimensional structure microscopic theories for energy-transfer reactions were developed. Data of high resolving X-ray analysis gave the essential details on an atomic scale needed as input quantities for microscopic theories of ET in them. This gave insight into the relation between the structure and function for the energy and particle transfer in biomolecules and it has been shown how their steric structure can affect electron tunneling. In particular, experiments indicate that the H-bridges and covalent bonds involved in the biomolecules secondary structure are vital for mediating ET. On the other hand under physiological conditions (ca. 300 K) the ET may be activated by couplings to vibrational motion as long ago advocated by Hopfield [4]. Furthermore, molecular dynamics simulations have predicted that global molecule motions are very important for biochemical reactions for instance in light-induced reactions of chromophores accompanied by nuclear motions and for the ET in pigment protein complexes. In reaction center proteins proceed the protein nuclear motions coherently along the reaction coordinate on the picosecond time scale of ET as femtosecond spectroscopy revealed. Thus the vibrational dynamics of biomolecules may serve as the driving force of ET in them. Therefore investigations of transport mechanisms relying on the *mutual coupling between the electron amplitude and intramolecular bond vibrations* in biomolecules are of paramount importance.

Studies of energy storage and transport in macromolecules on the basis of self-trapped states have a long history beginning with the work of Landau [24] and Pekar [25, 26]. They introduced the concept of polaron (or as earlier said electron self-trapping), i.e. an electron accompanied by its own lattice distortion (a few phonons in another language) forming a localized quasiparticle compound which becomes the true electric carrier. In this context an approximate Hamiltonian system is often used to model transport of such localized excitations [27, 28]. When the size of the polaron is large enough so that the continuum approximation can be applied to the underlying lattice system in a clever combination of physical insight and mathematical beauty Davydov showed that a mobile self-trapped state can travel as a solitary wave along the molecular structure and he coined the concept of electro-soliton as electrical carrier and the natural generalization of the polaron concept [29–32]. Since the work of Davydov the relevance of solitons for the energy and particle transport in biomolecules has been recognized [33–36]. Similar ideas to Davydov's were also advanced by Fröhlich [37–43]; the relationship between the two approaches was elucidated in [44]. Most of the studies of transport properties in biopolymers are based on one-dimensional nonlinear lattice models, and recent two- and three-dimensional extensions with respect to solitonic transport of vibrational energy can be found, e.g. in [45–47]. Recent findings suggest that supersonic acoustic solitons can capture and transfer self-trapping modes in anharmonic one-dimensional lattices [48]. Regarding the enforcing role played by soliton motion in the functional processes in biomolecules we note that recently it has been proposed that the folding and conformation process of proteins may be mediated by solitons traveling along the polypeptide chains while interacting with a field corresponding to the conformation angles of the protein [49]. Furthermore, in a nonlinear dynamics approach to DNA dynamics it has been suggested that solitons propagating along the DNA molecule may play an important role in the denaturation and transcription process [50–54].

Hence, for a theoretical understanding of ET mechanisms in biomolecule the models should not only incorporate the static aspect of the protein structure but also its dynamics [55]. In particular, it has been illustrated that the dynamical coupling of moving electrons to vibrational motions of the peptide matrix can lead to some biological reactions in an activationless fashion [56]. In this spirit the investigations in [57–65] have been devoted to bond-mediated biomolecule ET using the concept of breather solutions. The transfer of electrons along folded polypeptide chains arranged in three-dimensional conformations constituting the secondary helix structure of the proteins has been considered. It has been demonstrated that the coupling between the electron and the vibrations of the protein matrix can activate coherent ET.

In view of the above and to better place the work that follows here let us insist on the fact that it is the nonlinearity induced by the electron-(acoustic) phonon interaction that led Davydov to his electro-soliton concept for otherwise dynamically *harmonic* lattices. Davydov argued that these excitations could be stable at finite temperatures and could persist even at physiological or room temperatures. Several authors have checked this conjecture and have shown that Davydov's electro-solitons are destroyed already around 10 K lasting at most 2 ps [33–36]. We shall follow Davydov's line of thought here but rather than using a harmonic lattice we shall consider *anharmonic* lattice dynamics. It is now well established that if the underlying lattice dynamics involves *anharmonic* interaction this may result in the appearance of *supersonic* (acoustic) solitons running free along the lattice like in a Toda lattice and in some other cases [66–83]. We shall make use of the Morse potential [84] (akin to the Toda repulsive interaction and to the Lennard–Jones potential) together with the electron–(acoustic)soliton interaction. As shown in Fig. 1, these potentials can be scaled around the minimum in such a way that the first three derivatives are identical what guarantees a close relationship of their nonlinear (soliton) excitations when acting in a lattice. It is also known that these excitations bring a new form of dressed electrons or electro-soliton dynamic bound states [74–83]. They have been called solectrons to mark the difference with Davydov's electro-solitons (for further historical details see [85]). We shall show that due to the added lattice *anharmonicity* and the excitation of lattice solitons there is solectron stability well above 10 K, in fact up to the physiological or room temperature range (ca. 300 K).

After introducing the model lattice problem in Section 2, we develop in Section 3 a method of visualization of soliton excitations as well as estimation of their life times. Discussed also there are the processes of solectron formation, electron pairing and solectron pair formation. In Section 4 using the tight-binding approximation we further explore how lattice deformations (or relative displacements between lattice units) affect solectron evolution. We also return there to the question of soliton-mediated electron pairing. In Section 5 we explore in depth how Coulomb repulsion affects solectron formation and electron pairing. A summary of results and comments are given in Section 6.

2 Lattice Dynamics

2.1 *Lattice Anharmonicity and Temperature*

We shall consider in a mixed classical–quantum description a 1d nonlinear lattice with added (free) conduction electrons allowing donor-acceptor electron transfer (ET) or electric current in the presence of an external field. The system consists of N classical units (atoms or screened ion cores). We shall focus on the case of periodic boundary conditions on a lattice like a ring of length L. These electrons are allowed to occupy some 3d volume surrounding the 1d lattice. For the *heavier* lattice units (relative to the electrons) we shall consider that have all equal mass m, and are described by coordinates $x_n(t)$ and velocities $v_n(t), n = 1, \ldots, N$. We take

$$H_a = \frac{m}{2} \sum_n v_n^2 + \frac{1}{2} \sum_{n,j} V(x_n, x_j). \tag{1}$$

The subscripts locate lattice sites and the corresponding summations run from 1 to N. The mean equilibrium distance (lattice constant) between the particles in the lattice is σ ($\sigma = L/N$). We shall assume that the lattice particles repel each other with a strong Born repulsive force and attract each other with a weak dispersion force with a potential which depends on the relative distance $r = x_{n+1} - x_n$ between nearest-neighbors only. As earlier indicated we shall take the Morse function, one if not the earliest quantum-mechanics based interaction potential [84]. As Fig. 1 shows its repulsive core is to a good approximation like that of the Toda potential though the latter possesses an unphysical attractive component. As the Hamiltonian (Eq. (1)) with V taken as a Toda potential is integrable and we know analytically in compact form its exact solutions this is of interest to us as we shall be concerned with relatively strong lattice compressions where what really matters is atomic repulsion. On the other hand it also appears of interest that the Toda interaction yields the hard rod impulsive force in one limit (the fluid phase) while in another limit it becomes a harmonic oscillator (the solid lattice crystal-like phase). Thus we take

$$V = D \{\exp[2B(r - \sigma)] - 2\exp[-B(r - \sigma)]\}. \tag{2}$$

Fig. 1 Toda (*upper curve*),
Morse (*middle curve*) and
Lennard–Jones L-J(12-6)
(*lower curve*) potentials suitably scaled around their
minima to have identical
second and third derivatives.
Another L-J potential used by
chemists is the so-called standard screw L-J(32-6) potential
offering no added advantages
for the purposes of this report

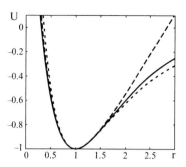

Exponentials are easily implemented in analog computers and they are also easier
to handle mathematically for our purposes here. For illustration in our computer
simulations we shall use $N = 200$ and $B = 1/\sigma$. B accounts for the stiffness along
the lattice and D provides an estimate of the binding/break-up energy of lattice
bonds.

Then in the presence of random forces (hence non zero temperature) and external
forces, H, the evolution of lattice particles is described by the (Newton-)Langevin
equations ($n = 1, 2, \ldots, N$) [86]

$$\frac{d v_n}{dt} + \frac{1}{m}\frac{\partial H}{\partial x_n} = -\gamma_0 v_n + \sqrt{2D_v}\,\xi_n(t), \tag{3}$$

where the stochastic force $\sqrt{2D_v}\,\xi_n(t)$ models a surrounding heat bath (Gaussian
white noise). The parameter γ_0 describes the common standard friction frequency
acting on the lattice units or atoms from the surrounding heat bath. The validity of
an Einstein relation is assumed $D_v = k_B T \gamma_0/m$, thus binging temperature T; k_B
in Boltzmann's constant. In most cases we shall use σ as the length unit (though in
occasions we may use $1/B$) and the frequency of oscillations around the potential
minimum ω_0^{-1} as the time unit. Typical parameter values for biomolecules are $\sigma \simeq$
$1 - 5\text{Å}$; $B \simeq 1\text{Å}^{-1}$; $D \simeq 0.1 - 0.5$ eV [87–89]. This means that $B\sigma \simeq 1$ (it
could take a higher value) and $1/\omega_0 \simeq 0.1 - 0.5 ps$. As the energy unit we shall use
$2D = m\omega_0^2\sigma^2/(B\sigma)^2$, that with $B\sigma = 1$ reduces to $m\omega_0^2\sigma^2$, commonly used by
most authors. This energy will be used also as the unit to measure the temperature
T ($k_B = 8.6 \cdot 10^{-5} \text{eV}/K$; $k_B T = 2D$).

The specific heat (at constant volume/length) of system Eqs. (1)–(3) is shown
in Fig. 2. Accordingly, the region where *anharmonicity* plays significant role is
$0.75 < C_v/k_B < 0.95$. This is the multiphonon range or highly deformed-phonons
domain on the way to melting in the system (recall that at high T, $C_v = 0.5$,
there is transition to a hard-sphere fluid phase). The corresponding temperatures
in our energy units are in the range $T \simeq 0.1$–0.5 (and even up to 1–2). Introducing
the binding strength of the Morse lattice, as the Morse potential can be suitably
adapted to the Toda interaction (Fig. 1), we foresee that solitonic effects are to be

Fig. 2 Toda–Morse lattice.
Specific heat at constant
volume/length (*upper curve*)
and ratio of *potential* energy,
U, to *kinetic* energy, T_{kin} of
the anharmonic lattice. Note
that we have only the "high"
temperature range

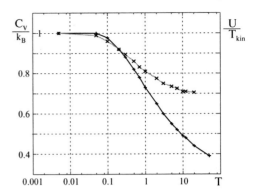

expected in the range $T_{sol}^M \simeq 0.2 - 1.0D$. In electron volts this would be the range $T_{sol}^M \simeq 0.01 - 0.1$ eV. This range of temperatures includes for biomolecules the range of physiological temperatures (ca. 300 K).

2.2 Lattice Units as Atoms and Lattice Solitons

We can visualize the time evolution of the lattice atoms and hence the lattice deformations or lattice excitations by representing the density of the *valence* electrons which are moving bound to the ion cores. This can be achieved by considering, for simplicity, that each lattice unit is surrounded by a Gaussian electron density (atomic density) of width, e.g. $s = 0.35\sigma$. Then the total atomic electron density is given by

$$\rho(x) = \sum_n \frac{1}{\sqrt{2\pi}s} \exp\left[-\frac{(x - x_n(t))^2}{2s^2}\right]. \tag{4}$$

Thus each lattice atom is like a spherical unit with continuous (valence) electron density concentrated around its center. In regions where the atoms overlap, the density is enhanced. This permits identifying solitonic excitations based on a color code in a density plot. This is of course a rough approximation which helps visualization of the location of dynamic excitations by using the (covalence) electrons density enhancements as an alternative to directly locating mechanical lattice compressions. For our purposes in Sections 2 and 3 this suffices. The mechanical approach is used in Sections 4 and 5. We show in Fig. 3 the results of computer simulations for three temperatures $T = 0.005$ (~ 10 K), $T = 0.1$ ($\sim 2 \cdot 10^2$K) and $T = 0.5$ ($\sim 10^3$K) with $D = 0.1$ eV. If we use $D = 0.05$ eV, then $T = 0.5$ corresponds to $T = 575$ K.

The diagonal stripes correspond to regions of enhanced density which are freely running along the lattice, this is the sign of solitonic excitations. Checking the slope we see that the excitations which survive more than 10 time units move with *supersonic* velocity. The pictures shown are quite similar to those described by Lomdahl and Kerr [33, 36] who gave a life-time of at most 2 ps and being stable

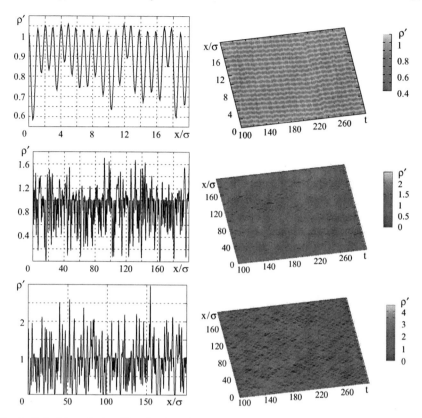

Fig. 3 Toda–Morse lattice. Visualization of running excitations (phonons and solitons) along the lattice. For convenience we use $\rho' = \sqrt{2\pi}s\rho$ to account for atomic core (valence) electrons density (the grey scale coding is in arbitrary units). We study three temperatures (given in units of $2D$): upper set of figures: $T = 0.005(\sim 10 \text{ K})$: we see only harmonic lattice vibrations or phonons and no evidence of strong (soliton-like) excitations; center two-figures: $T = 0.1(\sim 2.10^2 \text{ K})$: many density peaks show solitons (diagonal stripes). The strongest compressions move with velocity around $1.1v_{sound}$; lower two-figures: $T = 0.5(\sim 10^3 \text{ K})$: among the many excitations appearing we observe solitons running with velocity around $1.3v_{sound}$. Parameter values: $N = 200$ and $B\sigma = 1$

only up to 10 K. Ours, however, live about 10–50 time units that is for several picoseconds and survive even at $T = 1$ which is well above physiological temperatures. This confirms an earlier finding were at $T \simeq 300$ K stable solitons and solectrons could be identified [79, 80]. Recall that Davydov's electro-solitons and hence Lomdahl and Kerr's earlier mentioned work refer to solitons induced by the presence of originally free (conduction) electrons and subsequent electron–phonon (polaron-like states) whereas in the case described in this Section the conduction electrons are yet to be added as we shall do in the following section.

3 Local Electronic States

Let us now add to the system free electrons surrounding in 3d space the 1d lattice (Fig. 4).

3.1 Local Pseudo-Potentials and Classical Densities

The lattice creates a field which acts on the free electrons. In order to construct this field we need to evaluate the interaction between the lattice units and the surrounding electrons. The latter form a narrow 3d neighborhood of the lattice with diameter about the Bohr-radius a_B.

We can assume that all lattice atoms (with their valence electrons) which are near to each other by 1.5σ or less contribute to the local potential $V(x)$ acting on each conduction electron

$$V(x) = \sum_n V_n(x - x_n), \qquad r = |x - x_n| < 1.5\sigma. \tag{5}$$

The potential $V_n(x - x_n)$ created by the lattice particles n at the place of the electron x may be estimated by a pseudo-potential approach [90, 91]. One possible ansatz for the interaction of electrons with ions is

$$V_n(y) = -U_e \frac{h}{\sqrt{y^2 + h^2}}. \tag{6}$$

The value of the binding energy U_e is in the range $U_e \simeq 0.05 - 0.1$ eV. This is a second (independent) energy unit of the system, in general lower in value than the earlier mentioned binding energy between lattice units. Let us consider for numerical convenience $U_e \simeq 0.02 - 0.2D$ and $h = 0.3\sigma$. The choice $h = 0.3\sigma$ provides shallow minima at the location of the lattice atoms with significantly deep local minima at the location of lattice compressions. In view of the value U_e the electrons are only weakly bound to the atoms and may transit from one side to the other of a lattice unit. Accordingly the (free) conduction electrons are able to wander through the lattice eventually creating an electron current. To place a pair of such electrons between two lattice particles is in general not favorable in energetic terms, since the energy of repulsion $e^2/\varepsilon_0 r$ has to be overcome; ε_0 denotes dielectric constant. However the electron may bind to more than two lattice atoms thus forming a deep potential hole akin to a *polaron* state which is a *static* structure corresponding to

Fig. 4 Toda–Morse lattice. Sketch of lattice atoms or ion cores surrounded by added free electrons in 3d space

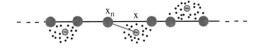

favorable energetic configurations. Here we are rather interested in the *dynamic* or time evolving phenomena initiated by solitonic excitations in the lattice. However we have to take into account that both of these items, the local compression by a static process (polaron formation) and by a running compression (soliton excitation) are intimately connected.

In the simplest entirely classical approximation we can assume that the evolution of the conduction electrons is very fast and the corresponding probability density follows locally a Boltzmann distribution. Note that when the electron density is sufficiently low, so that the electrons are still nondegenerated we may approximate the Fermi statistics by the Boltzmann statistics. In a heated lattice the units perform quite complex motions, we may expect therefore a rather complex structure of the field acting on the electrons. Let us give now examples of the fields created by the lattice atoms. The potential energy is given in units of the binding energy U_e. Taking into account the energy unit $2D(B\sigma)^2$ ($= m\omega_0^2\sigma^2$), the scale is set by the ratio $\eta = \frac{U_e}{2DB^2\sigma^2} = \frac{1}{2B^2\sigma^2}\frac{U_e}{D}$. For $B\sigma = 1$ the energy scale is therefore $\eta = \frac{U_e}{2D}$. To estimate any physical quantity the value of η is very important.

The potential $V(x, t)$ is time-dependent and gives at each time instant a snapshot of the actual situation. The potential changes quickly and the distribution of the electrons tries to follow it as fast as possible and hence the electrons are "slaved" accordingly, thus permitting an *adiabatic* approximation. We have a situation similar to that described for free electron statistics in semiconductor theory [92]. Then, we assume as a first approximation a Boltzmann distribution

$$n(x, t) = \frac{\exp[-\beta V(x, t)]}{\int dx' \exp[-\beta V(x', t)]}, \qquad (7)$$

with $\beta = 1/k_B T$. Here x denotes the coordinate along the lattice. An example of the estimated density Eq. (26) is shown in Fig. 5. The (relatively high) peaks correspond to the enhanced probability of a rather strong lattice compression, i.e., a soliton ready to meet and trap an electron. This defines the solectron as an electron "surfing" on a soliton for about 10–50 time units (i.e. a few picoseconds) then getting off it and eventually finding another soliton partner once more to surf-on and so on. For $T = 0.1$ we observe several rather stable running excitations (diagonal stripes) with velocities around $1.2v_{sound}$. For $T = 1$ (not shown in the figure) one can observe many weak and only a few very stable excitations moving with *supersonic* velocity $1.4v_{sound}$. The probabilities estimated from the Boltzmann distribution are strongly concentrated at the places of minima. This means that most of the electrons are concentrated near to solitonic compressions.

3.2   *Bound States of 3d Electrons in a Nonlinear Lattice Ring*

So far our estimates of the electronic states in the local potential were entirely classical. The Boltzmann distribution finds the deepest minima of the local potential

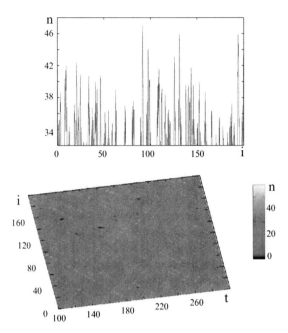

Fig. 5 Heated Toda–Morse lattice. Classical probability distribution of an electron in a heated anharmonic lattice in the *adiabatic* approximation according to local Boltzmann distribution. On the upper figure a snapshot of the distribution is given for a certain time instant. On the lower figure the actual time evolution of the distribution is displayed. The temperature is $T = 0.1$. Parameter values: $N = 200$, $h = 0.3$, $\sigma = 1$ and $B\sigma = 1$

$V(x, t)$ acting on an electron. The problem to find the quantum states for an electron in the anharmonic lattice is more difficult. There exist different situations depending on the relative values of the four length scales a_B, h, σ, d, where a_B is Bohr-radius, h is kind of softness scale of lattice particles (according to the pseudopotential (Eq. (6)), σ is the lattice spacing at equilibrium and d is the smallest interatomic lattice distance at solitonic compressions. As earlier noted, the character of the electron dynamics depends strongly on the value of h and on the distance σ. Figure 6 shows that the choice $h = 0.3\sigma$ depending on the distance between the neighbors in the lattice allows two kinds of minima. Accordingly, for a compressed lattice with $a_B \simeq h \simeq \sigma$ and $d < \sigma$, solectrons are to be expected.

Let us investigate now the conditions for possible formation of pairs of solectrons and under which conditions a solectron pair is more stable than a single solectron. As shown above, the electrons in soliton-bearing lattices move in a fastly changing potential landscape. The structure of this landscape is similar to the landscapes known from the theory of disordered systems [93–95]. At variance with the latter cases, our potential is time-dependent. In typical snapshots of the potential landscape acting on the electrons we see relatively flat normal parts showing only small oscillations of the potential and deep local minima which move approximately with soliton velocity. Certainly, the character of the bound states which may be formed depends on the depth of the potential U_{min}, on the temperature T and on the relation

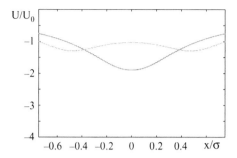

Fig. 6 Potential (in units of U_0) felt by an electron placed between two ions *versus* lattice spacing. If the ions are at equilibrium distance $r = \sigma$ the potential minima are at the center of two nearby ion cores. Between two compressed ions $r = \sigma/5$, a new potential minimum appears midway between two nearest-neighbor ions

between the characteristic quantum time $\tau_q \propto \hbar/U_{min}$ and the classical time scale $1/\omega_0$. Assuming that the classical time scale is much longer, we may work in an adiabatic approximation. Let the deep potential minimum (like a potential well) created by a soliton be approximated by a parabolic profile

$$U(r) = U_0 + \frac{a_0}{2}r^2 + \dots \tag{8}$$

where r denotes distance in 3d space. The second derivative is

$$a_0 = U''(r)|_{r=r_0} \simeq \frac{c}{\sigma^2}, \tag{9}$$

with $c \simeq 1$. A typical valley includes just a few lattice units. Then the bound states are approximately given by

$$\varepsilon_n = U_0 + 3\hbar\omega_0\left(n + \frac{1}{2}\right), \qquad n = 0, 1, 2, \dots \tag{10}$$

where $a = m\omega_0^2$. The ground state wave function is

$$\Psi_0(r) = (r_0)^{-3/2}\pi^{-3/4}\exp(-r^2/2r_0^2), \tag{11}$$

where $r_0^2 = (\hbar/m\omega_0)$. Note that this estimate is valid only for sufficiently deep minima. These states can in principle be filled by electrons albeit obeying Pauli's exclusion principle. In the ground state, if sufficient solitons are available, each of the solitons can capture two electrons with opposite spin or possibly more electrons as suggested by classical estimates. However higher occupation is less probable. Indeed a second electron with opposite spin may be placed on the same level as the first one, but a third electron in a potential valley cannot occupy the ground state level any more.

3.3 Binding Energy and Wave Functions of Solectron Pairs

As we have seen, the potential well created by a soliton may in principle be occupied by pairs of electrons with opposite spins satisfying Pauli's exclusion principle. At first sight, these electron pairs, which are Bosons, appear like "bipolarons" or "Cooper pairs". However looking at the details we see, that the solectron pairs are something new. The problem of pairs or clusters of quantum electrons in a parabolic trap is not new [96–98]. In the case of solectrons the width of the potential well is of the order of a few equilibrium inter-atomic lattice distances. In a first estimate the energy of a solectron is about

$$\varepsilon_n = U_0 + \frac{3}{2}\hbar\omega_{min}, \tag{12}$$

and correspondingly the ground state energy of a Coulomb pair is

$$\varepsilon_{0p} = 2\left[U_0 + \frac{3}{2}\hbar\omega_0\right] + \langle\frac{e^2}{\varepsilon_0 r_p}\rangle, \tag{13}$$

where r_p is the distance of the electrons in the pair and as earlier ε_0 is the dielectric constant of the medium. Due to the factor two this energy is in general lower than the energy of the state of one bound and one free electron. If the term arising from Coulomb repulsion is weak, pairing is favorable. An estimate follows from the condition that repulsion and attraction to the center of the well balance each other

$$m\omega_0^2 r_1 = \frac{e^2}{\varepsilon_0(2r_1)^2}. \tag{14}$$

This leads to a classical estimate of the Coulomb shift

$$\Delta\varepsilon_{cl} = \frac{e^2}{2\varepsilon_0 r_1} = \left[\frac{e^2}{\varepsilon_0}\right]^{2/3}(m\omega_0^2)^{1/3}. \tag{15}$$

Within quantum theory we may estimate the Coulomb shift by using perturbation theory as done in the study of the Helium atom. We take two electrons which are confined in the field of a spherical potential well given by Eq. (8). The symmetric ground state wave function of two electrons with opposite spin is given by

$$\Psi(r_1, r_2) = \frac{1}{\pi^3 r_0^6}\exp\left[-\frac{r_1^2 + r_2^2}{2r_0^2}\right]. \tag{16}$$

The mean energy calculated with these wave functions is then

$$\varepsilon_{p0} = 2U_0 + 3\alpha\hbar\omega_0 + \Delta\varepsilon_{qm}, \tag{17}$$

$$\Delta\varepsilon_{qm} = \frac{1}{\pi r_0^6} \int dr_1 \int dr_2 \frac{e^2}{\varepsilon_0 |r_1 - r_2|} \exp\left[-\frac{r_1^2 + r_2^2}{2r_0^2}\right]. \tag{18}$$

After doing the integral over the angles (Eq. (18)) becomes

$$\Delta\varepsilon_{qm} = \frac{16e^2}{\pi r_0^6} \int_0^\infty dr_1 r_1 \exp[-r_1^2/r_0^2] \int_0^{r_1} dr_2 r_2^2 \exp[-r_2^2/r_0^2], \tag{19}$$

or else

$$\Delta\varepsilon_{qm} = \frac{e^2}{\varepsilon_0 r_0} A_0, \tag{20}$$

where the constant is defined by the integral

$$A_0 = \frac{16}{\pi} \int_0^\infty dy y \exp[-y^2] \int_0^y dz z^2 \exp[-z^2] \approx 0.32. \tag{21}$$

In view of this estimate, the mean distance between the electrons in a solectron pair is around $3r_0$, i.e. three times the "size" of the wave function. We expect that the real Coulomb shift is between the classical and the quantum estimates. In order to find solectron pairs we need conditions where the Coulomb shift is much smaller than the gap to the next level which is $3\,\hbar\omega_0$. To be on the safe side we require conditions such that

$$max[\Delta\varepsilon_{cl}, \Delta\varepsilon_{qm}] < 3\hbar\omega_0. \tag{22}$$

Under these conditions the formation of a solectron pair is favored.

3.4   Soliton Mediated Electron Pairing

Let us further comment on how electron pairing could be influenced by the presence of solitons. If one could obtain Boson pairs with sufficient density, then interesting effects may be expected. Looking at the classical probability distributions in Figs. 3 and 5 we see that there are minima of different types. There are flat and narrow minima which carry just one electron as in a solectron. Further there are minima with two electrons and finally deep minima capable of carrying many electrons. In fact as Figs. 3 and 5 show most of the minima carry 3–10 electrons. However quantum-mechanical effects rather provide a new picture: (i) quantum solectrons are in energy just a bit higher than the classical solectrons due to the ground state energy shift $1.5\hbar\omega_0$; (ii) a second electron with opposite spin may be placed at the

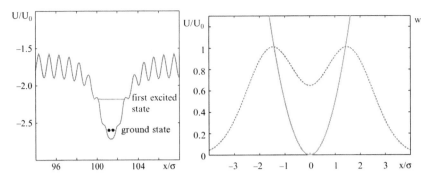

Fig. 7 Toda–Morse lattice. Left figure: formation of an electron pair in the potential minimum created by a solitonic excitation in the lattice. The formation of a trio requires a much higher energy. Therefore a solectron trio is ruled out except at very high temperatures. Right figure: shape of the pair wave function near to the minimum of the potential

same level (see Fig. 7). These solectron pairs are rather stable since the binding of a third electron needs a relatively high amount of energy, namely $3\hbar\omega_0$.

Let us estimate the chance to form trios. To place another electron into a solitonic well which is already occupied by an electron-pair needs the overcoming of a gap with the amount $3\hbar\omega_0$ between the ground state and the first excited level. Thus if

$$k_B T < 3\hbar\omega_0, \tag{23}$$

the occupation by trios, quartets, etc. (which is classically possible) is more or less prevented by quantum effects. Indeed we may assume that the extension of a solitonic minimum is about ten times wider than the Morse potential minimum (Fig. 1). Such minimum corresponds to a frequency about $3 \cdot 10^{12}\mathrm{s}^{-1}$. Then the frequency of oscillations around the minimum of the soliton potential is about $10^{12}\mathrm{s}^{-1}$. This is about $1 - 2$ eV. Accordingly, the inequality (Eq. (23)) implies $T < 10^3$K naturally fulfilled in all interesting cases.

Thus under special conditions, in certain windows of parameter values, the formation of pairs is more favorable than the single solectron. Under quasi-classical conditions however the system seems to favor electron clusters. In conclusion we may say that the Pauli exclusion principle has the consequence that instead of classical clusters we observe quantum–mechanical pairs of solectrons. This supports the soliton-mediated electron pairing mechanism proposed by Velarde and Neissner [99] (Fig. 8).

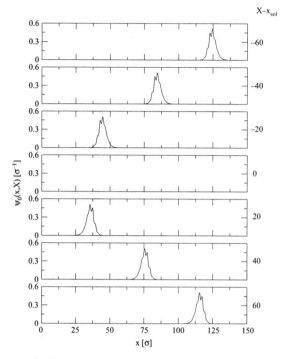

Fig. 8 Toda–Morse lattice. Evolution of the wave function of an electron-electron pair created by a moving solitonic lattice deformation [99]

4 Electron-Lattice Dynamics in Tight-Binding Approximation

Let us now go deeper into the question of solectron formation by describing the electrons on the lattice using the tight-binding approximation (TBA).

4.1 The Tight-Binding Approximation

The tight-binding approximation replaces the Schrödinger continuum dynamics by a hopping process along the discrete lattice sites. Assuming that there is only one atomic state per lattice unit we get for the electrons the following Hamiltonian in second-quantization formalism [100, 101]

$$H_{el} = \sum_n \left[E_n(..., x_{n-1}, x_n, x_{n+1}, ...) c_n^+ c_n \right.$$
$$\left. - V_{n,n-1}(x_n, x_{n-1})(c_n^+ c_{n-1} + c_n c_{n-1}^+) \right]. \qquad (24)$$

Recall that (24) refers to initially *free* or *excess* electrons added to the lattice with atoms assumed to be located at sites n. The quantities c_n, c_n^+ originally refer to Fermion destruction and creation operators, respectively, with appropriate anti-commutation relations but here they are just complex numbers. Purposely in this section we shall consider the non-uniformity of the on-site energy levels (diagonal elements, V_{nn}, of the transfer matrix). Further assuming that the interaction depends exponentially on the distance between the lattice units, we set

$$V_{n,n-1} = -V_0 \exp[-\alpha(q_n - q_{n-1})]. \tag{25}$$

Then the Hamiltonian (Eq. (24)) becomes

$$H_{el} = \sum_n \{(E_n^0 + \delta E_n)c_n^+ c_n$$
$$- V_0 \exp[-\alpha(q_n - q_{n-1})](c_n^+ c_{n-1} + c_n c_{n-1}^+)\}, \tag{26}$$

where, for convenience in notation, q_n denotes a lattice site spatial vibration (relative displacement) coordinate defined by $x_n = n\sigma + q_n/B$. The term E_n^0 denotes on-site energy levels of the unperturbed lattice and δE_n is the perturbation due to the lattice vibrations (harmonic as well as anharmonic modes may contribute). The simplest approximation is

$$\delta E_n = \chi(q_n/B), \tag{27}$$

where the "electron–phonon coupling constant", χ, indicates that the on-site energy level E_n, i.e. the local site energy, depends on the displacement of the unit at that site; q_n is dimensionless (unit: $1/B$). As shown e.g. in [87–89], this coupling between lattice deformations and electronic states, leads for large enough values of the parameter χ to the formation of *polarons*. In view of the above given parameter values, the value of the coupling constant is in the range $\chi \simeq 0.1 - 2$ eV/Å. We have to take into account that our model is translationally invariant and we are considering relative lattice displacements. Accordingly, we set

$$\delta E_n \simeq \frac{\chi_1}{2} [(q_{n+1} - q_n) + (q_n - q_{n-1})], \tag{28}$$

with $\chi_1 = \chi/B$ as a new constant. An alternative, using a pseudopotential like Eq. (6), is the approximation

$$E_n = E_n^0 - U_e \sum_{j \neq n}' \frac{h}{\sqrt{(x_n - x_j)^2 + h^2}}, \tag{29}$$

where the over-dash in the sum indicates that it is to be restricted in an appropriate way by introducing screening effects. For instance, as earlier done, we may cut the sum at a distance 1.5σ from the center of the ion core, or in other words include all terms corresponding to lattice units which are nearer than 1.5σ. Then we assume that the energy levels are shifted like the field created by the pseudopotentials acting on the electron from the side of the neighboring atoms. To linear approximation we get

$$\delta E_n \simeq \frac{hU_e\sigma}{B(\sigma^2 + h^2)^{3/2}} [(q_{n+1} - q_n) + (q_n - q_{n-1})]. \tag{30}$$

Comparing Eqs. (28) and (30) we find

$$\chi = U_e \frac{2\sigma h}{(\sigma^2 + h^2)^{3/2}} = \left(\frac{U_e}{\sigma}\right) \frac{2(h/\sigma)}{[1 + (h/\sigma)^2]^{3/2}}. \tag{31}$$

Then for $U_e = 0.1 - 1.0D$, $h = 0.3\sigma$, $D = 0.1 - 0.5$ eV, and $\sigma = 1 - 5$Å we obtain $\chi = 0.001 - 0.1$ eV/Å. As the parameter values in this approach are about one or two orders of magnitude below the earlier indicated values we expect that here polaron effects are rather weak and hence the system dynamics is dominated by solitons.

The probability to find the electron at the lattice site or atom located at x_n, i.e. the occupation number p_n, is $p_n = c_n c_n^*$. Solving the Schrödinger equation for the components of the wave function c_n we get

$$i\frac{dc_n}{dt} = \tau[E_n^0 + \delta E_n(q_{n+1}, q_{n-1})]c_n$$
$$-\tau \{\exp[-\alpha(q_{n+1} - q_n)]c_{n+1}$$
$$+ \exp[-\alpha(q_n - q_{n-1})]c_{n-1}\}, \tag{32}$$

where E_n^0 and δE_n are dimensionless (unit: 2D). The corresponding Newtonian equations for the lattice units are

$$\frac{d^2q_n}{dt^2} = -p_n \frac{\partial \delta E_n(q_{n+1}, q_{n-1})}{\partial q_n}$$
$$+ \{1 - \exp[-(q_{n+1} - q_n)]\}\exp[-(q_{n+1} - q_n)]$$
$$- \{1 - \exp[-(q_n - q_{n-1})]\}\exp[-(q_n - q_{n-1})]$$
$$-\alpha V_0 \{\exp[-\alpha(q_n - q_{n-1})](c_{n+1}^+ c_n + c_{n+1}c_n^+)$$
$$+ \exp[-\alpha(q_{n+1} - q_n)](c_n^+ c_{n-1} + c_n c_{n-1}^+)\}. \tag{33}$$

The role of temperature would be considered further below. The problem reduces, in principle, to solving both Eqs. (32) and (33) coupled together. It is not, however, the only possible approach to our problem as we shall see below.

4.2 Discussion About Solectronic Excitations and Expected Consequences

Let us consider one of the possible soliton-mediated processes: single electron transfer (ET) in a soliton-mediated hopping process along the lattice from a *donor* to an *acceptor*. When an added, excess electron is placed at a *donor* located at site

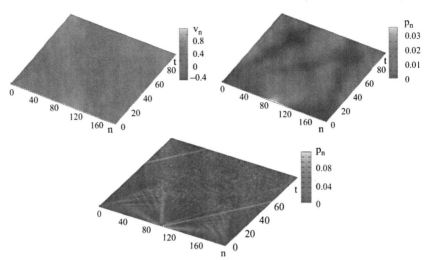

Fig. 9 Toda–Morse lattice. Soliton, electron and solectron. Results of numerical integration of Eqs. (32) and (33). Upper left figure: $\alpha = 0$, soliton alone; upper right figure: $\alpha = 0$, electron alone; bottom figure: $\alpha = 1,75$, solectron (electron dynamically bound to the soliton). The grey scales (velocity and probability density) are in arbitrary units, just for illustration

$n = 100$ at time $t = 0$, Fig. 9 shows our findings: (a) pure anharmonic lattice vibration without electron–lattice interaction ($\alpha = 0$): time evolution of one soliton as predicted by the Morse Hamiltonian (Eq. (33)), thus illustrating how little we depart from the Toda solitons; (b) free electron alien to lattice vibrations ($\alpha = 0$): spreading of the free electron probability density as a consequence of Schrödinger equation (32); and (c) electron-lattice interaction ($\alpha = 1,75$): soliton-mediated ET as predicted by Eqs. (32) and (33) coupled together. The electron is dynamically bound to the soliton which is the solectron excitation.

When the electron–lattice interaction is operating, we see that the electron moves with the soliton with a slightly supersonic velocity $v_{el} \sim \frac{100}{70} v_{sound}$ and is running to the right border of the square plot. Let us assume that there an acceptor is located. This means that the electron is guided by the soliton from *donor* to *acceptor*. In reality the electron cannot ride on just a single soliton from donor to acceptor. Several solitons should be involved in transport. We have already mentioned this kind of promiscuity of the electron. Therefore the above given soliton velocity is an upper bound for the ET process. In principle this effect may be used as a way to manipulate the transfer of electrons between donor and acceptor. Clearly in our case we may have a polaron effect due to the electron–phonon (or soliton) interaction in addition to the genuinely lattice soliton effect due to the anharmonicity of the lattice vibrations. Thus, from *donor* to *acceptor*, we have not just phonon-assisted ET but a much faster soliton-assisted ET.

4.3 Adiabatic Canonical Distributions in the Tight-Binding Approximation

In a first approximation with non-interacting electrons the canonical equilibrium distribution is

$$p_n^0 = \exp[\beta(\mu - E_n)], \tag{34}$$

where the chemical potential μ is given by the normalization. In the adiabatic approximation we assume that this distribution is reached in a very short time. Using the approximation (Eq. (28)) we get

$$p_n^0 \simeq \exp\left[-\frac{\delta E_n}{k_B T}\right] = \exp\left[-\frac{\chi(q_{n+1} - q_{n-1})}{B k_B T}\right]. \tag{35}$$

Suppose now that one big soliton is excited by appropriate heating of the lattice to the temperature T. We assume the following shape of the solution

$$\exp[-3(q_n - q_{n-1}) = 1 + \beta_0 \cosh^{-1}[\kappa n - \beta_0 t]. \tag{36}$$

Incidentally, the computations by Rice and collaborators [71, 72] show that for Morse or L-J [L-J(12-6) and L-J(32-6)] potentials a Gaussian profile could also be used as an reasonably valid approximation to the exact solution (Eq. (36)) of the Toda lattice.

By introducing this into Eq. (35) we find

$$p_n^0 \simeq [1 + \beta_0 \cosh^{-2}[\kappa n - \beta_0 t]]^\zeta [1 + \beta_0 \cosh^{-2}[\kappa(n+1) - \beta_0 t]]^\zeta, \tag{37}$$

where

$$\zeta = \frac{\chi}{6 B k_B T}. \tag{38}$$

We see that a thermally excited soliton is quite similar to a mechanically excited soliton except for some kind of a twin structure and a little deformation of the shape and the amplitude, both temperature-dependent. The velocity of such thermal soliton is the same as the standard soliton velocity.

Quantum mechanically the canonical equilibrium distribution is given by the time-dependent energy eigenvalues and hence rather than Eq. (35) we now get

$$p_n^0 \simeq \exp[-c(q_{n+1}(t) - q_{n-1}(t))], \tag{39}$$

with $c = \chi/B k_B T$. The displacements have to be taken from computer simulations of thermally excited solitons. The distribution is a quickly changing local function of the displacements. In the *adiabatic* approximation we assume that this distribution is reached in a very short time, as shown in Fig. 10. Noteworthy is that this picture of a canonical quantum distribution is qualitatively similar to the classical distributions shown in Fig. 5. We may estimate the soliton frequency from the thermal statistics

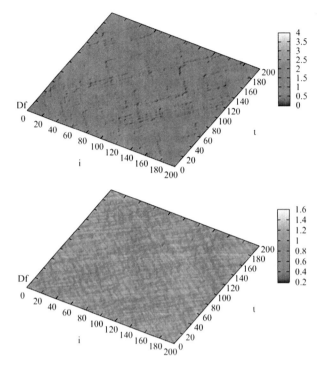

Fig. 10 Toda–Morse lattice. Probability distribution of an electron in a heated anharmonic lattice in the *adiabatic* approximation according to the quantum canonical distribution. The actual time evolution of the distribution is displayed. Upper figure: $T = 0.1$; lower figure: $T = 0.5$. Parameter values: $B\sigma = 1, \alpha = 1.75, V_0 = 1, \tau = 10$ and $\gamma = 0.002$

of the solitons in the lattice as done in Refs. [70, 73] for Toda interactions: (i) single solitons with parameter κ are described by Eq. (36); and (ii) the density of solitons depending on parameter κ is known. Following [73] we have

$$n(\kappa, T) = \frac{4a\kappa}{\pi k_B T} \exp(-\kappa) \exp[-(E(\kappa) - 2\kappa)/k_B T], \qquad (40)$$

where

$$E(\kappa) = \frac{2a}{b}[\sinh \cosh \kappa - \kappa]. \qquad (41)$$

Since the quantities p_n depend on κ we get this way the distribution of electron occupation numbers. In a thermally excited system, the number of solitons depends on the initial and boundary conditions. In an infinite Toda system (and the like for a Morse potential) the number of solitons can be approximated by

$$n(T) \simeq const \, T^{1/3}. \qquad (42)$$

Thus the number of solitons appears increasing with increasing temperature. On the other hand their contribution to macroscopic properties, as, e.g. the specific heat goes down as seen in Fig. 2. Therefore we expect that there exists a kind of "optimal temperature" where solitons have the strongest influence [70, 81].

5 Coulomb Repulsion and Electron-Lattice Dynamics in Hubbard Approximation

Let us complete our analysis by considering in more details the role of Coulomb repulsion between two added excess electrons thus supplementing our findings in Sections 3.3. and 3.4. We shall do it in the simplest possible way using Hubbard's model Hamiltonian [102–104]. Thus shall take the Coulomb repulsion when the electrons are at their shortest separation distance (local on-site repulsion).

5.1 The Hubbard Hamiltonian

When we add a spin variable and augment (Eq. (24)) with an on-site local (Coulomb)–Hubbard repulsion we get

$$H_{el} = -\sum_{n,\sigma} \left(V_{n\,n-1}\, \hat{a}_{n\sigma}^{+}\hat{a}_{n-1\sigma} + V_{n\,n+1}\hat{a}_{n\sigma}^{+}\hat{a}_{n+1\sigma} \right) + U \sum_{n} \hat{a}_{n\uparrow}^{+}\hat{a}_{n\uparrow}\hat{a}_{n\downarrow}^{+}\hat{a}_{n\downarrow}\,, \quad (43)$$

where the index n denotes the lattice site. Here σ accounts for the electron spin which can be up or down. For clarity we now have made explicit the Fermion operators $\hat{a}_{n\sigma}^{+}$ creates an electron with spin σ at site n and $\hat{a}_{n\sigma}$ annihilates the electron. The second term in Eq. (43) represents the on-site electron–electron interaction due to Coulomb repulsion of strength U (here it has positive values only). The transfer matrix is like Eq. (25). For $H_{lattice}$ we take Eq. (1) with Eq. (2).

5.2 Localized Paired Electron-Lattice Deformation States

We start with the exact two-electron wavefunction given by

$$|\psi(t)\rangle = \sum_{m,n} \phi_{mn}\left(\{p_m\}, \{q_m\}\right) \hat{a}_{m\uparrow}^{+}\hat{a}_{n\downarrow}^{+}\,|0\rangle\,, \quad (44)$$

where $|0\rangle$ is the vacuum state (containing no electrons) and ϕ_{mn} denotes the probability amplitude for an electron with spin up to occupy site m while an electron with spin down is at site n; $p_n = mv_n$. The symmetric $\phi_{mn} = \phi_{nm}$ probability amplitudes are normalized $\sum_{mn} |\phi_{mn}|^2 = 1$ and depend on the set of lattice variables $(\{p_n\}, \{q_n\})$.

To obtain the equations of motion for the probability amplitudes the wavefunction (Eq. (44)) is inserted into the corresponding Schrödinger equation and the evolution of the lattice variables is derived from Hamilton's variational principle with an energy functional $\mathcal{E}^2 = \langle \psi | H | \psi \rangle$. Suitable choice of scales permits rewriting the evolution equations in dimensionless form in a similar way as earlier done. Then we get:

$$i \frac{d\phi_{mn}}{dt} = -\tau \{ \exp[-\alpha (q_{m+1} - q_m)] \phi_{m+1n} + \exp[-\alpha (q_m - q_{m-1})] \phi_{m-1n}$$
$$+ \exp[-\alpha (q_{n+1} - q_n)] \phi_{mn+1} + \exp[-\alpha (q_n - q_{n-1})] \phi_{mn-1} \}$$
$$+ \bar{U} \phi_{mn} \delta_{mn}, \tag{45}$$

$$\frac{d^2 q_n}{dt^2} = [1 - \exp\{-(q_{n+1} - q_n)\}] \exp[-(q_{n+1} - q_n)]$$
$$- [1 - \exp\{-(q_n - q_{n-1})\}] \exp[-(q_n - q_{n-1})]$$
$$+ \alpha V \exp[-\alpha (q_{n+1} - q_n)]$$
$$\sum_m \{ [\phi^*_{mn+1} \phi_{mn} + \phi^*_{mn} \phi_{mn+1}] + [\phi^*_{n+1m} \phi_{nm} + \phi^*_{nm} \phi_{n+1m}] \}$$
$$- \alpha V \exp[-\alpha (q_n - q_{n-1})]$$
$$\sum_m \{ [\phi^*_{mn} \phi_{mn-1} + \phi^*_{mn-1} \phi_{mn}] + [\phi^*_{nm} \phi_{n-1m} + \phi^*_{n-1m} \phi_{nm}] \} . \tag{46}$$

Comparing with the equations in Section 4, Eq. (45) replaces Eq. (32), having assumed, for simplicity, that all on-site diagonal factors are equal and hence can be scaled away by suitable choice of the reference energy level. This suffices for our purpose in this Section. Correspondingly, Eq. (46) replaces Eq. (33). As in Eqs. (32) and (33) the parameter τ appearing in the R.H.S. of Eq. (45) determines the degree of time scale separation between the (fast) electronic and (slow) acoustic phonon or soliton processes. For computational illustration we shall use in what follows: $\tau = 10$, $V = 0.1$, and $\alpha = 1.75$. To obtain localized stationary solutions of the coupled system (Eqs. (45) and (46)) an energy functional is minimized yielding the lowest energy configuration.

The probability for one electron to be in site n with spin up, respectively spin down, is determined by

$$\rho_{n\uparrow} = \langle \psi | \hat{a}^+_{n\uparrow} \hat{a}_{n\uparrow} | \psi \rangle = \sum_k |\phi_{nk}|^2, \tag{47}$$

$$\rho_{n\downarrow} = \langle \psi | \hat{a}^+_{n\downarrow} \hat{a}_{n\downarrow} | \psi \rangle = \sum_k |\phi_{kn}|^2. \tag{48}$$

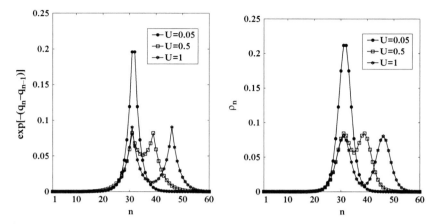

Fig. 11 Toda–Morse lattice. Lattice solitons and the role of Coulomb repulsion for an electron pair. Left figure: initial profile of the localized lattice deformation; right figure: electron probability distribution corresponding to a minimum of the variational energy for three different values of the Hubbard parameter U (values in insets). Other parameter values: $\alpha = 1.75$ and $V = 0.1$

Typical electron probability distributions and the corresponding profile of displacements of the molecules are depicted in Fig. 11 for three different values of U (because of symmetry $\rho_{n\uparrow} = \rho_{n\downarrow}$ and we plot half the electron density at a site n defined as $\rho_n = \frac{1}{2}\sum_k (|\phi_{kn}|^2 + |\phi_{nk}|^2)$). The corresponding localized compound comprises an exponentially localized two-electron state and the associated pair of kink-shape lattice deformations which represented as $\exp(-(q_n - q_{n-1}))$ are of bell-shape. These are the earlier introduced lattice solitons (Eq. (36)). Increasing the repulsive (Coulomb–) Hubbard-interaction has the impact that the inter-electron distance (and accordingly also the distance between the centers of the solitons) widens. At the same time the degree of localization reduces, i.e. broader profiles of lower peak values result. Notably, the localized solutions are fairly broad width and thus are expected to be mobile when appropriate kinetic energy is added. While for low values $U \lesssim 0.05$ the electron probability density is single-peaked increasing U causes a split up of ρ_n into a double-peak structure. For $U \gtrsim 0.9$ the inter-electron distance exceeds the width of either of the two peaks of the electron probability density. Therefore the two electrons can no longer be regarded as paired. Those features of the electron probability are equivalently exhibited by the soliton patterns, that is the stronger the repulsive interaction is, the less is the lattice compression reflected in the width and amplitude of the soliton patterns.

5.3 *Moving Electron-Pair Soliton Compounds*

Let us now see the evolution of the localized electrons coupled with the corresponding lattice deformations. The motion of the lattice soliton is achieved with the excitation of the soliton momenta according to

$$p_n = 2\sinh(\kappa)/\kappa \{\exp[2\kappa(n-1)]/(1+\exp[2\kappa(n-1)])$$
$$- \exp[2\kappa(n-l)]/(1+\exp[2\kappa(n-l)])\}$$
$$+2\sinh(\kappa)/\kappa \{\exp[2\kappa(n-l-1)]/(1+\exp[2\kappa(n-l-1)])$$
$$- \exp[2\kappa(n-l)]/(1+\exp[2\kappa(n-l)])\} .\tag{49}$$

One should bear in mind that while in this way the lattice is equipped with kinetic energy the electrons are presented as a standing state. To investigate whether a soliton-assisted transport is achievable for two correlated standing electrons in the lattice suffices to integrate the system (Eqs. (45) and (46)). For illustration this has been done with $N = 61$ lattice sites and as in all previous cases with periodic boundary conditions. The evolution of paired electrons and solitons for repulsive interaction strength, $U = 0.05$, is illustrated in Fig. 12. Noteworthy is that for such

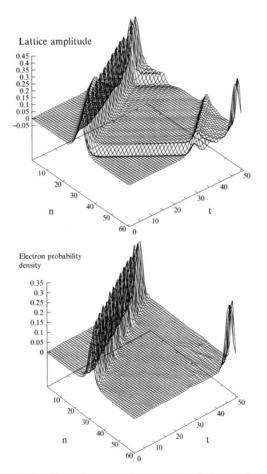

Fig. 12 Toda–Morse lattice. Upper figure: spatio-temporal evolution of a lattice soliton pair; and bottom figure: the electron probability distribution. Parameter values: $\alpha = 1.75$, $V = 0.1$ and $U = 0.05$

particular values the lattice solitons travel with *subsonic* velocity along the lattice retaining their localized profile save an early emission of small radiation to either side. Likewise the localized shape of the electron pair probability distribution as well as the inter-electron distance of a single site are maintained throughout the computation. Apparently part of the energy contained initially in the lattice deformation flows to the electronic degree of freedom with the result that the height of the electron probability density increases, with consequent lowering of the velocity of the corresponding solitons. For higher repulsion strengths, $U \gg 0.05$, *supersonic* moving paired electron lattice solitons compounds can be observed as it is the case for $\alpha = 2$ and $V = 0.25$ for which an inter-electron distance of a single site is attained.

6 Summary and Concluding Remarks

Davydov's approach to ET in biomolecules was a clever combination of physical insight and mathematical beauty. His electro-soliton concept was a fruitful step forward from the polaron concept due to Landau and Pekar. In both cases the underlying lattice dynamics is *harmonic* hence leading to phonons which are linear, infinitesimal excitations of the lattice crystal. The electro-soliton originates in the nonlinearity of the electron–lattice coupling. A natural generalization of the polaron and electro-soliton concepts is possible if consideration of lattice *anharmonicity* is added to the electron–lattice interaction. Indeed, if we focus first on the anharmonic lattice dynamics there are known Hamiltonian cases like the Toda one which being integrable possess as exact solutions, both solitons and solitonic periodic waves obtained in analytical compact form. Such lattice solitons are natural "carriers" of either matter or charge along the lattice crystal [105] and can trap excess, added electrons thus leading to dynamic bound states which have been called solectrons. There is a major component in the solectron concept that makes clear-cut distance with Davydov's electro-soliton. The underlying lattice excitations are of finite amplitude, and not merely infinitesimal.

Davydov's electro-solitons do not survive above 10 K and do this with just a few picoseconds lifetimes. In the present report we have shown that at variance with Davydov's electro-solitons, at least for Morse–Toda-like interactions thermally excited solectrons survive well above the physiological or room temperature range (ca. 300 K) with several picoseconds lifetimes. First we have shown that thermally excited solitons do survive at such temperatures. This was explored assuming that lattice units are atoms or screened ion cores and then tracking lattice compressions by the alternative offered by enhanced (covalent) electron densities. Then adding excess, free (conduction) electrons we have shown how solectrons are formed and survive at the physiological or room temperature range. Subsequently, we have considered pairs of electrons with added Coulomb repulsion albeit in the local, screened Hubbard approximation. The computer simulations have shown that electron pairs dynamically bound to solitons can travel along the charged lattice with speeds either subsonic or supersonic. It clearly appears that Coulomb repulsion does not alter the possibility of solectrons being ET carriers or at the origin of a new form of (non-Ohmic) electric conduction in the presence of an external field [75]. Furthermore, by allowing electron pairing the results here reported open the path to the study of solectron pairs as Bosons and whether or not such a system is prone to Bose–Einstein condensation is an appealing question.

Acknowledgements The authors are grateful to Professors J.J. Kozak, G. Nicolis and G. Tsironis, for fruitful discussions. This research has been sponsored by the EU under Grant SPARK II-FP7-ICT-216227 and by the Spanish Government under Grant MEC-VEVES-FIS2006-01305.

References

1. B. Alberts, D. Bray, J. Lewis, M.Raff, K. Roberts, J.D. Watson, *Molecular Biology of the Cell* (Garland, New York, 1983)
2. J.A. McCammon and S.C. Harvey, *Dynamics of Proteins and Nucleic Acids* (Cambridge University Press, Cambridge, 1987)
3. C. Branden and J. Tooze, *Introduction to Protein Structure* (Garland, New York, 1991)
4. J.J. Hopfield, Proc. Nat. Acad. Sci. USA **71**, 3649 (1974)
5. D.N. Beratan, J.N. Onuchic, J.J. Hopfield, J. Chem. Phys. **86**, 4488 (1987)
6. J.J. Hopfield, J.N. Onuchic, D.N. Beratan, Science **241**, 817 (1988)
7. J.N. Onuchic and D.N. Beratan, J. Chem. Phys. **92**, 722 (1990)
8. D.N. Beratan, J.N. Betts, J.N. Onuchic, Science **252**, 1285 (1991)
9. J.N. Onuchic, P.C.P. Andrade, D.N. Beratan, J. Chem. Phys. **95**, 1131 (1991)
10. K. Schulten, M. Tesh, Chem. Phys. **158**, 421 (1991)
11. J.N. Onuchic, D.N. Beratan, J.R. Winkler, H.B. Gray, A. Rev. Biophys. Struct. **21**, 349 (1992)
12. C. Turrò, C.K. Chang, G.E. Leroi, R.I. Cukier, D.G. Nocera, J. Am. Chem. Soc. **114**, 4013 (1992)
13. M.H. Vos, M.R. Jones, C.N. Hunter, J. Breton, J.-C. Lambry, J.-L. Martin, Biochem. **33**, 6759 (1994)
14. H.B. Gray, J.R. Winkler, Annu. Rev. Biochem. **65**, 537 (1996)
15. S.S. Skourtis, D.N. Beratan, J. Biol. Inorg. Chem. **2**, 378 (1997)
16. M. Bixon, B. Giese, S. Wessely, T. Langenbacher, M.E. Michel-Beyerle, J. Jortner, Proc. Natl. Acad. Sci. USA **96**, 11713 (1999)
17. V. Sartor, P.T. Henderson, G.B. Schuster, J. Am. Chem. Soc. **121**, 11027 (1999)
18. C. Wan, T. Fiebig, S.O. Kelley, C.R. Treadway, J.K. Barton, A.H. Zewail, Proc. Natl. Acad. Sci. USA **96**, 6014 (1999)
19. E.M. Conwell, S.V. Rakhmanova, Proc. Natl. Acad. Sci. USA **97**, 4556 (2000)
20. E.W. Schlag, D.-Y. Yang, S.-Y. Sheu, H.L. Selzle, S.H. Lin, P.M. Rentzepis, Proc. Natl. Acad. Sci. USA **97**, 9849 (2000)
21. C. Wan, T. Fiebig, O. Schiemann, J.K. Barton, A.H. Zewail, Proc. Natl. Acad. Sci. USA **97**, 14052 (2000)
22. S.-Y. Sheu, D.-Y. Yang, H.L. Selzle, E.W. Schlag, Eur. Phys. J. D **20**, 557 (2002)
23. H.B. Gray, J.R. Winkler, Proc. Natl. Acad. Sci. USA **102**, 3534 (2005)
24. L.D. Landau, Phys. Z. Sowjetunion **3**, 664 (1933)
25. S.I. Pekar, Sov. Phys. JETP, **16**, 335 (1946)
26. L.D. Landau, S.I. Pekar, Sov. Phys. JETP, **18** 419 (1948)
27. T.D. Holstein, Ann. Phys. NY **8**, 325, 343 (1959)
28. A.S. Alexandrov, N. Mott *Polarons and Bipolarons*, (World Scientific, Singapore 1995)
29. A.S. Davydov, J. Theor. Biol. **38**, 559 (1973)

30. A.S. Davydov, N.I. Kislukha, Sov. Phys. JETP **44**, 571 (1976)
31. A. S. Davydov, Sov. Phys. Rev. B **25**, 898 (1982)
32. A.S. Davydov, *Solitons in Molecular Systems*, 2nd edn. (Reidel, Dordrecht, 1991)
33. A.L. Christiansen, A.C. Scott (eds.), *Davydov's Soliton Revisited. Self-Trapping of Vibrational Energy in Protein* (Plenum Press, New York, 1983)
34. A.C. Scott, Phys. Rep. **217**, 1 (1992)
35. M. Peyrard (ed.), *Nonlinear Excitations in Biomolecules* (Springer-Verlag, Berlin, 1995)
36. P.S. Lomdahl, W.C. Kerr, Phys. Rev. Lett. **55**, 1235 (1985)
37. H. Fröhlich, Proc. R. Soc. London, Ser. A **215**, 291 (1952)
38. H. Fröhlich, Phys. Lett. **26A**, 402 (1968)
39. H. Fröhlich, Int. J. Quant. Chem. **2**, 641 (1968)
40. H. Fröhlich, Nature (London) **228**, 1093 (1970)
41. H. Fröhlich, Phys. Lett. **39A**, 153 (1972)
42. H. Fröhlich, Phys. Lett. **51A**, 21 (1975)
43. H. Fröhlich, Nuovo Cimento **7**, 416 (1977)
44. J.A. Tuszynski, R. Paul, R. Chatterjee, S.R. Sreenivasan, Phys. Rev. A **30**, 2666 (1984)
45. O.H. Olsen, M.R. Samuelsen, S.B. Petersen, L. Nørskov, Phys. Rev. A **38**, 5856 (1988)
46. A.V. Zolotaryuk, P.L. Christiansen, A.V. Savin, Phys. Rev. E **54**, 3881 (1996)
47. P.L. Christiansen, A.V. Zolotaryuk, A.V. Savin, Phys. Rev. E **56**, 877 (1997)
48. A.V. Zolotaryuk, K.H. Spatschek, A.V. Savin, Phys. Rev. B **54**, 266 (1996)
49. S. Caspi, E. Ben-Jacob, Europhys. Lett. **47**, 522 (1999)
50. S.W. Englander, N.R. Kallenbach, A.J. Heeger, J.A. Krumhansl, S. Litwin, Proc. Nat. Acad. Sci. USA **77**, 7222 (1980)
51. M. Peyrard, A.R. Bishop, Phys. Rev. Lett. **62**, 2755 (1989)
52. G. Gaeta, C. Reiss, M. Peyrard, T. Dauxois, Riv. Nuovo Cim. **17**, 1 (1994)
53. T. Dauxois, M. Peyrard, *Physics of Solitons* (Cambridge University Press, Cambridge, 2006)
54. A.C. Scott, *The Nonlinear Universe. Chaos, Emergence, Life* (Springer, Berlin, 2007)
55. Q. Xie, G. Archontis, S.S. Skourtis, Chem. Phys. Lett. **312**, 237 (1999)
56. E.S. Medvedev, A.A. Stuchebrukhov, J. Chem. Phys. **107**, 3821 (1997)
57. S. Yomosa, Phys. Rev. A **32**, 1752 (1985)
58. D. Hennig, Phys. Rev. E **64**, 041908 (2001)
59. N. Voulgarakis, D. Hennig, H. Gabriel, G.P. Tsironis, J. Phys.: Cond. Matter **13**, 9821 (2001)
60. D. Hennig, Eur. Phys. J. B **24**, 377 (2001)
61. D. Hennig, Phys. Rev. B **65**, 174302 (2002)
62. D. Hennig, Physica A **309**, 243 (2002)
63. D. Hennig, Eur. Phys. J. B **30**, 211 (2002)
64. S. Komarnicki, D. Hennig, J. Phys.: Cond. Matter **15**, 441 (2002)
65. D. Hennig, J.F.R. Archilla, J. Agarwal, Physica D **180**, 256 (2003)
66. M. Toda, *Theory of Nonlinear Lattices*, 2nd edn. (Springer-Verlag, New York, 1989) (and references therein)
67. T.P. Valkering, J. Phys. A: Math. Gen. **11**, 1885 (1978)
68. G. Friesecke, J.A.D. Wattis, Commun. Math. Phys. **161**, 391 (1994)
69. V.I. Nekorkin, M.G. Velarde, *Synergetic Phenomena in Active Lattices. Patterns, Waves, Solitons, Chaos* (Springer-Verlag, Berlin, 2002)
70. F.G. Mertens, H. Büttner, in *Solitons*, edited by S.E. Trullinger, V.E. Zakharov and V.L. Pokrousky (North-Holland, Amsterdam, 1986) ch. 15
71. J. Dancz, S.A. Rice, J. Chem. Phys. **67**, 1418 (1977)
72. T.J. Rolfe, S.A. Rice, J. Dancz, J. Chem. Phys. **70**, 26 (1979)
73. F. Marchesoni, C. Lucheroni, Phys. Rev. B **44**, 5303 (1991)
74. A.P. Chetverikov, W. Ebeling, M.G. Velarde, Eur. Phys. J. B **44**, 509 (2005)
75. M.G. Velarde, W. Ebeling, A.P. Chetverikov, Int. J. Bifurcation Chaos **15**, 245 (2005)
76. A.P. Chetverikov, W. Ebeling, M.G. Velarde, Eur. Phys. J. B **51**, 87 (2006)
77. A.P. Chetverikov, W. Ebeling, M.G. Velarde, Int. J. Bifurcation Chaos **16**, 1613 (2006)
78. V.A. Makarov, M.G. Velarde, A.P. Chetverikov, W. Ebeling, Phys. Rev. E **73**, 066626 (2006)
79. M. G. Velarde, W. Ebeling, D. Hennig, C. Neissner, Int. J. Bifurcation Chaos **16**, 1035 (2006)

80. D. Hennig, C. Neissner, M.G. Velarde, W. Ebeling, Phys. Rev. B **73**, 024306 (2006)
81. A.P. Chetverikov, W. Ebeling, G. Röpke, M.G. Velarde, Contr. Plasma Phys. **47**, 465 (2007)
82. D. Hennig, A. Chetverikov, M.G. Velarde, W. Ebeling, Phys. Rev. E **76**, 046602 (2007)
83. M.G. Velarde, W. Ebeling, A.P. Chetverikov, D. Hennig, Int. J. Bifurcation Chaos **18**, 521 (2008)
84. P. Morse, Phys. Rev. **34**, 57 (1929)
85. M.G. Velarde, Int. J. Comput. Appl. Math. DOI 10.10.16/j.cam.2008.07.058
86. W. Ebeling, I.M. Sokolov, *Statistical Thermodynamics and Stochastic Theory of Nonequilibrium Systems* (World Scientific, Singapore, 2005)
87. G. Kalosakas, S. Aubry, G.P. Tsironis, Phys. Rev. B **58**, 3094 (1998)
88. G. Kalosakas, K.O. Rasmussen, A.R. Bishop, J. Chem. Phys. **118**, 3731 (2003)
89. G. Kalosakas, K.O. Rasmussen, A.R. Bishop, Synthetic Metals **141**, 93 (2004)
90. V. Heine, D. Weaire, Pseudopotential Theory of Cohesion and Structure, in *Solid State Physics*, vol. 24, pp. 250–463, eds. H. Ehrenreich, F. Seitz, D. Turnbull (Academic, New York, 1970)
91. W.D. Kraeft, D. Kremp, W. Ebeling, G. Röpke, *Quantum Statistics of Charged Particle Systems* (Akademie-Verlag, Berlin, 1986)
92. J.S. Blakemore, *Semiconductor Statistics* (Pergamon Press, 1962); *Solid State Physics* (Cambridge University Press, Cambridge, 1985)
93. W.A. Harrison, *Solid State Theory* (Dover Ed., New York, 1979)
94. P.W. Anderson, Phys. Rev. **112**, 1900 (1958)
95. I.M. Lifshitz, S.A. Gredeskul, L.A. Pastur, *Introduction to the Theory of Disordered Systems* (in Russian) (Nauka, Moscow, 1982)
96. V.A. Schweigert, F.M. Peeters, Phys. Rev. B **51**, 7700 (1995)
97. V.M. Bedanov, F.M. Peeters, Phys. Rev. B **49**, 2667 (1994)
98. M. Bonitz, V. Golubichnyi, A.V. Filinov, Yu.F. Lozovik, Microelectron. Eng. **62**, 141 (2002)
99. M. G. Velarde, C. Neissner, Int. J. Bifurcation Chaos **18**, 885 (2008)
100. J.D. Patterson, B.C. Bailey, *Solid State Physics. Introduction to the Theory* (Springer-Verlag, Berlin, 2007)
101. H. Böttger, V.V. Bryksin, *Hoping Conduction in Solids* (Academie-Verlag, Berlin, 1985)
102. J. Hubbard, Proc. Roy. Soc. (London) A **276**, 238 (1963); A **277**, 237 (1964); A **281**, 401 (1964)
103. A. Montorsi (ed.), *The Hubbard Model. A Reprint Volume* (World Scientific, Singapore, 1992)
104. D. Hennig, M.G. Velarde, W. Ebeling, A.P. Chetverikov, Phys. Rev. E, **78**, 066606 (2008)
105. E. del Rio, M.G. Velarde, W. Ebeling, Physica A **377**, 435 (2007)

How Exponential Type Orbitals Recently Became a Viable Basis Set Choice in Molecular Electronic Structure Work and When to Use Them

Philip E. Hoggan

Abstract This paper advocates the use of the atomic orbitals which have direct physical interpretation, i.e. Coulomb Sturmians and hydrogen-like orbitals. They are exponential type orbitals (ETOs). Their radial nodes are shown to be essential in obtaining accurate nuclear shielding tensors for NMR work.

Until 2008, their products on different atoms were difficult to manipulate for the evaluation of two-electron integrals. The difficulty was mostly due to somewhat cumbersome orbital translations involving slowly convergent infinite sums. These are eliminated using Coulomb resolutions. Coulomb resolutions provide an excellent approximation that reduces these integrals to a sum of one-electron overlap-like integral products that each involve orbitals on at most two centers. Such two-center integrals are separable in prolate spheroidal co-ordinates. They are thus readily evaluated. Only these integrals need to be re-evaluated to change basis functions.

In this paper, a review of the translation procedures for Slater type orbitals (STO) and for Coulomb Sturmians follows that of the more recent application to ETOs of a particularly convenient Coulomb resolution.

Keywords Coulomb Sturmian basis · nodal structure · Coulomb resolutions · *ab initio* quantum chemistry

1 Introduction

The criteria for choice between Gaussian and exponential basis sets for molecules do not seem obvious at present. In fact, it appears to be constructive to regard them as being complementary, depending on the specific physical property required from molecular electronic structure calculations.

The present work describes a breakthrough in two-electron integral calculations, as a result of Coulomb operator resolutions. This is particularly significant in that

P.E. Hoggan
LASMEA, UMR 6602 CNRS, University Blaise Pascal, 24 avenue des Landais,
63177 Aubiere Cedex, France
e-mail: phhoggan@univ-bpclermont.fr

N. Russo et al. (eds.), *Self-Organization of Molecular Systems: From Molecules and Clusters to Nanotubes and Proteins*, NATO Science for Peace and Security Series A: Chemistry and Biology, © Springer Science+Business Media B.V. 2009

it eliminates the arduous orbital translations which were necessary until now for exponential type orbitals. The bottleneck has been eliminated from evaluation of three- and four- center integrals over Slater type orbitals and related basis functions.

The two-center integrals are replaced by sums of overlap-like one-electron integrals. This implies a speed-up for all basis sets, including Gaussians. The improvement is most spectacular for exponential type orbitals. A change of basis set is also facilitated as only these one-electron integrals need to be changed. The Gaussian and exponential type orbital basis sets are, therefore interchangeable in a given program. The timings of exponential type orbital calculations are no longer significantly greater than for a Gaussian basis, when a given accuracy is sought for molecular electronic properties.

Atomic orbitals are physically meaningful one-electron atom eigenfunctions for the Schrödinger equation. This gives them as well-known analytical expressions: hydrogen-like orbitals.

Boundary conditions allow the principal quantum number n to be identified as the order of the polynomial factor in the radial variable. It must therefore be positive and finite. It is also defined such that $n - l - 1$ is greater than or equal to 0. This gives the number of zeros of the polynomial (radial nodes). Here, $l = 0$, or a positive integer, which defines the angular factor of the orbital. (i.e. a spherical harmonic, or, more rarely, its Cartesian equivalent) The number n gives the energy of the one-electron atomic bound states. Frequently, basis set studies focus on the radial factor. That is, for our present purposes, the angular factor can be assumed sufficiently defined as a spherical harmonic.

The key issue is whether to choose basis sets with exponential or Gaussian asymptotic factors.

Certain physical properties, such as NMR shielding tensor calculations directly involve the nuclear cusp and correct treatment of radial nodes, which indicates that basis sets such as Coulomb Sturmians are better suited to their evaluation than Gaussians [36–38].

There is also evidence to suggest that CI expansions converge in smaller exponential basis sets compared to Gaussians [45, 46]. Benchmark overlap similarity work is available [46, 47].

2   Wave-Function Quality

The following quantity:

$$-1/2 \frac{\nabla \rho(r)}{\rho(r)}$$

is used to test wave-function quality. It is smooth, to varying degrees, in different basis sets. Atomic positions must give cusps. The importance for Quantum Monte Carlo work and DFT applications has been detailed elsewhere [61].

Much molecular quantum chemistry is carried out using Gaussian basis sets and they are indeed convenient and lead to rapid calculations. The essential advantage they had over exponential basis sets was the simple product theorem for Gaussians on two different atomic centers. This allows all the two-electron integrals, including three- and four-center terms to be expressed as single-center two-electron integrals.

The corresponding relationship for exponential type orbitals generally led to infinite sums and the time required, particularly for four-center integrals could often become prohibitive.

Recent work by Gill has, nevertheless been used to speed up all three- and four-center integral evaluation, regardless of basis using the resolution of the Coulomb operator [15–17]. This work by Gill is used here to reduce the three- and four- center two-electron integrals to a sum of products of overlap-like (one-electron) integrals, basically two-centered. This algorithm was coded in a Slater type orbital (STO) basis within the framework of the STOP package [4] (in fortran) during summer 2008. Note, however, that other exponential or Gaussian basis sets can readily be used. The set of one-electron overlap-like auxiliary integrals is the only calculation that needs to be re-done to switch basis functions. They may be re-evaluated for the basis set that the user selects for a given application. This procedure makes the approach highly versatile, since a change of basis set requires relatively few simple new evaluations. A modular or object-oriented program is being designed to do this efficiently [17, 48, 49].

The present article gives illustrative test results on molecular systems, e.g. the H_2 dimer.

The layout of this appraisal of recent work is as follows: the review begins with a brief recap of basis sets and programming strategy in the next two sections. Atom pairs are the physical entity used for integral evaluation, both in the Poisson equation technique and the Coulomb resolution. Two sections are devoted to these progressively more powerful techniques which both reduce two-electron to one-electron integrals. The overlaps required for the Coulomb resolution differ by a potential factor from orbital overlaps. Their evaluation is nevertheless analytic, using well-known techniques summarized in the subsequent section. Finally, to illustrate what can be gained by eliminating orbital translations, the translation of Slater type orbitals is reviewed briefly, from recent work on BCLFs. Translation of Coulomb Sturmians is briefly outlined to review work on he Shibuya–Wulfman matrix. Both these techniques have been studied by the present author. A few numerical results are given on the dimer of molecular hydrogen which show progressive speedup particularly for the Coulomb resolution given a pre-selected accuracy, which proves sufficient to provide satisfactory confirmation of experimental vibrational spectroscopy work on this dimer.

3 Basis Sets

Although the majority of electronic quantum chemistry uses Gaussian expansions of atomic orbitals [20, 21], the present work uses exponential type orbital (ETO) basis sets which satisfy Kato's conditions for atomic orbitals: they possess a cusp at the nucleus and decay exponentially at long distances from it [58–60]. It updates a 'real chemistry' interest beginning around 1970 and detailed elsewhere [22–33, 51].

Two types of ETO are considered here: Slater type orbitals (STOs) [56, 57] and Coulomb Sturmians, which may be written as a finite combination thereof [54]. Otherwise, STOs may be treated as multiple zeta basis functions in a similar way to the approach used with Gaussian functions.

Many exponential type functions exist [54]. Preferential use of Sturmian type functions is discussed [17].

Coulomb Sturmians have the advantage of constituting a complete set without continuum states because they are eigenfunctions of a Sturm–Liouville equation involving the nuclear attraction potential, i.e. the differential equation below.

$$\nabla_r^2 S_{nl}^m(\beta, \vec{r}) = \left[\beta^2 - \frac{2\beta n}{r}\right] S_{nl}^m(\beta, \vec{r}).$$

The exponential factor of Coulomb Sturmians; $e^{-\beta r}$ has an arbitrary screening parameter β. In the special case when $\beta = \zeta/n$ with n the principal quantum number and ζ the Slater exponent, we obtain hydrogen-like functions, which do not span the same space and require inclusion of continuum states to form a complete set [54]. Hydrogen-like functions are, however well known as atomic orbitals: the radial factor contains the associated Laguerre polynomial of order $2l + 1$ with suffix $n - l - 1$ and the exponential $e^{-\zeta r/n}$ as indicated above. The angular factor is just a spherical harmonic of order l. These functions are ortho-normal. The optimal values of the β parameters may be determined analytically by setting up secular equations which make use of the fact that the Sturmian eigenfunctions also orthogonalise the nuclear attraction potential, as developed by Avery [39].

$$\int S_{nl}^m(r, \theta, \phi) S_{n'l'}^{m'}(r, \theta, \phi) \frac{dr}{r} = \delta_{nn'll'mm'}.$$

Alternative ETOs would be Slater type orbitals and B-functions with their simple Fourier transforms. Strictly, they should be combined as linear combinations to form hydrogen-like or, better, Sturmian basis sets prior to use.

STOs allow us to use routines from the STOP package [55] directly, whereas Coulomb Sturmians still require some coding. The relationship to STOs is used to carry out calculations over a Coulomb Sturmian basis with STOP until the complete Sturmian code is available. The present state-of-the-art algorithms require at most twice as long long per integral than GTO codes but the CI converges with fewer functions and the integrals may be evaluated after Gaussian expansion or expressed as overlaps to obtain speed up [63]. Recent iterative procedures devised by Nakatsuji to be published in IJQC during 2009 and independent of basis prove that CI requires at least three times more Gaussians than Slater type orbitals for an electron pair.

After a suitably accurate electron density has been obtained for the optimized geometry over a Coulomb Sturmian basis set, the second-order perturbation defining the nuclear shielding tensor should be evaluated in a Coupled perturbed Hartree Fock scheme.

The integrals involved may conveniently be evaluated using B-functions with linear combinations giving the Coulomb Sturmians.

$$S_{nl}^m(r) = (2\alpha)^{3/2} \frac{2^{2l+1}}{2l+1)!!} \sum_{l=0}^{n-l-1} \frac{(-n+l+1)_t \, (n+l+1)_t}{t! \, (l+3/2)_t} B_{t+1,l}^m(r)$$

The techniques exploit properties of Fourier transforms of the integrand.

Note that either HF or DFT can serve as zero order for the present nuclear shielding tensor calculation over ETOs.

A full *ab initio* B-function code including nuclear shielding tensor work is expected to be complete shortly.

Some tests show that Slater type orbitals (STO) or B-functions (BTO) are less adequate basis functions that Coulomb Sturmians, because only the Sturmians possess the correct nuclear cusp and radial behavior.

4 Programming Strategy

Firstly, the ideal *ab initio* code would rapidly switch from one type of basis function to another.

Secondly, the chemistry of molecular electronic structure must be used to the very fullest extent. This implies using atoms in molecules (AIM) and diatomics in molecules (DIM) from the outset, following Bader (in an implementation due to Rico et al. [50] and Tully [41] implemented in our previous work [55], respectively. The natural choice of atomic orbitals, i.e. the Sturmians or hydrogen-like orbitals lend themselves to the AIM approach. To a good approximation, core eigenfunctions for the atomic hamiltonian remain unchanged in the molecule. Otherwise, atom pairs are the natural choice, particularly if the Coulomb resolution recently advocated by Gill is used. This leads us to products of auxiliary overlaps which are either literally one- or two- centered, or have one factor of the product where a simple potential function needs to be translated to one atomic center.

The Slater basis set nightmare of the Gegenbauer addition theorem is completely avoided. Naturally, the series of products required for, say a four-center two-electron integral may require 10 or even 20 terms to converge to chemical accuracy, when at least one atom pair is bound but the auxiliaries are easy to evaluate recursively and re-use. Unbound pairs may be treated using a smaller number of terms since the integrals can be predicted to be small, using a Schwarz inequality.

Now, the proposed switch in basis set may also be accomplished just by re-evaluating the auxiliary overlaps. Furthermore, the exchange integrals are greatly simplified in that the products of overlaps just involve a two-orbital product instead of a homogeneous density. The resulting cpu-time growth of the calculation is n^2 for SCF, rather than n^4. Further gains may be obtained by extending the procedure to post-Hartree-Fock techniques involving explicit correlation, since the r_{12}^{-1} integrals involving more than two electrons, that previously soon led to bottlenecks, are also just products of overlaps.

5   Atom Pairs: Solving Poisson's Equation

All the molecular integrals over CS required for standard SCF may be evaluated using analytical two-center terms based on the solution of Poisson's equation for the
Coulomb potential in an ETO basis. This uses the spectral forms (involving incom-
plete gamma functions and regular and irregular solid harmonics) defined initially
in [52, 53, 63] and subsequently generalized to ensure numerical stability as shown
in a brief summary below.

Recalling the definition of a Slater type orbital:

$$\chi_{n, l, m, \zeta}(r, \theta, \phi) = N_1 \, r^{n-1} e^{-\zeta r} \, Y_l^m(\theta, \phi) \, .$$

Define the radial factor $g(r)$:

$$g(r) = r^{n-1} e^{-\zeta r} \, .$$

Then, (from the spectral forms in [63]), the potential due to this distribution is
immediately written:

$$\Pi_l(g) = r^2 F(r) \, ,$$

Where g is short for g(r) and F(r) is given below, with a suitable variable of
integration; u:

$$F(r) = \int_0^1 du \, g(ru) \, u^{l+2} + \int_1^\infty du \, g(ru) \, u^{1-l} .$$

This expression is used to write all radially dependent one and two-center integrals in analytical closed form.

The next section describes a more profound advance, that reduces the atom-pair
evaluation to one-electron overlap-like integrals. It is related to the Poisson equation
technique, as detailed in [34, 63].

6   Avoiding ETO Translations for Two-Electron Integrals over 3 and 4 Centers

Previous work on separation of integration variables is difficult to apply, in contrast
to the case for Gaussians [43] cf. [40]. Recent work by Gill et al. [15] proposes a
resolution of the Coulomb operator, in terms of potential functions ϕ_i, which are
characterized by examining Poisson's equation. In addition, they must ensure rapid
convergence of the implied sum in the resulting expression for Coulomb integrals
J_{12} as products of "auxiliaries", i.e. overlap integrals, as detailed in [15]:

$$J_{12} = < \rho(r_1) \, \phi_i(r_1) > < \phi_i(r_2) \, \rho(r_2) > \quad \text{with implied sumation over i} \qquad (1)$$

This technique can be readily generalized to exchange and multi-center two-electron integrals.

Note, however, that the origin of one of the potential functions only may be chosen to coincide with an atomic (nuclear) position.

Define potential functions ϕ_i in the scope of a Coulomb operator resolution, as follows $\phi_i = 2^{3/2} Y_l^m(\theta, \phi) \phi_{nl}(r)$:

$$\phi_{nl}(r) = \int_0^{+\infty} h_n(x) j_l(rx) dx \quad \text{with} \quad j_l(x) \text{ denoting the spherical Bessel function} \quad (2)$$

Here, $h_n(x)$ is the nth member of any set of functions that are complete and orthonormal on the interval $[0, +\infty)$, such as the nth order polynomial function (i.e. polynomial factor of an exponential). The choice made in [15] is to use parabolic cylinder functions (see also another application [64]), i.e. functions with the even order Hermite polynomials as a factor. This is not the only possibility and a more natural and convenient choice is based on the Laguerre polynomials $L_n(x)$: Define:

$$h_n(x) = \sqrt{2} \, L_n(2\,x) e^{-x} \quad (3)$$

These polynomial functions are easy to use and lead to the following analytical expressions for the first two terms in the potential defined in Eq. (2):

$$V_{00}(r) = \sqrt{2} \, \frac{\tan^{-1}(r)}{r} \quad (4)$$

$$V_{10}(r) = \sqrt{2} \, [\frac{\tan^{-1}(r)}{r} - \frac{2}{(1 + r^2)}] \quad (5)$$

Furthermore, higher n expressions of $V_{n0}(r)$ all resemble Eq. (5) (see [16] Eq (23)):

$$V_{n0}(r) = \sqrt{2} \, \frac{1}{r}(1 + \sum_1^n (-1)^k \frac{\sin(2\,k\,\tan^{-1}(r))}{k}) \quad (6)$$

and analytical expressions of $V_{nl}(r)$ with non-zero l are also readily obtained by recurrence.

The auxiliary overlap integrals $< \rho(r_1)\ \phi_i(r_1) >$ and $< \phi_i(r_2)\ \rho(r_2) >$ will involve densities obtained from atomic orbitals centered on two different atoms in most multi-center two-electron integrals. The integrals required in an ETO basis are thus of the type:

$$< \psi_a(r_1)\ \psi_b(r_1)\ \phi_i(r_1) > \quad (7)$$

Such integrals appear for two-center exchange integrals and all three- and four center integrals. Note that exchange integrals require distinct orbitals ψ_a and ψ_b. In the atomic case, they must have different values for at least one of n, l, m or ζ. In the two-center case, the functions centered at a and b may be the same.

The product does not correspond to a single-center density: it is two-centered. The above equation then illustrates the relationship to the one-electron two-center overlap integral, although it clearly includes the extra potential term from the Coulomb operator resolution.

The overlap integrals may be evaluated by separating the variables in prolate spheroidal co-ordinates, following Mulliken and Roothaan [42] and using recurrence relations in [19]:

$$S(n_1, l_1, m, n_2, l_2, \alpha, \beta) = \alpha^{n_1+1/2} \beta^{n_2+1/2} \left[(2n_1)! \, (2n_2)!\right]^{-1/2} s(n_1 l_1 m n_2 l_2 \alpha \beta)$$

$$= N(n_1, n_2, \alpha\beta) s(n_1, l_1, m, n_2, l_2, \alpha\beta)$$

where: $\alpha = k_1 R$ and $\beta = k_2 R$. The k_1, k_2 are Slater exponents. The core overlaps are given by:

$$s(n_1, l_1, m, n_2, l_2, \alpha, \beta) = \int_1^\infty \int_{-1}^1 exp \left\{ -\frac{1}{2}(\alpha + \beta)\mu - \frac{1}{2}(\alpha - \beta)v \right\}$$
$$(\mu + v)^{n_1} (\mu - v)^{n_2} T(\mu, v) d\mu dv$$

$$\mu = \frac{r_a + r_b}{R}$$

$$v = \frac{r_a - r_b}{R}$$

r_a and r_b are the instantaneous position vectors of the electron from the two centers labeled a and b, respectively and separated by a distance R. We also define, using the normalised spherical tensors S:

$$T(\mu, v) = S_{l_1}^m(\mu, v)_a S_{l_2}^m(\mu, v)_b$$

The core overlaps then take the form:

$$s(n_1, l_1, m, n_2, l_2, \alpha, \beta) = D_{l_1, l_2, m} \sum_{ij}^\lambda Y_{ij}^\lambda A_i \left\{ \frac{1}{2}(\alpha + \beta) \right\} B_j \left\{ \frac{1}{2}(\alpha - \beta) \right\}$$

Y_{ij}^λ is a matrix with integer elements uniquely determined from n, l and m. It is obtained as a generalised binomial coefficient, in the expansion of:

$$(r_a - r_b)^n \, (r_a + r_b)^n$$

$D_{l_1, l_2, m}$ is a coefficient that is independent of the principal quantum number. It is obtained upon expanding the product of two Legendre functions in this co-ordinate system. Symmetry conditions imply that only $m_1 = m_2 = m$ lead to non-zero coefficients.

$$A_i \left\{ \frac{1}{2}(\alpha + \beta) \right\} = \int_1^\infty exp \left\{ -\frac{1}{2}(\alpha + \beta)\mu \right\} \mu^i d\mu$$

$$B_j \left\{ \frac{1}{2}(\alpha - \beta) \right\} = \int_{-1}^{1} exp \left\{ -\frac{1}{2}(\alpha - \beta)v \right\} v^j \, dv$$

Here, recurrence relations on the auxiliary integrals A and B lead to those for the requisite core integrals [18, 19].

This assumes tacitly that the potential obtained from the coulomb operator resolution be centered on one of the atoms. Whilst this choice can be made for one pair in a four-center product, it cannot for the second. There remains a single translation for this potential in one auxiliary of the two in a product representing a four-center integral and none otherwise. The structure of these potential functions obtained by recurrence from Eq. (6) shows that the translation may be accomplished readily in the prolate spheroidal co-ordinates. This point is addressed in detail in a recent publication [17].

This method obviates the need to evaluate infinite series that arise from the orbital translations efficiently. They have been eliminated in the Coulomb operator resolution approach, since only orbitals on two centers remain in the one-electron overlap-like auxiliaries. These can be evaluated with no orbital translation, in prolate spheroidal co-ordinates, or by Fourier transformation [16, 17].

7   How Slater Type Orbitals Were Translated

The Barnett–Coulson–Löwdin functions (BCLFs see [2]) arise as coefficients in the series expansion of a Slater type orbital centered at a distance a from the origin, placed on an atomic nucleus where a set of Slater type orbitals are centered [11, 14]. This allows the one- and two-electron multi-center integrals to be evaluated at a given origin in the molecule. The series expansion obtained is infinite, since the molecular geometry variable a (usually 1–20 a.u.) is fixed for an electronic structure calculation, whereas the instantaneous electron position variable r is independent of it and $0 < r < \infty$. They are both radial vectors and generally cannot be aligned.

Much work is already available on BCLFs [5, 7–10, 12, 35, 36] and references therein. Nevertheless, two bottlenecks are yet to be efficiently resolved when Slater type orbital translations are required:

(i) Rapid and accurate generation of the BCLFs themselves
(ii) Acceleration of the convergence of the infinite series generated, which typically do not converge quickly

In the present work, the first item, (i) is thoroughly addressed.

The value of the screening parameter ζ generally exceeds 1 and should not exceed the atomic number. In practice, the lower limit for ζ is related to the first ionization potential I in atomic units, i.e., ζ must not be less than $\sqrt{2I}$. These limitations are helpful in establishing the numerical behavior of the BCLFs.

In this work, we treat the problem of efficient computation of BCLFs. Our aim is to develop a computational procedure by which a whole sequence of BCLFs can be computed fast and accurately. In Section 8, we present an up-to-date review of properties of BCLFs. In more complete work on the subject [62], we discuss and evaluate possible strategies for computing them, and conclude that recursion relations can be used efficiently for this purpose, provided that the modified spherical Bessel functions $I_{n+1/2}(x)$ and $K_{n+1/2}(x)$ can be computed fast and accurately.

A method by which a whole array of BCLFs can be computed simultaneously, quickly and accurately is also detailed in previous work [62]. In this recent work, we also discuss the details of the programming of our method. It is important to note that, in our method, we do not compute $I_{n+1/2}(x)$ and $K_{n+1/2}(x)$ directly. Taking into account the asymptotics of $I_\nu(x)$ and $K_\nu(x)$ as $\nu \to \infty$, we compute some appropriately scaled versions of these functions instead. The scaling we use enables us to avoid the underflows and overflows that may occur in direct computation of $I_{n+1/2}(x)$ and $K_{n+1/2}(x)$ for large values of n; it is thus an important ingredient of our method. This also allows us to scale the BCLFs appropriately. In order to end up with BCLFs that have double-precision accuracy, in our method, we compute both the functions $I_{n+1/2}(x)$ and $K_{n+1/2}(x)$ and the BCLFs in extended precision arithmetic, the idea being that the quadruple-precision arithmetic is shown to suffice and it is offered with some high-level programming language compilers used for scientific applications, e.g. Fortran 77 and C. As the number of arithmetic operations required is very small (of the order of wN, where N is the number of BCLFs computed and w is a small integer), the use of quadruple-precision arithmetic cannot increase the cost of the computation time-wise. We provide an error analysis for the procedure we use to compute the scaled modified spherical Bessel functions, which shows that the procedure is indeed very accurate in previous work [62].

Finally, in [62], we also provide three appendices that contain several results that seem to be new. In the first, we analyze the asymptotic behavior of the modified Bessel functions $I_\nu(x)$ and $K_\nu(x)$ as $\nu \to \infty$. We derive two sets of full asymptotic expansions that have some quite interesting properties. The scalings we use in [62] are based on the results of this appendix. In the second, we obtain explicit power series expansions for products of modified spherical Bessel functions. In the third appendix of [62], we derive asymptotic expansions of BCLFs as their order tends to infinity.

8 Review of BCLFs

8.1 Definition and Properties of BCLFs

Let n be a non-negative integer, a and r two real positive numbers, ζ a real positive number. Of these, a and ζ are finite, while r assumes values from 0 to infinity. With R defined as in

$$R = \sqrt{a^2 + r^2 - 2ar\cos\theta} \tag{8}$$

consider the function $R^{n-1}e^{-\zeta R}$. Letting $x = \cos\theta$ so that $x \in [-1, +1]$, its expansion in Legendre polynomials $P_\lambda(x)$ may be expressed as

$$R^{n-1}e^{-\zeta R} = \frac{1}{\sqrt{ar}} \sum_{\lambda=0}^{\infty} (2\lambda + 1) A_{\lambda+1/2}^{n}(\zeta, a, r) P_{\lambda}(x), \quad -1 \le x \le 1, \quad (9)$$

$A_{\lambda+1/2}^{n}$ being the BCLFs. From this relation, it is seen that $R^{n-1}e^{-\zeta R}$ serves as a "generating function" for the BCLFs. Since

$$\int_{-1}^{+1} P_{\lambda}^{2}(x)\, dx = \frac{2}{2\lambda + 1}, \quad \lambda = 0, 1, \ldots, \quad (10)$$

we immediately deduce from Eq. 9 that

$$A_{\lambda+1/2}^{n}(\zeta, a, r) = \frac{\sqrt{ar}}{2} \int_{-1}^{+1} R^{n-1} e^{-\zeta R} P_{\lambda}(x)\, dx, \quad \lambda = 0, 1, \ldots . \quad (11)$$

Clearly, the $A_{\lambda+1/2}^{n}(\zeta, a, r)$ are symmetric functions of a and r, that is,

$$A_{\lambda+1/2}^{n}(\zeta, a, r) = A_{\lambda+1/2}^{n}(\zeta, r, a), \quad (12)$$

because the function $R^{n-1}e^{-\zeta R}$ is.

A simple expression for BCLFs with $n = 0$ and $\lambda = 0, 1, \ldots$, is known (see [1, p. 445, formula 10.2.35]):

$$A_{\lambda+1/2}^{0}(\zeta, a, r) = I_{\lambda+1/2}(\zeta\rho) K_{\lambda+1/2}(\zeta\rho'); \quad \rho = \min\{a, r\}, \quad \rho' = \max\{a, r\}. \quad (13)$$

Here, $I_{\lambda+1/2}(x)$ and $K_{\lambda+1/2}(x)$ are the modified spherical Bessel functions: [The functions $I_{\lambda+1/2}(x)$ and $K_{\lambda+1/2}(x)$ satisfy three-term recursion relations in λ that are given in this work, and are defined for *all* integer values of λ. Those $I_{\lambda+1/2}(x)$ with $\lambda \ge 0$ are called modified spherical Bessel functions of the first kind, while those with $\lambda < 0$ are called modified spherical Bessel functions of the second kind. The $K_{\lambda+1/2}(x)$ are called modified spherical Bessel functions of the third kind. Each of the two pairs $[I_{\lambda+1/2}(x)$ and $I_{-\lambda-1/2}(x)]$ and $[I_{\lambda+1/2}(x)$ and $K_{\lambda+1/2}(x)]$ is a linearly independent set of solutions of the modified spherical Bessel equation of order λ. See Abramowitz and Stegun [1, Chapter 10]]. of order λ, of the first and third kind, respectively. Because $I_{\lambda+1/2}(x)$ and $K_{\lambda+1/2}(x)$ are defined for *all* integer values of λ, we let (13) *define* $A_{\lambda+1/2}^{0}(\zeta, a, r)$ for $\lambda < 0$ as well. This is an important step that enables us to *define* $A_{\lambda+1/2}^{n}(\zeta, a, r)$ for $\lambda < 0$ as well, which is what we consider next (see [6]).

From the integral representation in Eq. (11), it follows that, for $n \ge 0$,

$$A_{\lambda+1/2}^{n+1}(\zeta, a, r) = -\frac{\partial}{\partial \zeta} A_{\lambda+1/2}^{n}(\zeta, a, r), \quad (14)$$

and hence

$$A^n_{\lambda+1/2}(\zeta,a,r) = (-1)^n \frac{\partial^n}{\partial \zeta^n} A^0_{\lambda+1/2}(\zeta,a,r) . \tag{15}$$

From Eq. (13), it is obvious that $A^0_{\lambda+1/2}(\zeta,a,r) = A^0_{\lambda+1/2}(1,\zeta a,\zeta r)$. By a simple manipulation of the integral representation in Eq. (11), it can be shown analogously that $A^n_{\lambda+1/2}(\zeta,a,r)$ satisfy the "homogeneity relation"

$$A^n_{\lambda+1/2}(\zeta,a,r) = \zeta^{-n} A^n_{\lambda+1/2}(1,\zeta a,\zeta r), \quad n \geq 0. \tag{16}$$

This relation shows that $A^n_{\lambda+1/2}(\zeta,a,r)$ are actually functions of two variables, namely, of ζa and ζr, and can be computed directly from the functions $\bar{A}^n_\lambda(a,r)$ that are defined as in

$$\bar{A}^n_\lambda(a,r) = A^n_{\lambda+1/2}(1,a,r). \tag{17}$$

From Eqs. (16) and (17), it follows that $A^n_{\lambda+1/2}(\zeta,a,r)$ can be computed from $\bar{A}^n_\lambda(a,r)$ via

$$A^n_{\lambda+1/2}(\zeta,a,r) = \zeta^{-n} \bar{A}^n_\lambda(\zeta a,\zeta r). \tag{18}$$

Invoking Eq. (18), it is easy to show that Eq. (14) can be rewritten as

$$\bar{A}^{n+1}_\lambda(a,r) = n\bar{A}^n_\lambda(a,r) - \left(a\frac{\partial}{\partial a} + r\frac{\partial}{\partial r}\right)\bar{A}^n_\lambda(a,r), \quad n \geq 0. \tag{19}$$

Translation of Coulomb Sturmians requires the procedures described in the next two sections which are given, for comparison. Much work by Avery and others is already available [70–73, 75, 77, 78].

9   Definition of the Shibuya–Wulfman Integrals

Coulomb Sturmian basis sets are sets of solutions to the one-electron wave equation

$$\left[-\frac{1}{2}\nabla^2 - \frac{nk}{r} + \frac{k^2}{2}\right]\chi_{nlm}(\mathbf{x}) = 0 \tag{20}$$

where k is held constant for the entire set. It can be seen that Eq. (20) is the same as the equation obeyed by hydrogen-like orbitals, except that Z/n has been replaced by the constant k. Thus Coulomb Sturmians have the same form as hydrogen-like orbitals, except that Z/n is everywhere replaced by k. A set of Coulomb Sturmian basis functions obey the potential-weighted orthonormality relations

$$\int d^3x\, \chi^*_{n'l'm'}(\mathbf{x})\frac{1}{r}\chi_{nlm}(\mathbf{x}) = \frac{k}{n}\delta_{n'n}\delta_{l'l}\delta_{m'm} \tag{21}$$

By projecting momentum-space onto the surface of a 4-dimensional hypersphere, V. Fock [76] was able to show that the Fourier-transformed Coulomb Sturmians

$$\chi_{nlm}^{t}(\mathbf{p}) = \frac{1}{\sqrt{(2\pi)^3}} \int d^3x \, e^{-i\mathbf{p}\cdot\mathbf{x}} \chi_{nlm}(\mathbf{x}) \tag{22}$$

can be very simply expressed in terms of 4-dimensional hyperspherical harmonics [65, 80] through the relationship

$$\chi_{n,l,m}^{t}(\mathbf{p}) = M(p)Y_{n-1,l,m}(\mathbf{u}) \tag{23}$$

where

$$M(p) \equiv \frac{4k^{5/2}}{(k^2 + p^2)^2} \tag{24}$$

and

$$\lambda_1 = \frac{2kp_1}{k^2 + p^2}$$

$$\lambda_2 = \frac{2kp_2}{k^2 + p^2}$$

$$\lambda_3 = \frac{2kp_3}{k^2 + p^2}$$

$$\lambda_4 = \frac{k^2 - p^2}{k^2 + p^2} \tag{25}$$

Extending Fock's method, Shibuya and Wulfman [79] found momentum–space solutions to the one-electron many-center wave equation

$$\left[-\frac{1}{2}\nabla^2 - \sum_a \frac{Z_a}{|\mathbf{x} - \mathbf{X}_a|} + \frac{k^2}{2} \right] \varphi(\mathbf{x}) = 0 \tag{26}$$

These authors started with the momentum-space counterpart of Eq. (26), and they showed that if the molecular orbitals are built up as superpositions of Coulomb Sturmians,

$$\varphi(\mathbf{x}) = \sum_{k,l,m,a} \chi_{n,l,m}(\mathbf{x} - \mathbf{X}_a)C_{n,l,m,a} \tag{27}$$

then the coefficients in the superposition are given by the solution of the secular equation

$$\sum_{n,l,m,a} \left[K_{n',l',m':n,l,m}(\mathbf{X}_{a'} - \mathbf{X}_a) - k\delta_{n',n}\delta_{l',l}\delta_{m',m}\delta_{a',a} \right] C_{n,l,m,a} = 0 \tag{28}$$

where

$$K_{n',l',m';n,l,m}(\mathbf{X}_{a'} - \mathbf{X}_a) \equiv \sqrt{\frac{Z_{a'}Z_a}{n'n}} S_{n',l',m';n,l,m}(\mathbf{X}_{a'} - \mathbf{X}_a) \tag{29}$$

and

$$S_{n',l',m';n,l,m}(\mathbf{X}_{a'} - \mathbf{X}_a) \equiv \int d\Omega e^{i\mathbf{p}\cdot\mathbf{R}} Y^*_{n'-1',l',m'}(\mathbf{u}) Y_{n-1,l,m}(\mathbf{u}) \tag{30}$$

with $\mathbf{R} \equiv \mathbf{X}_{a'} - \mathbf{X}_a$.

10 Evaluation of Shibuya–Wulfman Integrals

The structure of the matrix of Shibuya–Wulfman integrals is closely related to the composition properties of the 4-dimensional hyperspherical harmonics. When two of these hyperspherical harmonics are multiplied together, the result can be expressed as a sum over single harmonics:

$$Y^*_{n'-1',l',m'}(\mathbf{u}) Y_{n-1,l,m}(\mathbf{u}) = \sum_{n'',l''} Y_{n''-1,l'',m-m'}(\mathbf{u}) C[\{n'',l''\},\{n',l',m'\},\{n,l,m\}]$$

$$\tag{31}$$

where

$$C[\{n'',l''\},\{n',l',m'\},\{n,l,m\}] = \int d\Omega\, Y^*_{n''-1,l'',m-m'}$$

$$(\mathbf{u}) Y^*_{n'-1',l',m'}(\mathbf{u}) Y_{n-1,l,m}(\mathbf{u}) \tag{32}$$

In general, there can be several terms in such a sum. This is important for the structure of the Shibuya–Wulfman integrals because these integrals are defined by Eq. (30). It can be shown using a Sturmian expansion of a plane wave [74, 79, 81] and using generalized Wigner coefficients [66, 67, 80] that

$$\int d\Omega e^{i\mathbf{p}\cdot\mathbf{R}} Y_{n-1,l,m}(\mathbf{u}) = (2\pi)^{3/2} f_{nl}(s) Y_{lm}(\hat{\mathbf{s}}) \tag{33}$$

where $\hat{\mathbf{s}} \equiv k\hat{\mathbf{R}}$ is a unit vector and

$$k^{3/2} f_{nl}(s) \equiv R_{nl}(s) - \frac{1}{2}\sqrt{\frac{(n-l)(n+l+1)}{n(n+1)}} R_{n+1,l}(s)$$

$$-\frac{1}{2}\sqrt{\frac{(n+l)(n-l-1)}{n(n-1)}} R_{n-1,l}(s) \tag{34}$$

$$R_{nl}(s) \equiv \begin{cases} N_{nl}(2s)^l e^{-s} F[l+1-n; 2l+2; 2s] & n > l \\ \\ 0 & \text{otherwise} \end{cases} \tag{35}$$

with

$$\mathcal{N}_{nl} = \frac{2k^{3/2}}{(2l+1)!}\sqrt{\frac{(l+n)!}{n(n-l-1)!}} \tag{36}$$

while

$$F[a;b;x] \equiv 1 + \frac{a}{b}x + \frac{a(a+1)}{2b(b+1)}x^2 + \cdots \tag{37}$$

By comparing Eqs. (30), (31) and (33) we can see that the Shibuya–Wulfman integrals can be written in the form:

$$S_{n',l',m';n,l,m}(\mathbf{X}_{a'} - \mathbf{X}_a) = (2\pi)^{3/2} \sum_{n'',l''} f_{n'',l''}(s) Y_{l'',m-m'}(\hat{\mathbf{s}})$$

$$C[\{n'',l''\},\{n',l',m'\},\{n,l,m\}] \tag{38}$$

If we introduce the notation

$$Y_{l,m}(\hat{\mathbf{s}}) \equiv \mathcal{Y}_{l,m}(\theta_R)e^{im\phi_R} \tag{39}$$

where θ_R and ϕ_R are the angles associated with the vector $\mathbf{R} \equiv \mathbf{X}_{a'} - \mathbf{X}_a$, we can see that that $e^{i(m-m')\phi_R}$ can always be factored out. Thus we can write:

$$\frac{S_{n',l',m';n,l,m}(\mathbf{X}_{a'} - \mathbf{X}_a)}{(2\pi)^{3/2}} = e^{i(m-m')\phi_R} \sum_{n'',l''} f_{n'',l''}(s) \mathcal{Y}_{l'',m-m'}(\theta_R)$$

$$C[\{n'',l''\},\{n',l',m'\},\{n,l,m\}] \tag{40}$$

The coefficients $C[\{n'',l''\},\{n',l',m'\},\{n,l,m\}]$ can be evaluated by using the powerful angular and hyperangular integration theorem [68, 74]

$$\int d\Omega \, F_\lambda = \begin{cases} \dfrac{2\pi^{d/2}r^\lambda(d-2)!!}{\Gamma(d/2)\lambda!!(d+\lambda-2)!!}\Delta^{\frac{1}{2}\nu}F_\lambda & \lambda = \text{even} \\[4mm] 0 & \lambda = \text{odd} \end{cases} \tag{41}$$

Here F_λ is any homogeneous polynomial of degree λ, while d is the dimension of the space (in our case 4). The operator Δ is the generalized Laplacian operator

$$\Delta \equiv \sum_{j=1}^{d} \frac{\partial^2}{\partial x_j^2} \tag{42}$$

and $d\Omega$ is the generalized solid angle element. In order to evaluate the hyperangular integral that appears in Eq. (32), we convert the product of three 4-dimensional hyperspherical harmonics into a homogeneous polynomial in the coordinates of a 4-dimensional space by multiplying inside the integral by an appropriate power of the hyperradius and dividing by the same power outside. Then we simply apply the hyperangular integration theorem (Eq. 31). Mathematica and Maple programs which calculate the Shibuya–Wulfman integrals using this method may be found on the website http://sturmian.kvante.org

11 Numerical Results Compared for Efficiency

Consider the H_2 molecule and its dimer/agregates. In an s-orbital basis, all two-center integrals are known analytically, because they can be integrated by separating the variables in prolate spheroidal co-ordinates. A modest s-orbital basis is therefore chosen, simply for the demonstration on a rapid calculation, for which some experimental data could be corroborated.

The purpose of this section is to compare evaluations using the translation of a Slater type orbital basis to a single center (STOP) [55] with the Poisson equation solution using a DIM (Diatomics in molecules or atom pair) strategy and finally to show that the overlap auxiliary method is by far the fastest approach, for a given accuracy (the choice adopted is just six decimals, for reasons explained below).

H_2 molecule with interatomic distance of 1.402d0 atomic units (a.u.) (Table 1) assembles the full set of all Coulomb integrals; with one and two-centers evaluated using STOP, Poisson and overlap methods. Exponents may be found from the atomic integrals which do not include the constant factor (5/8 here).

The two-center integrals are dominated by an exponential of the interatomic distance and thus all ave values close to 0.3. The table is not the full set. All '15' terms, involving $1s_{a1}(1) \, 1s_{b1}(2)$ are given, to illustrate symmetry relations.

Note that this is by no means the best possible basis set for H_2, since it is limited to $l = 0$ functions (simply to ensure that even the two-center exchange integral has an analytic closed form).

The total energy obtained for the isolated H_2 molecule is -1.1284436 Ha as compared to a Hartree-Fock limit estimate of -1.1336296 Ha. Nevertheless, the Van der Waals well, observed at 6.4 a.u. with a depth of 0.057 kcal/mol (from Raman studies) is quite reasonably reproduced [82].

Table 1 Coulomb integrals in H_2

AOs (zeta)	$1s_{a1}$	$1s_{a2}$	$2s_{a1}$	$2s_{a2}$
$1s_{a1}$ 1.042999	1.042999	0	0	0
$1s_{a2}$ 1.599999	0.934309	1.599999	0	0
$2s_{a1}$ 1.615000	0.980141	0.870304	1.615000	0
$2s_{a2}$ 1.784059	0.901113	0.923064	1.189241	1.784059
$1s_{b1}$ 1.042999	3.455363	0.364117	0.659791	1.621644
$1s_{b2}$ 1.599999	0.433097	0.332887	0.635867	1.541858
$2s_{b1}$ 1.615000	0.323691	0.248050	0.529300	1.276630
$2s_{b2}$ 1.784059	0.402387	0.324872	0.636877	2.014196

Table 2 Atomic exchange integrals (six distinct single center values between pairs of different AOs)

AOs (zeta)	Label	[a(1)b(2)a'(2)b'(1)]	Value	Comment
$1s_{a1}$ 1.042999	1	1212	0.720716	–
$1s_{a2}$ 1.599999	2	1313	0.585172	–
$2s_{a1}$ 1.615000	3	1414	0.610192	–
$2s_{a2}$ 1.784059	4	2323	0.557878	–
$1s_{b1}$ 1.042999	5	2424	0.607927	–
$1s_{b2}$ 1.599999	6	3434	0.602141	–
$2s_{b1}$ 1.615000	7	2121	0.720716	= 1212
$2s_{b2}$ 1.784059	8	3232	0.585172	= 2323

Table 3 Two-center exchange integrals. All pair permutations possible. Some are identical by symmetry

Labels	Value	Comment
1515	0.319902	–
1516	0.285009	= 1525
1517	0.325644	= 1535
1518	0.324917	= 1545
1527	0.291743	= 1536
1528	0.293736	= 1547
1538	0.329543	= 1548
2525	0.260034	–
2516	0.254814	–
2517	0.290533	–
2518	0.290149	–

Dimer geometry: rectangular and planar. Distance between two hydrogen atoms of neighboring molecules: 6 au. Largest two-center integral between molecules: $4.162864 \ 10^{-5}$. (Note that this alone justifies the expression dimer—the geometry corresponds to two almost completely separate molecules; however, the method is applicable in any geometry).

Timings on an IBM RS6000 Power 6 workstation, for the dimer (all 4-center integrals in msec): STOP: 12 POISON: 10 OVERLAP: 2.

Total dimer energy: -2.256998 Ha. This corresponds to a well-depth of 0.069 kcal/mol, which may be considered reasonable in view of the basis set. The factor limiting precision in this study is the accuracy of input. The values of Slater exponents and geometric parameters are required to at least the accuracy demanded of the integrals and the fundamental constants are needed to greater precision.

12 Conclusions

A remarkable gain in simplicity is provided by Coulomb operator resolutions [15], that now enables the exponential type orbital translations to be completely avoided in *ab initio* molecular electronic structure calculations, although some mathematical structure has been emerging in the BCLFs used to translate Slater type orbitals and even more in the Shibuya–Wulfman matrix used to translate Coulomb Sturmians.

This breakthrough that Coulomb resolutions represent (in particular with the convenient choice of Laguerre polynomials) in the ETO algorithm strategy stems from a well-controlled approximation, analogous to the resolution of the identity. The convergence has been shown to be rapid in all cases [16].

The toy application to H_2 dimer Van der Waals complexes uses a general code within the STOP package [55]. Numerical vales for the geometry and interaction energy agree well with complete *ab initio* potential energy surfaces obtained using very large Gaussian basis sets and data from vibrational spectroscopy [82].

Acknowledgment The author would like to thank Peter Gill for helpful discussions at ISTCP-VI (July 2008) and rapidly transmitting the text [16]. The section on the Shibuya–Wulfman matrix is drawn from a collaboration with John Avery. Didier Pinchon, as usual, contributed to the scientific debate surrounding this work.

References

1. M. Abramowitz and I.A. Stegun. *Handbook of Mathematical Functions with Formulas, Graphs, and Mathematical Tables.* Number 55 in Nat. Bur. Standards Appl. Math. Series. US Government Printing Office, Washington, D.C., 1964.
2. M.P. Barnett. Some elementary two-center integrals over Slater orbitals. 1998. Preprint. Available at http://www.princeton.edu/~allengrp/ms/other/ajcat.pdf.
3. A. Bouferguène. Addition theorem of Slater type orbitals: a numerical evaluation of Barnett-Coulson/Löwdin functions. *J. Phys. A: Math. Gen.*, 38:2899–2916, 2005.
4. A. Bouferguène, M. Fares, and P.E. Hoggan. STOP: A Slater-type orbital package for molecular electronic structure determination. *Int. J. Quantum Chem.*, 57(4):801–810, 1996.
5. A. Bouferguène and D. Rinaldi. A new single-center method to compute molecular integrals of quantum chemistry in Slater-type orbital basis of functions. *Int. J. Quantum Chem.*, 50(1):21–42, 1994.
6. I.S. Gradshteyn and I.M. Ryzhik. *Table of Integrals, Series, and Products.* Academic Press, New York, 1994. Fifth printing.
7. H.W. Jones. Analytic Löwdin alpha-function method for two-center electron-repulsion integrals over Slater-type orbitals. In C.A. Weatherford and H.W. Jones, editors, *Int. Conf. on ETO Multicenter integrals*, page 53, Tallahasse, Florida 32307, 1981.
8. H.W. Jones. Analytical evaluation of multicenter molecular integrals over Slater-type orbitals using expanded Löwdin alpha functions. *Phys. Rev. A*, 38(2):1065–1068, 1988.
9. H.W. Jones. Analytic Löwdin alpha-function method for two-center electron-repulsion integrals over Slater-type orbitals. *J. Comput. Chem.*, 12(10):1217–1222, 1991.
10. H.W. Jones and J. Jain. Computer-generated formulas for some three-center molecular integrals over Slater-type orbitals. *Int. J. Quantum Chem.*, 23(3):953–957, 1983.
11. H.W. Jones and C.A. Weatherford. Modified form of Sharma's formula for STO Löwdin alpha functions with recurrence relations for the coefficient matrix. *Int. J. Quantum Chem.*, 12:483–488, 1978. (*International Symposium on Atomic, Molecular, and Solid-State Theory, Collision Phenomena and Computational Methods,* Flagler Beach, Florida, 1978.)
12. H.W. Jones and C.A. Weatherford. The Löwdin α-function and its application to the multi-center molecular integral problem over Slater-type orbitals. *J. Mol. Struct. : THEOCHEM*, 199:233–243, 1989.
13. F.W.J. Olver. *Asymptotics and Special Functions.* Academic Press, New York, 1974.

14. R.R. Sharma. Expansion of a function about a displaced center for multicenter integrals: A general and closed expression for the coefficients in the expansion of a slater orbital and for overlap integrals. *Phys. Rev. A*, 13(2):517–527, 1976.

15. S. A. Varganov, A. T. B. Gilbert, E. Duplazes and P. M. W. Gill. *J. Chem. Phys.* 128:201104, 2008.

16. P. M. W. Gill and A. T. B. Gilbert. Resolutions of the Coulomb Operator. II The Laguerre Generator. *Chem. Phys.* 356 (2009) 86–90.

17. P. E. Hoggan. Four center ETO integrals without orbital translations. *Int. J. Quantum Chem.* **109** (2009)

18. I. I. Guseinov, A. Ozmen, U. Atav, H. Uksel. Computation of overlap integrals over Slater Type Orbitals, using auxiliary functions. *Int. J. Quantum Chem.* 67:199–204 1998.

19. P. E. Hoggan. DSc Thesis, 1991; Appendix 2

20. S. F. Boys. *Electronic wave functions. I. A general method of calculation for the stationary states of any molecular system. Proc. Roy. Soc. [London]* **A (200)** (1950) 542.

21. S. F. Boys, G. B. Cook, C. M. Reeves and I. Shavitt. *Automated molecular electronic structure calculations Nature* **178** (1956) 1207.

22. V. N. Glushov and S. Wilson. Distributed Gaussian basis sets: Variationally optimized s-type sets for H2, LiH, and BH. Int. J. Quant. Chem, 89 (2002) 237–247; Distributed Gaussian basis sets: Variationally optimized s-type sets for the open-shell systems HeH and BeH. 99 (2004) 903-913; V. N. Glushov and N. Gidopoulos. Constrained optimized potential method and second-order correlation energy for excited states. Int. J. Quant. Chem, 107 (2007) 2604–2615. Adv. Quant. Chem. 39 (2001) 123.

23. I. Shavitt. *Methods in Computational Physics*, volume 2. Academic Press, New York, edited by B. Alder, S. Fernbach, M. Rotenberg, 1963. p. 15.

24. E. Clementi and D. L. Raimondi. Atomic screening constants from SCF functions. *J. Chem. Phys.* **38** (1963) 2686–2689.

25. S. J. Smith and B. T. Sutcliffe. *The development of computational chemistry in the United Kingdom* in Reviews in computational chemistry, edited by K. B Lipkowtz and B. D Boyd. VCH Academic Publishers, New York, 1997.

26. I. G. Csizmadia, M. C. Harrison, J. W. Moskowitz, S. Seung, B. T. Sutcliffe and M. P. Barnett. *POLYATOM: Program Set for Non-Empirical Molecular Calculations.* Massachusetts Institute of Technology Cambridge, 02139 Massachusetts. QCPE No 11, Programme 47 and M. P. Barnett, *Rev. Mod. Phys.* **35** (1963) 571.

27. H. O. Pritchard. Computational chemistry in the 1950s and 1960s. *J. Mol. Graphics and Mod.* **19** (2001) 623.

28. W. J. Hehre, W. A. Lathan, R. Ditchfield, M. D. Newton and J. A. Pople. *GAUSSIAN 70: Ab Initio SCF-MO Calculations on Organic Molecules* QCPE 11, Programme number 236 (1973).

29. R. Bonaccorsi, E. Scrocco and J. Tomasi. Molecular SCF Calculations for the Ground State of Some Three-Membered Ring Molecules: (CH2)3, (CH2)2NH, (CH2)2NH, (CH2)2O, (CH2)2S, (CH)2CH2, and N2CH2. *J. Chem. Phys.* **52** (1970) 5270–5284.

30. R. M. Stevens. *Geometry Optimization in the Computation of Barriers to Internal Rotation The POLYCAL program. J. Chem. Phys.* **52** (1970) 1397–1402.

31. E. J. Baerends, D. E. Ellis and P. Ros. Self-consistent molecular Hartree-Fock-Slater calculations I. The computational procedure. *Chem. Phys.* **2** (1973) 41–51.

32. A.D. McLean, M. Yoshimine, B. H Lengsfield, P. S. Bagus and B. Liu, *ALCHEMY II.* IBM Research, Yorktown Heights, MOTECC 91 1991.

33. J. A. Pople and D. L. Beveridge. *Approximate molecular orbital theory.* McGraw Hill, New York 1970.

34. D. Rinaldi and P. E. Hoggan. Evaluation of two-electron integrals over spd bases of STO. *Theo. Chim. Acta.* **72** (1987) 49.

35. A. Bouferguène. PhD thesis, Nancy I University, France, 1992.

36. A. J. Cohen and N. C. Handy. Density functional generalized gradient calculations using Slater basis sets. J. Chem. Phys. 117 (2002) 1470-1478. M. A. Watson, N. C. Handy and A. J. Cohen. Density functional calculations, using Slater basis sets, with exact exchange. J. Chem. Phys.

119 (2003) 6475–6481. M. A. Watson, N. C. Handy, A. J. Cohen and T. Helgaker. Density-functional generalized-gradient and hybrid calculations of electromagnetic properties using Slater basis sets. J. Chem. Phys. 120 (2004) 7252–7261.

37. P. E. Hoggan. Choice of atomic orbitals to evaluate sensitive properties of molecules. An example of NMR chemical shifts. *Int. J. Quantum Chem.* 100 (2004) 218.

38. L. Berlu. PhD thesis, Université Blaise Pascal, Clermont Ferrand, France, 2003.

39. J. Avery, *Hyperspherical Harmonics and Generalized Sturmians*, Kluwer, Boston 2000.

40. Y. Shao, C. A. White and M. Head-Gordon. Efficient evaluation of the Coulomb force in density-functional theory calculations. J. Chem. Phys. 114 (2001) 6572–6577.

41. J. C. Tully. Diatomics-in-molecules potential energy surfaces. I. First-row triatomic hydrides. J. Chem. Phys. 58 (1973) 1396–1410.

42. C. C. J. Roothaan. A Study of two-center integrals useful in calculations on molecular structure. J. Chem. Phys. 19 (1951) 1445–1458.

43. J. C. Cesco, J. E. Perez, C. C. Denner, G. O. Giubergiaand and Ana E. Rosso. Rational approximants to evaluate four-center electron repulsion integrals for 1s hydrogen Slater type functions. Applied Num. Math. 55 (2) (2005) 173–190 and references therein.

44. H.-J. Werner and P. J. Knowles and R. Lindh and F. R. Manby and M. Schütz and P. Celani and T. Korona and G. Rauhut and R. D. Amos and A. Bernhardsson and A. Berning and D. L. Cooper and M. J. O. Deegan and A. J. Dobbyn and F. Eckert and C. Hampel and G. Hetzer and A. W. Lloyd and S. J. McNicholas and W. Meyer and M. E. Mura and A. Nicklass and P. Palmieri and R. Pitzer and U. Schumann and H. Stoll and A. J. Stone and R. Tarroni and T. Thorsteinsson. *MOLPRO, version 2006.1 a package of ab initio programs*, www.molpro.net

45. J. Fernández Rico, R. López, A. Aguado, I. Ema, and G. Ramírez. Reference program for molecular calculations with Slater-type orbitals. J. Comp. Chem. 19(11) (1998) 1284–1293

46. R. Carbó, L. Leyda and M. Arnau. How similar is a molecule to another? An electron density measure of similarity between two molecular structures *Int. J. Quantum Chem.* 17 (1980) 1185–1189.

47. L. Berlu and P.E. Hoggan. Useful integrals for quantum similarity measurements *ab initio* over Slater type orbitals. *J. Theo. and Comp. Chem.* 2 (2003) 147.

48. D. Pinchon and P. E. Hoggan. Rotation matrices for real spherical harmonics: general rotations of atomic orbitals in fixed space axes. *J. Phys. A* 40 (2007) 1597–1610.

49. D. Pinchon and P. E. Hoggan. New index functions for storing Gaunt coefficients. *Int. J. Quantum Chem.* 107 (2007) 2186–2196.

50. J. Fernández Rico, R. López, G. Ramírez, I. Ema and E. V. Ludeña. *J. Comp. Chem.* 25 (2004) 1355.

51. H. W. Jones. *International Conference on ETO Multicenter Integrals*. (Tallahassee, USA 1981) Edited by C. A. Weatherford and H. W. Jones. Reidel, Dordrecht 1982. and International Journal of Quantum Chemistry 100 (2) (2004) pp 63-243. Special Issue in memory of H. W Jones. Edited by C. A. Weatherford and P. E. Hoggan.

52. C. A. Weatherford, E. Red and P. E. Hoggan. Solution of Poisson's equation using spectral forms over Coulomb Sturmians. *Mol. Phys.* 103 (2005) 2169.

53. F. R. Manby, P. J. Knowles and A. W. Lloyd. The Poisson equation in density fitting for the Kohn-Sham Coulomb problem. J. Chem. Phys. 115 (2001) 9144–9148.

54. E. J. Weniger. Weakly convergent expansions of a plane wave and their use in Fourier integrals. J. Math. Phys. 26 (1985) 276–291.

55. A. Bouferguène and P. E. Hoggan. STOP: A Slater Type Orbital Package. *QCPE*. Programme number 667, 1996.

56. J. C. Slater. Atomic shielding constants. Phys. Rev. 36 (1930) 57–64.

57. J. C. Slater. Analytic atomic wave functions. Phys. Rev. 42 (1932) 33–43.

58. T. Kato. On the eigenfunctions of many-particle systems in quantum mechanics. Commun. Pure Appl. Math. 10 (1957) 151–177. *Phys. Rev.* 36 (1930) 57.

59. T. Kato. *Schrödinger Operators*. Springer Verlag, Berlin, Edited by S. Graffi, 1985. p. 1–38.

60. S. Agmon. *Lectures on Exponential Decay of Solutions of Second Order Elliptic Equations: Bounds on Eigenfunctions of N-Body Schrödinger Operators*. Princeton University, Princeton, NJ, 1982.

61. P. E. Hoggan. Trial wavefunctions for Quantum Monte Carlo simlations over ETOs AIP Proceedings of ICCMSE 2007, Vol II 963.

62. A. Sidi, D. Pinchon and P. E. Hoggan. Fast and accurate evaluation of Barnett–Coulson–Löwdin functions. *Technical note, UBP.*

63. C. A. Weatherford, E Red, D Joseph and P. E. Hoggan. Solution of Poisson's equation: application to molecular Coulomb integrals. Mol. Phys. **104** (2006) 1385.

64. D. Pinchon and P. E. Hoggan. Gaussian approximation exponentially decaying functions: B-functions. *Int. J. Quantum Chem* **109** (2009) 135–148.

65. Aquilanti, V., Cavalli, S., Coletti, C. and Grossi, G., *Alternative Sturmian bases and momentum space orbitals; an application to the hydrogen molecular ion*, Chem. Phys. **209** 405, 1996.

66. Multicenter sturmians for molecules. Aquilanti, V. and Caligiana, A., Chem. Phys. Letters, **366** 157 2002.

67. Aquilanti, V., and Caligiana, A., in **Fundamental World of Quantum Chemistry: A Tribute to the Memory of P.O. Löwdin, I**, E.J. Brändas and E.S. Kryachko, Eds., Kluwer, Dordrecht, 297, 2003.

68. Avery, J., **Hyperspherical Harmonics; Applications in Quantum Theory**, Kluwer Academic Publishers, Dordrecht, 1989.

69. Avery, J., *Hyperspherical harmonics; Some properties and applications*, in **Conceptual Trends in Quantum Chemistry**, Kryachko, E.S., and Calais, J.L., Eds, Kluwer, Dordrecht, 1994.

70. Avery, J., *A formula for angular and hyperangular integration*, J. Math. Chem., **24** 169, 1998.

71. Avery, J., **Hyperspherical Harmonics and Generalised Sturmians**, Kluwer Academic Publishers, Dordrecht, Netherlands, 196 pages, 2000.

72. Avery, J., *Sturmians*, in **Handbook of Molecular Physics and Quantum Chemistry**, S. Wilson, ed., Wiley, Chichester, 2003.

73. Avery, J., *Many-center Coulomb Sturmians and Shibuya-Wulfman integrals*, Int. J. Quantum Chem., **100** 2004 121–130.

74. Avery, James. and Avery, John., **Generalised Sturmians and Atomic Spectra**, World Scientific, 2007.

75. Caligiana, Andreia, **Sturmian Orbitals in Quantum Chemistry**, Ph.D. thesis, University of Perugia, Italy, October, 2003.

76. Fock, V.A., *Hydrogen atoms and non-Euclidian geometry*, Kgl. Norske Videnskab Forh, **31** 138, 1958.

77. Koga, T. and Matsuhashi, T., *Sum rules for nuclear attraction integrals over hydrogenic orbitals*, J. Chem. Phys., **87** (8) 4696–9, 1987.

78. Koga, T. and Matsuhashi, T., *One-electron diatomics in momentum space. V. Nonvariational LCAO approach*, J. Chem. Phys., **89** 983, 1988.

79. Shibuya, T. and Wulfman, C.E., *Molecular orbitals in momentum space*, Proc. Roy. Soc. A, **286** 376, 1965.

80. Wen, Z.-Y. and Avery, J., *Some properties of hyperspherical harmonics*, J. Math. Phys., **26** 396, 1985.

81. Weniger, E.J., *Weakly convergent expansions of a plane wave and their use in Fourier integrals*, J. Math. Phys., **26** 276, 1985.

82. R. J. Hinde, *Six dimensional potential energy surface of H_2-H_2. J. Chem. Phys.* **128** 2008.

Cation Hydrolysis Phenomenon in Aqueous Solution: Towards Understanding It by Computer Simulations

M. Holovko, M. Druchok, and T. Bryk

Abstract Molecular dynamics studies of the influence of ionic charge on hydrated-hydrolyzed structure and dynamic properties of highly charged cations are reviewed. In order to clarify the influence of ion charge a simple model cation M^{Z+} called primitive cation was introduced. The investigations demonstrate a wide variety of hydrated-hydrolyzed forms of primitive cation, including aquo, hydroxo-aquo, hydroxo, oxo-hydroxo, and oxo forms. A transition between these forms is regulated by a value of cation charge Z. For correct description of cation hydrolysis we also modeled effects of charge redistribution on hydrolysis reaction products, that essentially modifies hydrated–hydrolyzed structures and initiates a partial dehydration of cation. Self-diffusion coefficients and spectral densities of hindered translation motions of primitive cation and oxygens of first hydration shell demonstrate a strong correlation with hydrated–hydrolyzed structure of cations.

Aqueous solutions of real cations Al^{3+} and UO_2^{2+} are modeled too. The aluminium ions are characterized by a strict octahedral arrangement of neighbors in hydration shell with tendency to hydrolysis. For the case of uranyl solution a bipyramidal pentacoordinated arrangement of UO_2^{2+} ion is found. It is shown that the influence of pH level can modify uranyl hydration shell from $UO_2^{2+}(H_2O)_5$ to $UO_2^{2+}(OH^-)_5$. Finally, uranyl–uranyl distribution functions are obtained and the mechanisms of formation of polynuclear ions are discussed.

Keywords Molecular dynamics · Water · Cation hydrolysis · Hydration · Polynuclear ion formation · pH influence

1 Introduction

Cation hydrolysis reactions in aqueous solutions of metal ions are important in many natural and industrial processes. The initial steps involved in a cation hydrolysis in aqueous solutions lead to a decay of water molecules in the hydration shell of

M. Holovko (✉), M. Druchok, and T. Bryk
Institute for Condensed Matter Physics, National Academy of Sciences of Ukraine,
1 Svientsitskii Str., 79011 Lviv, Ukraine
e-mail: holovko@icmp.lviv.ua

N. Russo et al. (eds.), *Self-Organization of Molecular Systems: From Molecules and Clusters to Nanotubes and Proteins*, NATO Science for Peace and Security Series A: Chemistry and Biology, © Springer Science+Business Media B.V. 2009

a cation M^{Z+} and can be considered as a chain of H^+-eliminating (acid dissociation) reactions, which turn hydrated cation complexes $[M(H_2O)_n]^{Z+}$ into new ionic species $[MO_nH_{2n-h}]^{(Z-h)+}$. Here n is a number of water molecules in the cation hydration shell, so-called hydration number, while h is a number of protons lost by water molecules from the cation hydration shell, so-called the molar ratio of hydrolysis [1, 2]. Usually the process does not stop at this stage and as a result of condensation reaction, polynuclear ions appear [2, 3]. The nature of these hydrated–hydrolyzed ionic species is of fundamental interest in chemistry of inorganic solutions and of particular importance in many areas ranging from nuclear technology to environmental chemistry [4].

For the last decades our knowledge of ionic hydration of metal cations has been essentially increased due to availability of more accurate experimental data [5–7]. The impressive progress has been also achieved in a computer modeling of the structure of hydration shells of monovalent alkaline and divalent alkaline earth cations [8]. It was shown that for divalent cations the electrostatic repulsion between cation and protons of water molecules can cause notable modification of an intramolecular geometry of water molecules in hydration shell. One can suppose that it can be sufficiently strong to repel protons of water molecules from the hydration shell causing the hydrolysis effect. However, despite their considerable importance, computer studies of hydration of cations with hydrolysis effect are scarce. This can be explained by inapplicability of standard water models for hydrolysis effect treatment and by difficulties of quantum mechanical calculations of hydrated–hydrolyzed complexes with a sufficient degree of reliability for further use in computer simulations [9]. In computer simulations, in order to treat a cation hydrolysis effect explicitly, water should be considered in the framework of a non-constrained flexible model such as the central force (CF) model [10, 11] or polarizable model [12, 13].

The modeling of cation–water interaction is more complex and is rather computationally intensive task even for the case of systems with only metal ion and water molecules. It should contain numerous details connected with the treatment of a many-body cation–water interaction, the covalent bond effects caused by specific features of cation electronic configuration etc. During the last years the interaction potentials between highly charged cations and water molecules have been largely developed by *ab initio* calculations. It was shown that they provide correct description of ionic hydration shell and dynamical properties for cations Al^{3+} [14, 15], Zn^{2+} [16], Cu^{2+} [17], Pb^{2+} [18], Cr^{3+} [19–21], Ti^{3+}, Co^{3+} [21], and Tl^{3+} [22]. Due to strong hydration of the highly charged cations in some of these investigations the hydrated complexes $[M(H_2O)_n]^{Z+}$ were considered as new species and computer modeling was performed in the form of hybrid quantum mechanical/molecular mechanical (QM/MM) simulations [16–20]. In this technique a cation with its hydration shell is treated quantum mechanically, while the rest of the system is described by classical potentials. Nevertheless in spite of large efforts to describe ion–water interactions more correctly and applications of enough sophisticated models no hydrolysis effects are taken into account directly in relevant simulations. We can mention only the success in the treatment of the problem of the hydrolysis for trivalent cation Fe^{3+} [23, 24]. In [23] molecular dynamics (MD) simulations were performed for a combined model: the pair potential for iron–water interaction was derived from *ab initio* calculations, while water was described in the framework of a polarizable model. In [24] the density functional technique for electronic structure of hydrated-hydrolyzed complexes $[Fe(H_2O)_{n-h}(OH)_h]^{Z-h}$ ($n = 6$; $h = 0, 1, 2$) in combination with a dielectric continuum model was used. In both cases adequate results for the first hydrolysis constant ($h = 1$) were obtained, but the second hydrolysis constant ($h = 2$) was highly overestimated.

The driving force of cation hydration and hydrolysis is a strong cation–water interaction, in which the electrostatic interaction dominates and increases with increasing of the ion valency Z and/or decreasing of ion size. The prevalent role of electrostatic cation–water interaction in stability of hydrated–hydrolyzed complexes of cations in aqueous solutions is confirmed by the existence of a linear relation between the first hydrolysis constant pK_a and a ratio of cation charge to cation–oxygen interatomic distance for metal ions of main and transition groups [1]

$$pK_a = A - 11.0 \cdot (Z/d) \tag{1}$$

where d is the metal–oxygen interatomic distance (in Å). Typical values for the constant A are 22.0 for metal ions with inert electronic configuration and $A = 19.8$ for transition and post-transition metal ions. Recently such a linear correlation between pK_a and binding energy of hydrated cations was also derived from electronic density calculations [25]. Consequently, the cations with small ionic radius and high charge are extensively hydrolyzed in aqueous solutions. Most of the trivalent and quadrivalent cations hydrolyze water molecules in their hydration shells, while this occurs only for Be^{2+}, the smallest of the divalent cations.

However, the hydrolysis also involves a bond-breaking process, where quantum mechanical effects are important [26–28], because the proton dynamics should be treated within quantum theory rather than by classical Newtonian equations. Very impressive achievements were recently reported in path integral molecular dynamics modeling of proton dynamics [26].

The complexity of problems appeared in microscopic modeling of hydrated–hydrolyzed forms of highly charged cations suggests the idea of formulation of a simplified model, in which a contribution of electrostatic interaction between cation and water molecules will be separated from other contributions, which are connected with the influence of ionic sizes, non-additivity of ion–water interactions, the peculiarities of cation electronic configuration etc. Theoretical study of such model can give us the information about the role of electrostatic interaction in molecular mechanism of creation of hydrated-hydrolyzed forms of highly charged cations in dependence on cationic charge. For this aim in this study we performed the molecular dynamics simulation for the model of primitive cation M^{Z+}, which was introduced by us previously [29–34]. In this model for the description of the influence of cation charge Ze (e is the elementary electric charge) onto intramolecular structure of water molecules we use the non-rigid model of water CF1 [35, 36], which is a slightly modified version of the original CF model in order to ensure more realistic value of pressure. The interaction of primitive cation M^{Z+} with water molecules is similar to the interaction of cation Na^+ with water [37] with assumption that the ion can possess the different valency Z. Since the cation Na^+ has the hydration structure close to the octahedral one can expect that with increasing of valency Z the octahedral hydration structure of primitive cation will became stronger. We hope that the use of the term "primitive cation" will not evoke any misleading due to specific meaning of the word "primitive" in the theory of electrolyte solutions, where this word is connected with the charged hard sphere model, the so-called primitive model of electrolyte (look, for e.g. [38]). Here we use the term "primitive cation" only in the meaning that we do not attempt to reproduce any real ions for $Z > +1$.

In our previous investigations [29–34] an effect of proton loss by some water molecules of hydration shell was noticed, which was treated as the hydrolysis of water molecules caused by highly charged cations. In Section 2 a molecular dynamics study of the influence of ion charge on hydrated–hydrolyzed structure and dynamical properties of multivalent cations in aqueous solutions is reported. Our investigation demonstrates a wide variety of hydrated–hydrolyzed forms around primitive cation, including aquo, hydroxo-aquo, hydroxo, oxo-hydroxo, and oxo forms. The transition between these forms is defined by a value of cation charge. In order to take into account the effect of charge redistribution between hydrolysis products the model of primitive cation M^{Z+} is improved by introduction of effective charges for oxygens and hydrogens dependent on O–H distance of dissociated water molecules, which roughly takes into account quantum mechanical effects in cation hydrolysis. It is shown that the charge redistribution between hydrolysis products reduces the number of exchanges between the bulk and coordination shell and leads to a partial dehydration of cation.

In Section 3 we report the results of a molecular dynamics investigation of structural and dynamic properties of hydrated–hydrolyzed forms of aluminium Al^{3+} [39] and uranyl $(UO_2)^{2+}$ [40] in aqueous solutions. For the uranyl case we extended our investigation by a study of the influence of pH level on uranyl hydration shell. We also discuss the possibility of creation of polynuclear uranyl complexes.

2   Primitive Model for Cation Hydrolysis

2.1   Model and Computer Simulation Details

This study of cation charge influence on the structure of cation hydration shell is performed in the framework of a model of a primitive cation M^{Z+} [29–34]. Water is described by the CF1 model [35, 36]. Within the CF1 model the water molecules are treated as a binary mixture of oxygen and hydrogen atoms bearing effective charges $Z_H = 0.32983$, $Z_O = -2Z_H$ and interacting via pairwise potentials

$$U_{OO}(r) = \frac{331.671Z_O^2}{r} + \frac{26758.2C_1}{r^{8.8591}} - 0.25e^{-4(r-3.4)^2} - 0.25e^{-1.5(r-4.5)^2},$$

$$U_{HH}(r) = \frac{331.671Z_H^2}{r} + \frac{18}{1 + e^{40(r-2.05C_2)}} - 17e^{-7.62177(r-1.45251)^2},$$

$$U_{OH}(r) = \frac{331.671Z_O Z_H}{r} + \frac{6.23403}{r^{9.19912}} - \frac{10}{1 + e^{40(r-1.05)}}$$

$$- \frac{4}{1 + e^{5.49305(r-2.2)}} \tag{2}$$

where $C_1 = 0.9$, $C_2 = 1/1.025$. Distances are measured in Å, and energies in kcal/mol.

The interaction between the model cation M^{Z+} with water molecules is presented by the following potentials

$$U_{MO}(r) = \frac{331.671Z_M Z_O}{r} - \frac{36.677}{r^2} + 116862\,e^{-4.526r}$$

$$U_{MH}(r) = \frac{331.671Z_M Z_H}{r} + \frac{7.479}{r^2} + 99545\,e^{-7.06r} \tag{3}$$

where Z_M is the cation valency.

As it was mentioned above in the case $Z_M = 1$ interaction potentials coincide with the ones of Na$^+$ ion in CF1 water and were successfully used for computer modeling of aqueous solution of NaCl salt in [37]. They were derived from quantum chemical calculations [41] by subtracting the coulombic contributions from the energy values for a greater number of energetically favorable and a smaller number of less probable ion-water arrangements and fitting the remainder to the analytical expression (3) [42]. For $Z_M > 1$ the model does not correspond to any real cation and is rather a toy model, which permits one to study an effect of increasing electrostatic cation–water interaction with a fixed non-Coulomb short-range interaction on the formation of hydrated-hydrolyzed complexes of cations in water. In contrast to Ref. [37], where the water–water interaction is described by the Bopp–Jancso–Heinzinger (BJH) model [43], in this study we used the CF1 model. Both models are identical for description of the intermolecular interaction but differ in the intramolecular part of interaction. The intramolecular terms in BJH are represented by anharmonic expansion, which would never permit the hydrolysis effect.

At the first step [29, 30] it was assumed the charges on products of hydrolysis reaction

$$H_2O \rightarrow H^{Z^*+} + (OH)^{Z^*-} \tag{4}$$

are equal to effective charges of CF1 model. Since in the CF1 model hydrogens and oxygens in water molecules carry effective charges due to an intramolecular screening by electronic density one has to take into account that during the hydrolysis, when proton leaves the "native" water molecule, it should change its charge from $Z^* = 0.32983$ to $Z^* = 1$. Correspondingly the charge on $(OH)^{Z^*-}$ group should be of the same value but of opposite sign. In order to take into account this phenomena at the second step we take into account the charge redistribution effect between the hydrolysis reaction products by using the functional dependence of proton charge on proton–oxygen distance OH_i ($i = 1,2$) in the following form:

$$Z^*_{H_i} = Z^{CF}_H + (1 + tanh((r_{OH_i} - r_0)/w)) \cdot (1 - Z^{CF}_H)/2 \tag{5}$$

which is characterized by two parameters r_0 and w. The r_0 localizes a charge redistribution region relatively to the proton–oxygen distance, w defines a width of this region. $Z^{CF}_H = 0.32983$ – hydrogen charge within the water molecule in CF1 model. According to Eq. (5) after the hydrolysis reaction (at high r_{OH} distances) lost protons will possess a charge $+1$, while the charge of oxygens of their native water molecules will be changed too. The electroneutrality condition is:

$$Z^*_O = -(Z^*_{H_1} + Z^*_{H_2}) \tag{6}$$

In this model the short-range part of the interaction potentials does not change during the charge redistribution process. One also should note that the mechanism of dissociation of water molecules is rather complex and consists of a few steps. In accordance with *ab initio* molecular dynamics simulations [44] it includes the stretching of O–H bond at $r_{OH} = 1 - 1.1$ Å, following conversion from O–H bond to hydrogen bond at $r_{OH} = 1.1 - 1.3$ Å, separation of dissociation products at $r_{OH} = 1.4 - 1.5$ Å and finally the formation of free dissociation products at $r_{OH} = 1.7 - 1.8$ Å. It is clear that the description of charge redistribution between cation hydrolysis products according to Eq. (5) is rather a simplification of dissociation process. We should also take into account that the processes of dissociation of free water molecules and in the presence of highly charged cation are not identical. The parameters of charge redistribution function (Eq. (5)) were set equal to $r_0 = 1.7$ Å, $w = 0.25$ Å. Such a choice of parameters provides the CF1 effective charge for hydrogen in water molecule and the value $Z_H = 1$, when it leaves the molecule and hydration shell.

We should note that the CF1 as a non-constrained flexible model possesses such properties as the proton affinity of water and the hydroxide affinity of water. Likewise in the case of Stillinger–David polarizable model [12] the proton and hydroxide affinities of CF1 model leads to a creation of complexes H_3O^+, $H_5O_2^+$ and $H_5O_3^-$ or more complex clusters. In the presence of the cation M^{Z+} due to the repulsion between the hydrolyzed protons and cation the proton affinity of water will be intensified. The attraction between OH^- group and a cation can form the "hydroxo-aquo" $[M(OH)_n(H_2O)_{n-h}]^{(Z-Z^*h)+}$ ($h < n$) or "hydroxo" complexes $[M(OH)_n]^{(Z-Z^*n)+}$ ($h = n$). Moreover a strong electrostatics field of cation M^{Z+} can initiate the reaction

$$(OH)^{Z^*-} \rightarrow H^{Z^*+} + O^{2Z^*-} \tag{7}$$

and formation of "oxo-hydroxo" $[MO_{h-n}(OH)_{2n-h}]^{(Z-Z^*h)+}$ ($n < h < 2n$) and "oxo" complexes $[MO_n]^{(Z-Z^*2n)+}$ ($h = 2n$).

After improvement of the CF1 model by the charge redistribution procedure the absolute values of charges on hydrolyzed proton and oxygen in considered complexes increase and these species principally differ from the non-hydrolyzed ones. Due to this the hydrolyzed proton attracts oxygens and repels protons of water molecules in complexes H_3O^+, $H_5O_2^+$, etc. stronger. It is also repelled more strongly by the cation. Similarly the hydrolyzed oxygen is attracted more strongly by the cation. One can expect that taking into account of the charge redistribution effect will modify the hydrated–hydrolyzed structure of cations.

The MD simulation was performed for a system of 1,727 water molecules and a single cation of different charge Z_M at normal conditions: pressure 1 atm and temperature 298 K, which were controlled by means of Nose–Hoover barostat and thermostat in isotropic NPT ensemble [45]. All the particles were placed into rectangular box with periodic boundary conditions with sides length $L_x : L_y : L_z = 1 : 1 : 1.375$. Production runs were performed over 2×10^5 time steps for each cation charge. We used the velocity Verlet algorithm with a time step $\tau = 10^{-16}$ s to integrate the classical equations of motion of the system [46]. The correct treatment of the long-range interactions is very important for the description of ion hydration. For example, in aqueous solution the Born correction for free energy of ion hydration of monovalent ions for 10 Å cutoff is about 10 kcal/mol [47]. The role of long-range interactions increases for multivalent cations. In this study long-range interactions extend beyond the box were they are treated by Ewald summation technique with a convergence parameter $\eta = 0.372$ Å^{-1} and maximal summation parameters in reciprocal space $|n_x| = |n_y| = 11$ and $|n_z| = 15$ [48] in the framework of standard DL_POLY package [49]. Although this treatment is time-consuming it provides almost exact accounting of electrostatic contributions for ion hydration.

2.2 Structural Properties

The effect of cation charge on the hydrated-hydrolyzed structure of cations was examined by the comparison of the radial distribution functions (RDF) cation–oxygen $g_{MO}(r)$ and cation–hydrogen $g_{MH}(r)$ and corresponding running coordination numbers $n_{MO}(r)$ and $n_{MH}(r)$

$$n_{M\alpha}(r) = 4\pi\rho_\alpha \int_0^r g_{M\alpha}(r')r'^2 dr' \qquad (8)$$

for primitive cations M^{Z+} with $Z_M = 1, 2, 3, 4, 5, 6, 7$. $\alpha = O, H$, ρ_α is the number density of atoms of species α. As a starting point of our investigation we consider the influence of hydrated–hydrolyzed structure of cation M^{Z+} neglecting the effects of charge redistribution on hydrolysis products. The effects of charge redistribution will be discussed later.

Table 1 Estimated parameters of hydration structure of cations M^{Z+}

	r_{max_1}	$g(r_{max_1})$	r_{min_1}	$n(r_{min_1})$	r_{min_2}	$n(r_{min_2})$
M^+O	2.33	8.11	3.03	6.08	5.28	22.73
M^+H	2.98	3.30	3.63	14.69	6.23	69.45
$M^{2+}O$	2.08	16.75	2.48–2.83	6.05	5.23	24.45
$M^{2+}H$	2.88	5.73	3.28	12.12	5.83	58.97
$M^{3+}O$	2.03	18.52	2.68	6.65	4.98	23.52
$M^{3+}H$	2.78	6.97	3.38	13.34	5.78	55.08
$M^{4+}O$	1.93	21.21	2.43	7.07	4.93	25.55
$M^{4+}H$	2.77	8.53	3.18	13.76	5.83	57.95
$M^{5+}O$	1.77	44.14	1.98–2.97	6.00	5.08	27.43
$M^{5+}H$	2.73	8.12	2.98–3.38	6.00	5.74	56.60
$M^{6+}O$	1.63, 1.73	21.99, 36.52	1.88–2.28	6.00	3.87	13.31
$M^{6+}H$	2.71	6.49	2.90–3.28	4.00	4.45	23.41
$M^{7+}O$	1.64	59.52	1.85–2.73	6.00	3.98	13.88
$M^{7+}H$	0	0	3.28	0	4.60	15.68

For all considered cations the M-O radial distribution functions reveal two well defined hydration shells. The M-H radial description functions also show two clearly distinguishable peaks. The characteristic data of hydration structure of primitive cation from $Z_M = 1$ up to $Z_M = 7$ are summarized in Table 1. It includes the positions of the first maxima r_{max_1}, the values of the radial distribution functions $g(r_{max_1})$, the first and second hydration numbers $n(r_{min_1})$, $n(r_{min_2})$, which were drawn by integration of $g_{MO}(r)$ and $g_{MH}(r)$ up to the first and second minima r_{min_1} and r_{min_2} correspondingly. The obtained results demonstrate the general tendency of the influence of cation charge on the hydration structure of cation. With the increase of cation charge the first peaks of $g_{MO}(r)$ and $g_{MH}(r)$ functions shift towards smaller distances and become more legible and narrow. The cation hydration structure becomes more stable. Its size decreases as a consequence of growing electrostatic interaction between cation and water molecules. The influence of cation charge on the hydrated–hydrolyzed structure of multivalent cations is also illustrated by Fig. 1, where the instantaneous configuration of hydration shells of primitive cation M^{Z+} for seven different charge states are shown.

In the monovalent case $Z_M = 1$ the model of the primitive cation M^+ reduces to the model for cation Na^+. A non-strict first peak of $g_{MO}(r)$ and $g_{MH}(r)$ for this case demonstrates a possibility of exchange of water molecules between the first and second hydration shells. For divalent cation M^{2+} the first and second hydration shells are well separated so the function $n_{MO}(r)$ has a plateau at the value 6, while $n_{MH}(r)$ has a plateau at 12. It means that for cation M^{2+} the strong cation–water electrostatic interaction leads to formation of stable structure constituted by six water molecules octahedrally arranged around the cation.

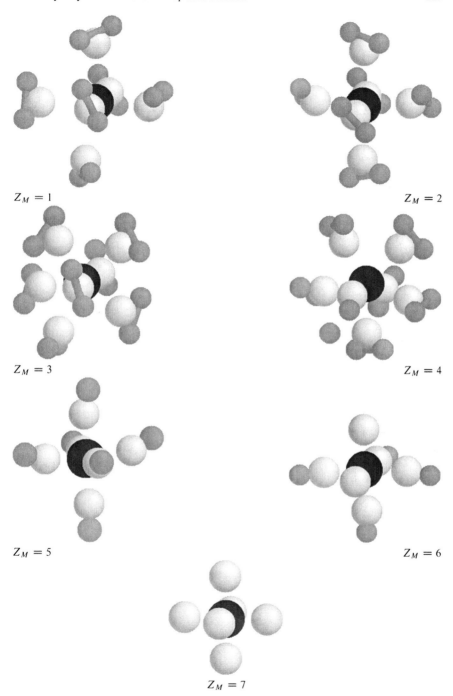

$Z_M = 1$

$Z_M = 2$

$Z_M = 3$

$Z_M = 4$

$Z_M = 5$

$Z_M = 6$

$Z_M = 7$

Fig. 1 Snapshots of instantaneous configurations of hydration structure of cations M^{Z+}

For cations M^{3+} and M^{4+} the balance between cation–oxygen attraction and cation–hydrogen repulsion slightly modifies octahedral configuration of cation hydration shell. For these cations the running coordination numbers $n_{MO}(r)$ and $n_{MH}(r)$ do not have a plateau like in the case of M^{2+}. The cation charge for M^{3+} and M^{4+} is high enough to attract the seventh water molecule but there is not enough space to form stable symmetric structure. That is why the hydration shell of cation M^{3+} is constituted by six or seven closest neighbors with possibility of exchange with bulk water molecules. In order to reduce these exchanges and stabilize the octahedral configuration of the hydration shell of cations M^{3+} and M^{4+} one should additionally take into account non-additive ion–water interaction [50].

A new phenomenon was observed for cation M^{4+}. Due to increasing of repulsion between M^{4+} and protons of water molecules one proton of water molecule can leave the hydration shell. The corresponding snapshot of hydration structure of cation M^{4+} is shown in Fig. 1. The hydration shell of cation M^{4+} is consisted of six water molecules and one $OH^{Z^{*-}}$ group, where Z^{*} is the effective charge of hydrolyzed protons in the framework of considered model. We interpret this phenomenon as the cation hydrolysis effect. It is convenient to describe the hydrolysis effect by the quantity

$$h = 2n_{MO}(r_{min}) - n_{MH}(r_{min}) \tag{9}$$

More strong hydrolysis effect was observed for cations M^{5+}, M^{6+} and M^{7+}. For these cations the octahedral configuration of the first hydration shell is stabilized again. The first and second hydration shells are strongly separated and the function $n_{MO}(r)$ is characterized by legible plateau with the value 6. However due to the cation hydrolysis the function $n_{MH}(r)$ has a plateaus at values smaller than 12. As we can see from Fig. 1 and Table 1 for cation M^{5+} six protons leave cation hydration shell and the hydration shell is constituted by six $OH^{Z^{*-}}$ groups. For cation M^{6+} eight protons leave the hydration shell and the cation is surrounded by four $OH^{Z^{*-}}$ groups and two $O^{2Z^{*-}}$ ions. Due to this the first maximum of $g_{MO}(r)$ has two peaks at 1.63 and 1.73 Å. The first of them corresponds to $O^{2Z^{*-}}$ ions and other one is connected with $OH^{Z^{*}}$ - groups. Finally, for cation M^{7+} all 12 protons leave the hydration shell and cation is surrounded by six $O^{2Z^{*-}}$ ions. As a result the $g_{MH}(r)$ is zero in the region of first hydration shell.

The second hydration shell due to the increase of cation charge shifts to smaller distances similar as the first one. But in contrast to the first hydration shell the second one is not so strongly separated from the bulk. With the increase of cation charge the number of water molecules in second hydration shell slightly increases from ≈ 23 for monovalent cation M^{+} to ≈ 27 for cation M^{5+}. However, for cations M^{6+} and M^{7+} the number of water molecules in the second hydration shell decreases by about two times comparatively to the cations with lower charges. This is a consequence of strong repulsion between hydrolyzed $O^{2Z^{*-}}$ ions in the first hydration shell and oxygens of water molecules in the second one.

2.3 Dynamical Properties

The influence of cation charge on dynamical properties of multivalent cations and oxygens in their hydration shells was investigated via an analysis of the velocity autocorrelation functions (VACF) of cations and oxygens in cation first hydration shell:

$$\Psi_\alpha(t) = \; < \vec{v}_\alpha(0) \vec{v}_\alpha(t) >, \tag{10}$$

where $v_\alpha(t)$ is the velocity of a particle $\alpha = M, O$ at time t.

The self-diffusion coefficients have been calculated by integrating the VACFs in accordance with the Green–Kubo relation:

$$D_\alpha = \frac{1}{3} \int_0^\infty < \vec{v}_\alpha(0) \vec{v}_\alpha(t) > dt, \; \alpha = M, O. \tag{11}$$

Two opposite tendencies determine the dependence of self-diffusion coefficient of ion D_M on cation charge. The first one is connected with the stabilization of the first and second hydration shells due to increasing of electrostatic ion–water interaction. It leads to a decrease of self-diffusion coefficient. The second tendency is connected with destabilization of the cation hydration structure and decreasing of the number of water molecules in the second hydration shell due to increasing of the repulsion between hydrolyzed oxygens of the first and oxygens of water molecules in the second hydration shells. These two effects lead to increasing of self-diffusion coefficient D_M. The dependence of the self-diffusion coefficients of cation M^{Z+} and oxygens in cation first hydration shell on cation charge Z_M is presented in Fig. 2. As we can see the self-diffusion coefficient of cation M^{Z+} is almost two times smaller than for M^+. This is a consequence of stabilization of hydration structure. The slight increase of D_M for M^{3+} is caused by destabilization of octahedral hydration structure before hydrolysis effect. A decrease of D_M for M^{4+} and M^{5+} is connected with a strengthening of cation hydration structure due to the increase of

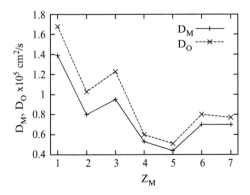

Fig. 2 Dependence of the self-diffusion coefficients of cation M^{Z+} (D_M) and oxygens in cation hydration shell (D_O) on cation charge Z_M

cation charge. Finally, the increase of D_M for M^{6+} and M^{7+} is a result of decreasing of the number of water molecules in the second hydration shell. The dependence of the self-diffusion coefficient D_O of oxygens in the cation first hydration shell on cation charge Z_M is strongly correlated with D_M dependence, being by about 10–20% higher than D_M. Such a behavior of D_O confirms the strong stability of hydrated-hydrolyzed structure of cations. However the oxygens have more degrees of freedom in the hydration shell than the cation.

Another important dynamical property we consider is the spectral density of hindered translation motions of cation $f_M(\omega)$ and oxygens in the first hydration shell $f_O(\omega)$. $\omega = 2\pi\nu$ is a frequency of hindered translation motions. These spectra are calculated by Fourier transform of normalized VACFs:

$$f_\alpha(\omega) = \int_0^\infty \frac{<\vec{v}_\alpha(0)\vec{v}_\alpha(t)>}{<(\vec{v}_\alpha(0))^2>}cos(\omega t)dt, \ \alpha = M, O \qquad (12)$$

The functions $f_M(\omega)$ and $f_o(\omega)$ for cases of cation M^{Z+} with $Z_M = 1, 2, 3, 4, 5, 6$ were presented by us previously in [33], and $Z_M = 7$ in [34]. The functions $f_M(\omega)$ and $f_O(\omega)$ are characterized by a broad distribution of frequencies. With increasing of cation charge both these distributions are extended to higher frequencies region. Both functions $f_M(\omega)$ and $f_O(\omega)$ have a couple of characteristic peaks, which become more pronounced with the increase of cation charge. The values of these characteristic frequencies on spectral densities $f_M(\omega)$ for different cation charges are collected in Table 2. For monovalent cation M^+ $f_M(\omega)$ has only one peak at about 30 ps^{-1}. The appearance of three spectral maxima for cation M^{2+} is a consequence of stabilization of its octahedral configuration. The destabilization of hydration structure by cation M^{3+} provokes the modification of $f_M(\omega)$, which has two smaller and wider peaks, the third one is not well distinguishable. The functions $f_O(\omega)$ for cations M^+, M^{2+} and M^{3+} have the form of vibrational spectrum for pure water but they are partially modified by the presence of cations.

For highly charged cations M^{4+}, M^{5+}, M^{6+} and M^{7+} the functions $f_M(\omega)$ and $f_O(\omega)$ in high frequency region have two common characteristic peaks ω_2 and ω_3, which can be attributed to the normal modes of hydrated cation complex. The values of these characteristic frequencies increase with the increase of cation charge.

Table 2 Characteristic frequencies of the spectral densities of the hindered translation motions of cations M^{Z+}

	ω_1, ps^{-1}	ω_2, ps^{-1}	ω_3, ps^{-1}
M^+	30.28	–	–
M^{2+}	41.16	82.31	109.75
M^{3+}	59.14	88.70	–
M^{4+}	14.72	73.57	161.86
M^{5+}	28.37	85.10	184.38
M^{6+}	24.56	125.10	223.70
M^{7+}	–	159.28	245.62

The functions $f_M(\omega)$ for cations M^{4+}, M^{5+} and M^{6+} in low frequency region are characterized by a peak, which for the cation M^{7+} is not well defined. Such a behavior of $f_M(\omega)$ agrees with the conclusion of our investigation of cation hydration structure about the formation of very stable hydrated-hydrolyzed aggregates around cations. The persistence of the hydration shells and existence of local hydration structures around cations was confirmed recently by investigation of Raman spectra and also by inelastic neutron scattering experiments for aqueous solutions of aluminium halides [51]. The measurements show three peaks, which had been assigned to the three predicted Raman modes of Al-O vibrations in octahedral $Al^{3+}(H_2O)_6$ clusters. The cation with its hydration shell is highly stable aggregate, whose internal vibrations can be considered independently from the rest of solution, anions, ion concentration and even water molecules in second hydration shell. The Raman spectra measured for concentrated aqueous solutions of $AlCl_3$ and $AlBr_3$ demonstate three characteristic frequencies $\omega_1 = 66$ ps^{-1}, $\omega_2 = 88$ ps^{-1}, and $\omega_1 = 105$ ps^{-1} [51], which are comparable with characteristic frequencies for primitive cations M^{2+} and M^{3+} (Table 2). The cation hydrolysis phenomenon observed in the case of M^{4+} leads to the essential increase of ω_3 and decrease of ω_1. For M^{6+} and M^{7+} the frequency ω_2 increases too, that can be explained by hydrolysis of second protons from oxygens in hydration shell. Thus the behavior of frequencies ω_1, ω_2, ω_3 could be an indirect proof of hydrolysis and its intensity.

2.4 Effects of Charge Redistribution Between the Cation Hydrolysis Products

We have shown above that the model cations M^{4+}, M^{5+}, M^{6+}, and M^{7+} have the hydration shell in form $M^{4+}(H_2O)_6OH^-$, $M^{5+}(OH^-)_6$, $M^{6+}(OH^-)_4O_2^{2-}$, and $M^{7+}O_6^{2-}$ correspondingly. It means that the charges of these cations are high enough to repel some protons of water molecules from the first hydration shell of the cation. We interpreted this phenomenon as the cation hydrolysis effect. However since in CF1 model the hydrogen and oxygen of water molecules carry effective charges one has to take into account that during the hydrolysis, when one of the protons leaves the "native" water molecule it should change its charge from $Z_H = Z_H^{CF} = 0.32983$ to $Z_H = 1$. Correspondingly the charge of OH^{Z_H-} group should be of the same value but of opposite sign. To improve the description of hydrolysis reaction products we expanded the regular CF1 model by inclusion of charge redistribution effects due to Eq. (5)–(6). Such redistribution procedure leads to difference between hydrogens H^{Z^*+} as hydrolysis products and hydrogens of non-hydrolyzed water molecules. There is also non-equivalence of oxygens in OH^{Z^*-} groups obtained in result of hydrolysis and oxygens in water molecules. This causes sufficient modifications of structure of cation hydration shell comparing to the results discussed above.

For cation M^{4+} only one proton leaves the first hydration shell. More strong hydrolysis effect was observed for cation M^{5+}, so we carried out more detailed study how the effect of charge redistribution changes the structural and dynamical

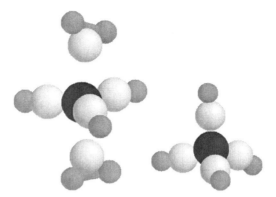

Fig. 3 Snapshots of instantaneous configurations of hydration shells of cation $Z_M = +5$ with five (*left*) and four (*right*) neighbors after charge redistribution

Table 3 The characteristics of hydration shell of cation M^{5+} in the framework of the model with charge redistribution between hydrolysis products. "5O", "4O" sections denotes configurations with five and four neighbors

		r_{max_1}	$g(r_{max_1})$	r_{min_1}	$n(r_{min_1})$	r_{min_2}	$n(r_{min_2})$
	$M-O_{OH}$	1.48	36.09	1.58–1.82	3.00	3.65	8.93
"5O"	$M-O_{H_2O}$	2.02	7.10	2.38–2.48	5.00		
	$M-H_{OH}$	2.38	6.74	2.53	3.00	4.13	14.89
	$M-H_{H_2O}$	2.78	2.46	3.12	7.00		
"4O"	$M-O_{OH}$	1.47	69.03	1.62–2.38	4.00	3.68	7.63
	$M-H_{OH}$	2.42	8.70	2.62	4.00	4.09	13.53

properties of cation M^{5+} and its hydration shell. Several MD runs yielded two types of stable hydration shell instead of octahedral configuration. The corresponding snapshots are presented in Fig. 3. One of these configurations demonstrates five neighbors in the shell: three OH^- groups in equatorial plane and two axial H_2O molecules. The second configuration is characterized by tetrahedral arrangement of four OH^- groups. During long simulation run we noticed an attempt of transition from configuration $M^{5+}(OH^-)_3(H_2O)_2$ to $M^{5+}(OH^-)_4$, but not the opposite transition. It implies that the second configuration is more stable.

Structural data for two hydration structures of M^{5+} cation are summarized in Table 3. Data in "5O" and "4O" sections correspond to configurations $M^{5+}(OH)_3^-(H_2O)_2$ and $M^{5+}(OH)_4^-$. Comparing Tables 2 and 3 one can see that the increase of charge on the oxygen of $(OH)^{Z^*-}$ group results in appearance of two peaks (instead of one peak for model with no charge redistribution) on $g_{MO}(r)$ at distances 1.48 and 2.02 Å for $M^{5+}(OH^-)_3(H_2O)_2$ configuration. The first peak of $g_{MO}(r)$ corresponds to three oxygens of $(OH)^{Z^*-}$ groups. Second maximum stands for two oxygens of non-hydrolyzed water molecules. Due to higher charge on oxygen the $(OH)^{Z^*-}$ group can approach the cation closer, while oxygens of non-hydrolyzed water molecules are attracted weaker and the octahedral geometry of hydration shell is broken. Total number of neighbors is equal to five instead of six neighbors for the regular model with no charge redistribution. For

Table 4 Self-diffusion coefficients of cation M^{5+} (D_M) and oxygens in cation hydration shell (D_O) for the model with charge redistribution between hydrolysis products. "5O" and "4O" correspond to configurations $M^{5+}(OH)_3^-(H_2O)_2$, $M^{5+}(OH)_4^-$

Configuration	$D_M \times 10^5$, cm²/s	$D_O \times 10^5$, cm²/s
"5O"	0.77	0.88
"4O"	1.08	1.32

the configuration $M^{5+}(OH^-)_4$ there is only one peak on $g_{MO}(r)$ at the distance 1.47 Å and one peak on $g_{MH}(r)$ at 2.42 Å. These peaks correspond to four $(OH)^{Z^*-}$ groups. Non-hydrolyzed water molecules are absent in the first hydration shell for this configuration.

The charge redistribution between the hydrolysis products leads also to a strong modification of second cation hydration shell. The second hydration shell shifts to the smaller distances and number of water molecules in it decreases significantly. This is the result of the increased charge on the hydrolysis products. After the charge redistribution the charge on oxygen in a group OH^- increases significantly. This implies the growth of repulsion between OH^- group and water molecules, which in turn leads to reduction of a number of water molecules in second hydration shell. Due to a higher repulsion between hydrolyzed hydrogen and cation the hydrogen moves to the bulk water.

The charge redistribution between the hydrolysis products is also important for dynamical properties of hydrated–hydrolyzed cation complexes. The influence of the charge redistribution on the self-diffusion coefficients of cation M^{5+} D_M and oxygens from its hydration shell D_O is presented in Table 4. As we can see (comparing with results presented in Fig. 2) the self-diffusion coefficient of cation M^{5+} in configurations $M^{5+}(OH^-)_3(H_2O)_2$ and $M^{5+}(OH^-)_4$ increases by two and three times respectively. Very similar increase of the self-diffusion coefficients of oxygens in cation hydration shell was also found. This result can be explained by reformation of the second hydration shell after charge redistribution on hydrolysis products. A strong decrease of number of water molecules in a second hydration shell is responsible for such strong increasing of diffusion coefficients D_M and D_O. The spectral densities of the hindered translation motions of cation M^{5+} for both configurations $M^{5+}(OH^-)_3(H_2O)_2$ and $M^{5+}(OH^-)_4$ are presented in Fig. 4. In these figures for comparison we also shown by dashed line the $f_M(\omega)$ obtained without consideration of the charge redistribution effects. The values of characteristic frequencies on spectral densities of cation M^{5+} for both configurations are collected in Table 5. As one can see for both configurations the function $f_M(\omega)$ has two parts. One of them is characterized by a broad distribution of frequencies up to 200–250 ps^{-1} with three characteristic peaks. The peak at 13.7 ps^{-1} is approximately of the same magnitude for both configurations and characterizes a "breathing" of hydration complex of cation. Vibrations between the cation and OH^- groups and between the cation and water molecules in configuration $M^{5+}(OH^-)_3(H_2O)_2$ lead to a splitting of the second peak of $f_M(\omega)$. It is interesting to note that the peak, which corresponds to the vibration between the cation and oxygens of OH^- groups is located almost at

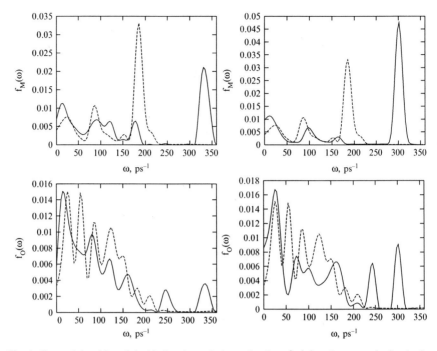

Fig. 4 Spectral densities of the hindered translations of cation $f_M(\omega)$ and oxygens in first hydration shell $f_O(\omega)$ for cation charge $Z_M = +5$ with inclusion of charge redistribution effect (solid line) compared to the case $Z_M = +5$ but without charge redistribution: configuration with five neighbors (*left*), configuration with four neighbors (*right*)

Table 5 Characteristic frequencies of the spectral densities $f_M(\omega)$ of the hindered translation motions of cation M^{Z+} for the model with charge redistribution between hydrolysis products. "5O" and "4O" denote configurations $M^{5+}(OH^-)_3(H_2O)_2$, $M^{5+}(OH^-)_4$

Configuration	ω_1, ps^{-1}	ω_2, ps^{-1}	ω_3, ps^{-1}	ω_4, ps^{-1}
"5O"	13.66	95.61	177.57	327.82
		122.94		
"4O"	13.69	95.82	164.27	301.16

the same frequencies (95.6 and 95.8 ps^{-1}) for both configurations. The third peak of $f_M(\omega)$ is located at $\omega = 177.6$ ps^{-1} for configuration $M^{5+}(OH^-)_3(H_2O)_2$ and at $\omega = 164.3$ ps^{-1} for configuration $M^{5+}(OH^-)_4$. The separated peak at frequency $\omega_4 = 327.8$ ps^{-1} for configuration $M^{5+}(OH^-)_3(H_2O)_2$ and at $\omega_4 = 301.2$ ps^{-1} for configuration $M^{5+}(OH^-)_4$ is the second part of function $f_M(\omega)$. This new feature of $f_M(\omega)$ describes the cation motion between two oxygens in hydration shell and is connected with strong stability of considered complexes due to charge redistribution between the hydrolysis products.

Similarly to $f_M(\omega)$ the function $f_O(\omega)$ has a part with a broad distribution of frequencies up to 200–250 ps^{-1} and second part with separated peaks. The broad part of $f_O(\omega)$ has four peaks which characterize different motion of the oxygen strongly coordinated by cation and other oxygens. The second part of $f_O(\omega)$ includes two separated peaks. The first of them is located at $\omega = 250$ ps^{-1} for configuration $M^{5+}(OH^-)_3(H_2O)_2$ and at $\omega = 240$ ps^{-1} for configuration $M^{5+}(OH^-)_4$. This peak characterizes the influence of cation on O–O stretching motion of the oxygens in hydration shell. The second separated peak exactly coincides with position of the separated peak of $f_M(\omega)$. This again confirms the strong stability of considered hydrated–hydrolyzed complexes. The existence of common separated peaks at a high frequency region for $f_O(\omega)$ and $f_M(\omega)$ was also observed in MD simulation of Cr^{3+} hydration [19].

Due to a charge redistribution effect the charge of hydrolyzed protons H* changes from the value $+0.32983e$ in CF model to $+e$ and the charge of hydrolyzed group OH$^-$ changes from the value $-0.32983e$ to $-e$. The OH$^-$ groups are strongly bonded to the cation, while hydrolyzed protons left cation hydration shell and formed some complexes with water molecules. In particular in our simulation we observed two types of aqua-complexes, with protons H*, which were created after hydrolysis. The first complex $(H_5O_2)^+$ is linear and consists of two water molecules and a proton between them. The H*–O lengths are 0.9 and 1.36 Å, which slightly differ from corresponding values from ab initio MD calculations: 0.98 and 1.23 Å [28]. The second configuration $(H_7O_3)^+$ is planar and consists of three water molecules and a proton in the center. The angles between H*–O arms are close to 120°, while all three lengths are equal to 1.36 Å. In addition one should note that the single proton being a result of hydrolysis has a small size and charge $+1$ can be a source of hydrolysis of another water molecule. After that the hydrolyzed proton changes its charge from $Z_H = 1$ to $Z_H = Z_H^{CF}$ while a new hydrolyzed proton changes its charge from $Z_H = Z_H^{CF}$ to $Z_H = 1$. This process can be repeated many times and leads to a proton relay transfer in aqueous solution.

3 More Realistic Models for a Cation Hydrolysis

The study of the primitive model of cation hydrolysis gives us a possibility to understand the role of electrostatic ion–water interaction in the formation of hydrated-hydrolyzed forms of highly charged cations in water. This approach utilizes a short-range part of the interaction potential, which was previously developed for Na$^+$ ion in water. The increase of a charge on real cations leads to shortening of distances between cation and closest water molecules, this in turn intensifies the hydrolysis tendency and stabilizes hydrated–hydrolyzed form of highly charged cations. This means that the primitive cation M^{Z+} even underestimates the electrostatic interactions in comparison with Na$^+$, Mg^{2+}, Al^{3+}, ... in a row of the periodical table of elements, since the sizes of these cations decrease through the sequence. So the role of the electrostatic interaction in a formation of hydrated–hydrolyzed complexes of real cations in the considered sequence will be even intensified. In this section we will analyze the hydrated–hydrolyzed structure and dynamical properties of one of the simplest realistic cations — Al^{3+} and more complex one

— the uranyl ion $(UO_2)^{2+}$. Like in the case with primitive cation in previous section water molecules are considered in the framework of the flexible CF1 model [35, 36].

3.1 A Molecular Dynamics Study of Al^{3+} Hydration

Ion Al^{3+} is one the simplest realistic trivalent cations. Due to a comparatively small size and high valency of Al^{3+} the electrostatic cation–water interaction is strong enough for hydrolysis triggering. The aqueous speciation of Al^{3+} is very important for many industrial processes such as waste water treatment, pharmaceutical design, catalysis optimization, remediation of wastes from plutonium production, etc. [52]. However, in spite of large efforts to describe the cation–water interaction correctly no hydrolysis effects are taken into account directly in previuos computer simulations of Al^{3+} in water [14, 15, 53]. Our results of molecular dynamics study of Al^{3+} in water were reported in [39]. In this investigation the description of the interaction between Al^{3+} and water molecules is chosen in the form

$$U_{AlO}(r) = -\frac{656.95}{r} - \frac{596.09}{r^2} + 63533.01 \cdot e^{-3.89948r}$$
$$U_{AlH}(r) = \frac{328.47}{r} + \frac{38.37}{r^2} + 68.66 \cdot e^{-0.35461r} \quad (13)$$

which was drawn from quantum chemical calculations [54]. In Eq. (13) the energies are given in kcal/mol, distances are in Å. For simplification we neglect the effects of charge redistribution on hydrolysis products.

In Fig. 5 we present the Al^{3+}-water RDFs and coordination numbers. The main peak of $g_{Al-O}(r)$ distribution is located at 1.78 Å. This result slightly underestimates the values range 1.8–1.97 Å yielded from calculations at various levels of theory in [14, 15, 55], while the experimental results collected in [6] are within the range 1.87–1.9 Å. However the number of oxygens in the first hydration shell in

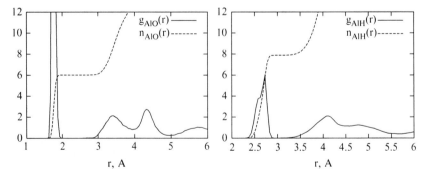

Fig. 5 Radial distribution functions (solid lines) $g_{Al-O}(r)$, $g_{Al-H}(r)$ and corresponding running coordination numbers (dashed lines) $n_{Al-O}(r)$, $n_{Al-H}(r)$

our calculations coincides with the Al–O coordination numbers reported from all mentioned investigations and is equal to 6. A well defined plateau on the $n_{Al-O}(r)$ indicates that no exchange processes between the first and second hydration shells were observed during simulation time.

The Al–H distribution is characterized by $g_{Al-H}(r)$, $n_{Al-H}(r)$ functions. The position of the first peak of Al–H RDF is at 2.71 Å. Different theoretical approaches [14, 15] yielded for r_{max} a range of values 2.47–2.59 Å. The deviation of our result can be easy explained if one will take into account that our water model is non-rigid (in contrast to the ones used in above references). The strong electrostatic repulsion between Al^{3+} and hydrogens stretches the O–H bonds of the aluminium cation hydration shell neighbors. In [14] stretching of O–H bonds in water molecules in the first hydration shell was noticed too. The corresponding running coordination number $n_{Al-H}(r_{min})$ indicates that the number of protons in the first hydration shell is equal to 8, instead of 12. This result we treat as the cation hydrolysis: strong electrostatic repulsion between Al^{3+} and protons pushes four of them outside the first hydration shell. Due to separation of the first hydration shell from the second one both first peaks of $g_{Al-O}(r)$ and $g_{Al-H}(r)$ are separated by gaps making the well-defined plateaus on the profiles of running coordination numbers. The structural details of hydration shell of Al^{3+} are collected in Table 6. A snapshot of aluminium cation first hydration shell is presented in Fig. 6. One can clearly see two water molecules, four OH^- groups and four protons appeared as a result of hydrolysis. Four protons are located in the second hydration shell and are strongly bonded to the first one. This is a consequence of neglect of the charge redistribution

Table 6 Structure of Al^{3+} hydration shell

	r_{max1}, Å	$g(r_{max1})$	r_{min1}, Å	$n(r_{min1})$
Al-O	1.78	40.99	2.02–2.52	6
Al-H	2.71	5.95	2.98	8

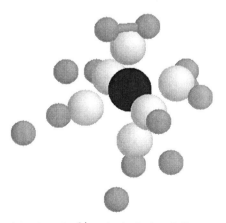

Fig. 6 Instantaneous configuration of Al^{3+} cation hydration shell

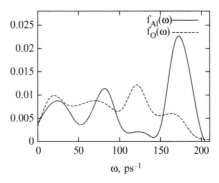

ω, ps^{-1}

Fig. 7 Spectral densities of hindered translation of aluminium cation and oxygens of hydration shell

effects, since non-redistributed charges are too weak to repell protons from cation, like it happens in case of primitive cation. All the neighbors are octahedrally arranged around the aluminium ion.

Another aspect of our investigation is connected with dynamical properties of Al^{3+} and oxygens in first hydration shell. The spectral densities $f_{Al}(\omega)$, $f_O(\omega)$ of hindered translational motions are presented in Fig. 7. The Al^{3+} ion distribution of frequencies is characterized by three main peaks approximately at: 25, 80 and, 170 ps^{-1}; one additional peak is located near 120 ps^{-1}. In the spectrum of oxygens one can see four peaks approximately at: 20, 70, 120, and 165 ps^{-1}. Obviously there is a correlation between the spectra of central cation and oxygens, furthermore the characteristic frequencies coincide, that allows us to conclude that the motion of Al ion is strongly coupled with the motion of water molecules and OH^- groups in hydration shell.

As we noted in previous section the observable characteristic peaks in Raman spectra [51] are located far from the values yielded from MD simulation. They are more closer to the values obtained for primitive cation M^{3+}. The reason for such a difference is in very strong hydrolysis effects found in case of Al^{3+} in contrast to experimental study where aluminium salts where chosen to avoid hydrolysis effect. The self-diffusion coefficient of aluminium ion, calculated in [39] is equal to 0.46×10^{-5} cm^2/s. In [15] a set of results for the self-diffusion coefficient of aluminium ion for different models is reported. It varies within the range $0.17 - 0.47 \times 10^{-5}$ cm^2/s, while the experimental value taken from [56] is 0.6×10^{-5} cm^2/s.

3.2 A Molecular Dynamics Study of Uranyl Hydration

In this subsection we study the aspects of cation hydrolysis in aqueous solutions of actinides such as uranium U, plutonium Pt, neptunium Np, etc. This investigation has connection with the problem of contact of nuclear fuel wastes with water. Understanding of complexation processes in such systems can provide one with important information for chemical technology, medicine, environmental ecology [57]. The actinides An in water usually have a valency $Z = 6$ and easily form a corresponding complex cation $(AnO_2)^{Z+}$ called actynil. Both An=O bond lengths are usually 1.7–1.8 Å and O=An=O angle is close to 180°. Actynil is usually hydrated by five water molecules, so-called ligands, located in equatorial plane (normal to O=An=O axis) at the distances 2.5–2.6 Å. This finding was confirmed by quantum chemical investigations. In particular in [58, 59] it is shown that the number of ligands of uranyl, neptunyl and plutonyl is equal to five. It was also found that hydrolysis reaction with one proton loss from one of five uranyl ligands is energetically favorable. However the quantum chemical calculations can not give a detailed understanding of a role of environment beyond the hydrated-hydrolyzed complex, which is essential for correct interpretation of many structural, dynamic, and thermodynamic properties of these complexes [60]. In Refs. [61–63] molecular dynamics simulations were carried out for aqueous solutions of actinides using rigid TIP3P or SPC water models. Neither deformation of water molecules nor hydrolysis effects in hydration shell were observed because of the constrained rigid model of water molecules. No hydrolysis evidence was also found in recent investigation of uranyl hydration by ab initio molecular dynamics simulation [64].

Next we review our recent molecular dynamics study of uranyl hydration performed in the framework of CF1 model for water [40]. For the uranyl–water interaction we made use of potentials from Ref. [63]. These potentials are of "1-12-6" type:

$$E_{ij} = \frac{Z_i Z_j}{r} + \frac{A_i A_j}{r^{12}} - \frac{B_i B_j}{r^6} \tag{14}$$

The corresponding potential parameters are listed in Table 7.

In order to preserve uranyl intramolecular geometry additional constraints are also used for U=O bond length in the form $E_{ij} \sim (r - 1.75)^2$ and for O=U=O angle in the form $E_{ij} \sim (\theta - 180)^2$. The distances are measured in Å, angles—in degrees, energies—in kcal/mol. We also take into account the effect of charge redistribution between hydrolyzed products utilizing Eqs. (5) and (6).

For convenience the oxygens in uranyl we note as O*. The results for radial distribution functions $g_{UO}(r)$, $g_{UH}(r)$, $g_{O^*O}(r)$, and $g_{O^*H}(r)$ are presented in Figs. 8. From Fig. 8 it can be clearly seen that the first peak of $g_{UO}(r)$ function is divided into two ones at positions 2.275 and 2.582 Å. These distances are close to the value 2.42 Å obtained from the X-ray data [65–67]. One should note that the experiments were performed under conditions excluding hydrolysis effect. The absence

Table 7 Potential parameters for uranyl-water interaction

	Z	A (kcal Å^{12}/mol)$^{1/2}$	B (kcal Å^6/mol)$^{1/2}$
O in H$_2$O	−0.65966	793.322	25.010
H	0.32983	0.1	0.0
U	2.50	629.730	27.741
O in UO$_2$	−0.25	793.322	25.010

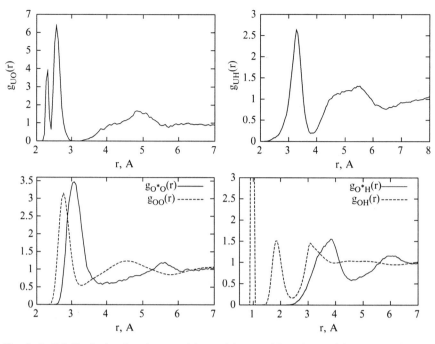

Fig. 8 Radial distribution functions $g_{UO}(r)$, $g_{UH}(r)$, $g_{O*O}(r)$, and $g_{O*H}(r)$. For a comparison we also show $g_{OO}(r)$ and $g_{OH}(r)$ of pure CF1 model (dashed lines)

of hydrolysis effect in results of ab initio simulation [64] explains the difference in results with [40]. Near $r_{UO} = 3.175$ Å there is a region where $g_{UO} = 0$. Due to this the $n_{UO}(r)$ function demonstrates at corresponding distances a shoulder and plateau with values approximately equal to 1 and 5. This means that the uranyl first hydration shell is separated from the bulk and consists of five neighbors, one of these is positioned closer to uranium than the others. The U–H distribution demonstrates a well-pronounced first peak at 3.281 Å and minimum at 3.712 Å. Comparing the first peak positions of $g_{UO}(r)$ and $g_{UH}(r)$ functions one can conclude that water molecules have their hydrogens oriented in direction opposite to uranium. Integration over the first peak of $g_{UH}(r)$ function yields the number of closest neighbors equal to 9. This indicates that in the first hydration shell there are four H_2O molecules and one $(OH)^-$ group. Obviously repulsion of proton from uranium overcomes attraction of proton in its native water molecule. We treat this phenomenon as hydrolysis reaction. After such a short analysis it is also easy to explain two peaks of $gUO(r)$. During the hydrolysis process new ions $((OH)^-$ and $H^+)$ appear. The distance between $(OH)^-$ and H^+ increases and when proton leaves the uranium first hydration shell because of charge redistribution effects the charges on ions are -1 and $+1$ respectively. This in turn differs the oxygens of $(OH)^-$ group from the oxygens of regular water molecules described in the framework of CF1 model. The increase of the oxygen charge makes the uranium–oxygen attraction

Table 8 Structure of uranyl hydration shell

	r_{max1} Å	$g(r_{max1})$	r_{min1} Å	$n(r_{min1})$
U–O	2.28	3.30	2.41	1.00
	2.58	6.37	3.18	5.00
U–H	3.28	2.64	3.71	9.13
O*–O	3.07	3.39	3.79	8.20
O*–H	3.84	1.56	4.56	21.02

stronger and as a consequence the $(OH)^-$ group becomes located closer to uranium than the other four molecules. In Table 8 we present a few most important results of radial distribution functions and running coordination numbers. Two lines for the U–O distribution correspond to two peaks of $g_{UO}(r)$ function.

In Fig. 8 we also present the radial distribution functions for coordination of water around uranyl oxygens O* - $g_{O*O}(r)$ (bottom left) and $g_{O*H}(r)$ (bottom right). For the comparison we also show the coordination of water around water oxygens $g_{OO}(r)$ and $g_{OH}(r)$. Wider and higher first peak of $g_{O*O}(r)$ (solid line) at ≈ 3.1 Å, comparatively to $g_{OO}(r)$ (dashed line), is shifted to larger distances. This implies some weaker coordination of water oxygens around uranyl oxygens. The corresponding running coordination numbers $n_{O*O}(r)$ demonstrate a smeared plateau at the value ≈ 8 in the region $3.5 - 3.7$ Å. Five equatorial uranyl neighbors and two or three non-equatorial water molecules contribute to the first peak of $g_{O*O}(r)$ distribution. The first peak is not separated from the rest of distribution by a zero-gap with $g_{O*O}(r) = 0$, that means that non-equatorial neighbors can be exchanged by bulk water molecules.

Similar analysis can be drawn for the uranyl oxygen–hydrogen distribution $g_{O*H}(r)$ (solid line) in comparison with water oxygen–hydrogen $g_{OH}(r)$ (dashed line). The wide first peak of $g_{O*H}(r)$ is located in the region from 3 to 4 Å. Partial contribution to this peak comes from hydrogens of five equatorial neighbors, while the rest is due to waters from non-equatorial plane. Exchanges between non-equatorial neighbors and bulk prevent from zero-gap appearance on $g_{O*H}(r)$ and plateau on the running coordination numbers $n_{O*H}(r)$. These findings allow one to conclude that non-equatorial neighbors of uranyl are not bound tight with hydrated–hydrolyzed complex.

The instantaneous configuration of uranyl and its first hydration shell will be presented below in Fig. 10 (top right configuration). All the ligands in the first hydration shell are located in a plane equatorial to the uranyl O=U=O axis. Such a bipyramidal pentacoordinated structure of hydration shell is in agreement with the data from quantum chemical and experimental studies [58, 59, 65].

In Fig. 9 the spectral densities of hindered translation motions $f_U(\omega)$, $f_{O*}(\omega)$, $f_O(\omega)$ are presented. One can see all spectral functions are characterized by a common peak at ≈ 26 ps^{-1}, which implies common vibrations of the particles within hydrated-hydrolyzed complex. Another common peak at ≈ 60 ps^{-1} in spectra of uranium and water oxygens indicates correlated vibration of these particles. The high-frequency peak at $\approx 115–120$ ps^{-1} in spectra of uranium and uranyl oxygens is

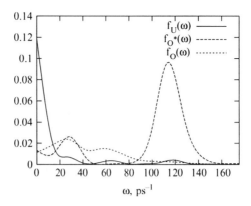

Fig. 9 Spectral densities of hindered translation motions of uranium $f_U(\omega)$, uranyl oxygens $f_{O^*}(\omega)$, and water oxygens $f_O(\omega)$ from uranyl first hydration shell

Table 9 Characteristic frequencies of spectral densities of uranyl and oxygens of first hydration shell; corresponding self-diffusion coefficients

	ω_1, ps^{-1}	ω_2, ps^{-1}	ω_1, ps^{-1}	$D \times 10^5$ cm^2/s
U	26.38	65.20	118.91	0.58
O*	28.02	–	114.22	0.69
O	25.98	59.88	–	0.91

a consequence of additional interaction terms used for preserving uranyl intramolecular geometry. Characteristic frequencies of spectral densities $f_U(\omega)$, $f_{O^*}(\omega)$, $f_O(\omega)$ and self-diffusion coefficients D_U, D_{O^*}, D_O are collected in Table 9.

3.3 Influence of the pH Level on Uranyl Hydration

The tendency for cation hydrolysis also depends on the pH value of aqueous solution. Intuitively it is clear that in alkaline solution the cation hydrolysis will be stronger than in acidic solution. However the investigations of pH influence on cation hydrolysis on the molecular level are very scarce. One should mention a recent study of Al^{3+} hydrolysis under neutral, alkaline, and acidic conditions by constrained molecular dynamics [68]. An analytical semiempirical approach was also used for a description of pH influence on a formation of hydroxocomplexes of actinides [69].

Above we reported the hydrolysis effects only in solutions with neutral pH level, in other words, with no additional extra H$^+$ or (OH)$^-$ ions. On the model level a problem of investigation of a pH influence can be formulated in the following manner: one needs to study time-dependent changes in uranyl hydration shell if the solution will include different amount of H$^+$ or (OH)$^-$ ions. Interaction potentials can be taken in the form as in previous subsection, but for simplification we will neglect the effect of charge redistribution between hydrolysis products.

Molecular dynamics simulations were performed for a few alkaline and acidic solutions, which differed by number of $(OH)^-$ or H^+ ions. Every considered alkaline system consisted of 1,600 water molecules, 1 uranyl ion, and 20/40/60/80 $(OH)^-$ groups. Analogically a series of acidic solutions were modeled too, but with H^+ ions instead of $(OH)^-$ groups.

In previous subsection we have shown that in the solution with no additional H^+ or $(OH)^-$ ions uranyl hydration shell consists of five neighbors: four water molecules and one hydrolyzed $(OH)^-$ group. As a result the whole complex of uranyl and its hydration shell changes into $(UO)_2^{2+}(H_2O)_4OH^-$. Our studies of alkaline and acidic solutions indicate a strong dependence of the composition of uranyl hydration shell on pH level. The modeling of acidic solution have shown that presence of extra H^+ ions in the bulk prevents hydrolysis reaction in uranyl shell and stabilizes the complex $(UO_2)^{2+}(H_2O)_5$. As one can expect the modeling of alkaline solutions demonstrated an opposite effect: water molecules in first hydration shell of uranyl are exchanged by $(OH)^-$ groups from bulk, changing the complex $(UO_2)^{2+}(H_2O)_5$ into $(UO_2)^{2+}(OH^-)_5$ form. In Fig. 10 we present snapshots of these configurations of uranyl and its hydration shell.

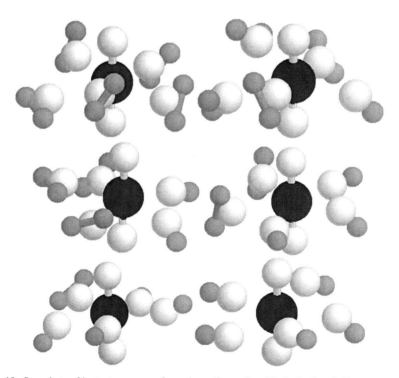

Fig. 10 Snapshots of instantaneous configurations of uranyl and its hydration shell

It is also convenient to analyze structural results in terms of radial distribution functions. In Fig. 11 (left panel) the RDFs for a series of configurations from $(UO_2)^{2+}(H_2O)_5$ to $(UO_2)^{2+}(OH^-)_5$ are shown. One can see that the first peak of $g_{UO}(r)$ becomes narrower and shifts to smaller distances for configurations with more $(OH)^-$ groups in the uranyl hydration shell. This can be explained by a stronger attraction between uranyl and $(OH)^-$ groups in comparison with the uranyl-water interaction. A zero-gap between the first peak the rest of distribution becomes wider indicating increased stability of configurations with more $(OH)^-$ groups. Our conclusion about increased stability of configurations with more $(OH)^-$ groups is confirmed by widening of plateau on $n_{UO}(r)$ for these complexes. In Fig. 11 (right panel) an evolution of radial distribution functions $g_{UH}(r)$ for a set of configurations is shown. It can be seen that first peak of $g_{UH}(r)$ becomes narrow and lower, which reflects a decrease of number of hydrogens in uranyl neighborhood. A shift of the first peak of $g_{UH}(r)$ to larger distances indicates that $(OH)^-$ groups prefer to be oriented with hydrogens outside the hydration shell.

In Fig. 12 we present a time dependence of composition of uranyl hydration shell for different numbers of $(OH)^-$ groups in solution where Shell N_{OH} is the number of $(OH)^-$ groups in uranyl hydration shell. It is clear that the more $(OH)^-$ groups

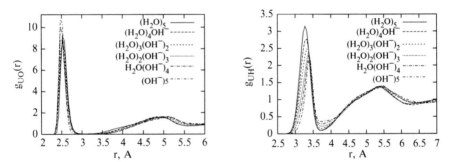

Fig. 11 Radial distribution functions uranyl–oxygen $g_{UO}(r)$ and uranyl–hydrogen $g_{UH}(r)$

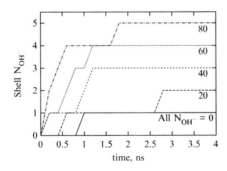

Fig. 12 Time dependence of composition of uranyl hydration shell for different numbers of $(OH)^-$ groups in solution

added to a solution the faster uranyl shell is being modified and filled by (OH)'s. For the case with out $(OH)^-$ groups we yield previous result for neutral solution: uranyl can hydrolyze one water molecule. The result for acidic solutions is not shown here since extra H^+ ions prevent hydrolysis reaction and no $(OH)^-$ groups were found in the uranyl hydration shell.

3.4 Formation of Polynuclear Complexes

Creation of hydrated–hydrolyzed cation structure is only the first step, however usually the process does not stop at this stage and at the finite ionic concentrations polynuclear ions can appear as a result of condensation reaction [2, 3]:

$$p[M(H_2O)n]^{Z+} + qH_2O \rightarrow M_p[(OH)_q H_2 O_{pn-q}]^{(pZ-q)+} + qH_3O^+ \quad (15)$$

The process of polynuclear ion formation can be described in the framework of polymer associative theory operating with a parameter m. It has a meaning of mean chain length defined by the equation [3, 70–72]:

$$m(m-1) = 2\pi c K_{++} g_{00}^{++}(d^+) \quad (16)$$

where K_{++} is the parameter of intercation associative interaction, c – ionic concentration, $g_{00}^{++}(d^+)$ is the contact value of the cation–cation distribution function, d is the cation size. For low ionic concentrations according to Eq. (16) $m \sim \sqrt{c}$ and has concentration dependence similar to inverse Debye length $\kappa \sim \sqrt{c}$ [38]. Depending on the value of m the effect of polynuclear ions formation can be significant and can lead to a strong deviation from the Debye–Huckel limiting behavior for the osmotic coefficients and activity coefficients [3]. As it was shown by Ramsay [73] from light and neutron scattering measurements on relatively concentrated solutions (more than 0.1 mol/l) Al^{3+} ions create chains with mean length ≈ 30 ions, while Zr^{4+} ions can create chains consisting of ≈ 200 ions. Considering Eq. (16) one can show that $m_{Al^{3+}} \simeq 3$ at the concentration 0.001 mol/l and $m_{Zr^{4+}} \simeq 2$ at the concentration 10^{-6} mol/l. For a more dilute solutions, according to Eq. (16), $m \rightarrow 1$ and only mononuclear ions exist.

In order to understand the mechanisms of polynuclear ion formation we performed molecular dynamics simulations of uranyl aqueous solution at finite ionic concentration with the same interaction potentials. The MD cell consisted of 1,600 water molecules and 16 uranyl ions, modeling a solution with neutral pH. To study pH influence we also considered a system with 100 extra $(OH)^-$ groups.

In this particular problem uranium–uranium, uranium–oxygen, and oxygen-oxygen radial distribution functions are of special interest since they provide the information about most favorable distances between these particles during formation of polynuclear complexes. These functions are collected in Fig. 13. The uranium–uranium RDF (top) demonstrates first peak in the region 5.4–7.5 Å with

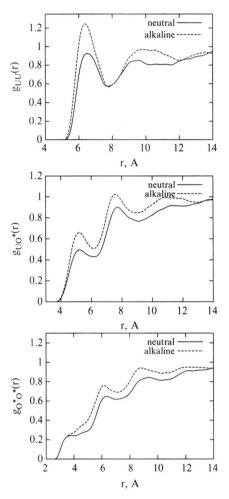

Fig. 13 Uranium–uranium (*top*), uranium-oxygen (*center*), and oxygen-oxygen (*bottom*) radial distribution functions for neutral (solid line) and alkaline (dashed) solutions. Only intermolecular part of RDFs is shown

the maximum at 6.3 Å. The function which describes distribution of uranyl oxygens around uranium (center) has two peaks at 5.3 and 7.4 Å. Distribution of uranyl oxygens (bottom) has two peaks at distances 6.3 and 8.7 Å. In such a way we have estimated the most probable interparticle distances when two uranyl ions approach. One should also note that the RDFs for alkaline solutions are higher than the RDFs for neutral conditions. This indicates that in alkaline solutions uranyl ions can approach one another easier than under neutral conditions.

One of the important quantities describing formation of polynuclear complexes is the mutual orientation of ions and their hydration shells. For such an orientational

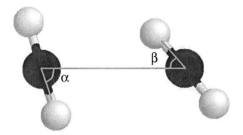

Fig. 14 Mutual orientation of two uranyl ions and characteristic angles

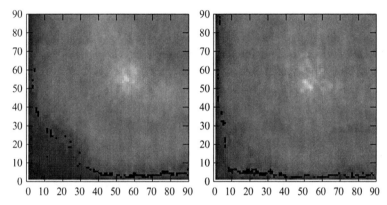

Fig. 15 Probability distribution for uranyl–uranyl mutual orientation under alkaline (left panel) and neutral (right panel) conditions

analysis we store the coordinates of all uranyl ions and oxygens in their hydration shells. At every simulation step one can assign a vector along uranyl axis O–U–O. We also consider a vector between two uranium ions when the distance between them is less than 7.5 Å (position of the first minimum of U–U RDF). Description of mutual uranyl orientations can be treated in terms of two angles α, β between each of O–U–O axis and U–U vector (see Fig. 14). These angles were collected during simulation time. The corresponding symmetrized probability distributions are shown in Fig. 15. Left panel stands for the alkaline solution, right—for the neutral one. Both distributions demonstrate very low probability (dark regions in the distribution maps) for configurations with two uranyl ions oriented axially or configurations with both angles less than 25°. More probable are the orientations with one of the angles within the region 10–25°, while the second angle is in the region of 40–80°. Most favorable configurations (bright regions) appear when both angles are close to 55°. One should also note that "neutral" distribution is more smeared, indicating less strict orientation of uranyl ions.

Besides the structural properties we built a probability distribution $f_{UU}(t)$ of lifetimes of polynuclear complexes, since this quantity allows one to judge how stable in time these complexes are. The same condition was used to decide whether a pair of uranyl ions is in "bonded" state: distance between uranium ions should be less than 7.5 Å. The shape of both distributions is almost the same, except $f_{UU}(0)$, which has a meaning of the probability that in any moment an arbitrary uranyl ion is in "bonded" state. In neutral solution it is equal to 40%, for "alkaline" solution it is $\approx 30\%$. This probability decays comparatively fast: a half of uranyl complexes lives during 0.7 ps, a quarter of them survives till 2 ps, and only 3% of ionic pairs reach 10 ps. Maximal lifetimes are ≈ 70 ps for neutral solution and ≈ 80 ps for alkaline one. The mean lifetime t_{UU} over the distribution $f_{UU}(t)$ is:

$$t_{UU} = \int_0^\infty t f_{UU}(t) dt \tag{17}$$

For neutral solution the mean lifetime is ≈ 10 ps, while in case of alkaline solution ≈ 12 ps. In other words, in neutral solution more uranyl complexes can form pairs, but in alkaline solution pairs live longer.

4 Summary

In this chapter molecular dynamics studies of an influence of ion charge on hydration structure and dynamic properties of highly charged cations in aqueous solutions are reviewed. A special attention is focused on the investigation of cation hydrolysis phenomena. Through this study the water molecules are described in the framework of the flexible non-constrained model CF1 [35, 36]. For a clear understanding of the influence of cation charge on the hydrated–hydrolyzed structure of cation we introduced a simple "toy" model of so-called primitive cation M^{Z+} in water, which in case of $Z = 1$ describes Na^+ ion in water [37]. It is shown that the increase of cation charge significantly decreases the size of the first hydration shell and stabilizes its octahedral arrangement.

The electrostatic repulsion between highly charged cation and protons of water molecules leads to a notable modification of intramolecular geometry of water molecules in hydration shell. Further increase of ion–water electrostatic interaction causes the loss of some protons from hydration shell. This is interpreted as the cation hydrolysis effect. Such treatment formally coincides with the concept of "aquo-acidity", which describes the cation hydrolysis as a stepwise sequence of proton removals from water molecules of cation hydration shell [1, 2]. As a result a hydrated aquo-ion form $[M(H_2O)_n]^{Z+}$ is transformed into hydrated–hydrolyzed form $[MO_nH_{2n-h}]^{(Z-Z^*h)+}$. Only one proton is lost from the hydration shell of cation M^{4+} and it has a typical hydroxo-aquo form $[M(H_2O)_6OH]^{(4-Z^*)+}$. The cations M^{5+}, M^{6+} and M^{7+} have an octahedral hydrated-hydrolyzed configuration but with different degrees of hydrolysis. In particular, six protons are lost from the hydration shell of cation M^{5+} and it is characterized by a hydroxo form $[M(OH)_6]^{(5-6Z^*)+}$. Eight protons are lost by the hydration shell of cation M^{6+} and it has a typical oxo-hydroxo form $[M(OH)_4O_2]^{(6-8Z^*)+}$. All 12 protons are lost from the hydration shell of cation M^{7+} and it has an oxo form $[MO_6]^{(7-12Z^*)+}$.

Stabilization of the cation hydration structure through the sequence M^{3+}, M^{4+}, M^{5+} leads to the decreasing of self-diffusion coefficients D_M and D_O. However for the cases of cations M^{6+} and M^{7+} due to formation of the oxo-hydroxo and oxo ionic forms the number of water molecules in the second hydration shell decreases significantly and self-diffusion coefficients D_M and D_O increase. We observed for cases of M^{4+} and higher charged cations that the spectral density of hindered translation motions of cations and oxygens in the first hydration shell have common characteristic frequencies in the high frequency region, which was interpreted as the presence of normal modes of highly stable hydrated–hydrolyzed complexes.

For correct description of cation hydrolysis the considered model is improved by taking into account charge redistribution effects between oxygens and hydrogens of hydrolyzed water molecules. A functional dependence of proton charge on proton-oxygen distance was chosen, which provides a monotonic change of hydrogen charge from CF1 value (when water molecule is not hydrolyzed) to the value $Z_H = 1$ (when proton leaves its "native" molecule). According to the electroneutrality condition the charge of oxygen is changed consistently. It is shown that the charge redistribution between hydrolysis products essentially modifies and stabilizes the hydrated–hydrolyzed structure of cation. In particular, the hydration shell of cation M^{5+} demonstrates two different types of configurations. First of them consists of two axial non-hydrolyzed water molecules and three OH^- groups in equatorial plane. The second configuration is characterized by the tetrahedral arrangement of four OH^- groups.

For real cations an increase of charge leads to a shortening of distances between cation and neighboring water molecules. Due to this a role of electrostatic ion-water interaction in the formation of hydrated–hydrolyzed structures of real cations is more important than for the primitive cation. This is demonstrated by a molecular dynamics study of Al^{3+} hydration. The results of this study show a strict octahedral arrangement of the aluminium hydration shell with tendency to cation hydrolysis.

Another realistic cation – anisotropic uranyl UO_2^{2+} ion – is considered. The simulations show that the uranyl hydration shell is characterized by a bipyramidal pentacoordinated arrangement of neighbors and a hydrolysis of one of the water molecules is observed. It is shown that the influence of pH level can modify the uranyl hydration structure from $UO_2^{2+}(H_2O)_5$ to $UO_2^{2+}(OH^-)_5$. In particular the hydrolysis reaction is inhibited under acidic conditions, while the alkaline conditions intensify it. Finally, the intercation correlations between uranyl ions are considered and the possibilities of a formation of polynuclear ions are discussed.

We have to mention here that the CF1 model, although was not specially designed for hydrolysis treatment, provided qualitatively correct results. In order to obtain good quantitative agreement with experimental data on the hydration shell of highly charged cations, the CF1 model must be revised towards adjusting the depth of potential wells responsible for confining oxygens and hydrogens in the flexible molecules to be in agreement with the hydrolysis energy data. It is hoped, that this kind of improvement of the potential model will be a step forward from the actual description of cation hydrolysis.

Acknowledgement The authors thank I.R. Yukhnovskii, K. Heinzinger, P. Bopp, H. Krienke, V. Vlachy, Yu.V. Kalyuzhnyi, and A.M. Gaspar for useful comments at different stages of this project. This work was supported by National Academy of Sciences of Ukraine in the framework of common Ukr–Rus scientific collaboration (project No.2-02-a).

References

1. C.F. Baes, R.E. Mesmer, *The Hydrolysis of Cations* (Wiley, New York, 1976)
2. J. Livage, M. Henry, C. Sanchez, Progr. Solid State Chem. **18**, 259 (1988)
3. M.F. Holovko, Condens. Matter Phys. **12**, 57 (1997)
4. W. Stumm, J.J. Morgan, *Aquatic Chemistry* (Wiley, New York, 1981)
5. Y. Marcus, *Ionic Solvation* (Wiley, New York, 1985)
6. H. Ohtaki, T. Radnai, Chem. Rev. **93**, 1157 (1993)
7. J.E. Enderby, Chem. Soc. Rev. **24**, 159 (1995)
8. K. Heinzinger, in *Computer Modelling of Fluids, Polymers and Solids*, edited by C.R.A. Catlow, S.C. Parker, M. Allen (Kluwer, Dordrecht, 1990), p. 357
9. K. Heinzinger, H. Schafer, Condens. Matter Phys. **2(18)**, 273 (1999)
10. H.L. Lemberg, F.H. Stillinger, J. Chem. Phys. **62**, 1677 (1975)
11. F.H. Stillinger, A. Rahman, J. Chem. Phys. **68**, 666 (1978)
12. F.H. Stillinger, C.W. David, J. Chem. Phys. **69**, 1474 (1978); **73**, 3384 (1980)
13. J.W. Halley, J.R. Rustad, A. Rahman, J. Chem. Phys. **98**, 4110 (1993)
14. E. Wasserman, J.R. Rustad, S. Xantheas. J. Chem. Phys. **106**, 9796 (1997)
15. D. Spangberg, K. Hermansson. J. Chem. Phys. **120**, 4829 (2004)
16. A.M. Mohammed, H.H. Loeffler, Y. Inada, K. Tanada, S. Funahashi. J. Mol. Liq. **119**, 55 (2005)
17. C.F. Schwenk, B.M. Rode, J. Chem. Phys. **119**, 9523 (2003)
18. T.S. Hofer, B.M. Rode, J. Chem. Phys. **121**, 6406 (2004)
19. J.M. Martinez, R.R. Pappalardo, E.S. Marcos. J. Chem. Phys. **109**, 1445 (1998)
20. J.M. Martinez, J. Hernandez-Cobos, H. Saint-Martin, R.P. Rappalardo, E.S. Marcos, J. Chem. Phys. **112**, 2339 (2000)
21. C. Kritayakornupong, J.I. Yague, B.M. Rode, J. Phys. Chem. **A106**, 10584 (2002)
22. V. Vchirawongkwin, T.S. Hofer, B.R. Randolf, B.M. Rode, J. Comp. Chem. **28**, 1006, 1057 (2007)
23. J.R. Rustad, B.P. Hay, J.W. Halley, J. Chem. Phys. **102**, 427 (1995)
24. R.L. Martin, P.J. Hay, L.R. Pratt, J. Phys. Chem. **A102**, 3565 (1998)
25. C.M. Chang, M.K. Wang, Chem. Phys. Lett. **286**, 46 (1998)
26. M.E. Tuckerman, D. Marx, M.L. Klein, M. Parrinello, Science **275**, 817 (1997)
27. D. Marx, M.E. Tuckerman, J. Hutter, M. Parrinello, Nature **397**, 601 (1999)
28. D. Wei, D.R. Salahub, J. Chem. Phys. **106**, 6086 (1997)
29. M.F. Holovko, Yu.V. Kalyuzhnyi, M.Yu. Druchok, J. Phys. Studies **4**, 100 (2000)
30. M.Yu. Druchok, T.M. Bryk, M.F. Holovko, J. Phys. Studies **7**, 402 (2003)
31. M. Holovko, M. Druchok, T. Bryk, Curr. Opin. Colloid Interface Science **9**, 64 (2004)
32. M. Holovko, M. Druchok, T. Bryk, J. Electroanal. Chem. **582**, 50 (2005)
33. M. Holovko, M. Druchok, T. Bryk, J. Chem. Phys. **123**, 154505 (2005)
34. M. Holovko, M. Druchok, T. Bryk, J. Mol. Liq. **131–132**, 65 (2007)
35. A. Nyberg, A.D.J. Haymet, in *Structure and Reactivity in Aqueous Solutions*, edited by D. Trular, C. Kramer, (American Chem. Soc., New York, 1994)
36. D.M. Duh, D.N. Perera, A.D.J. Haymet, J. Chem. Phys. **102**, 3736 (1995)
37. G. Jancso, K. Heinzinger, P. Bopp, Z. Naturforsch **A40**, 1235 (1985)

38. J. Barthel, H. Krienke, W. Kunz, *Physical Chemistry of Electrolyte Solutions: Modern Apsects* (Springer, New York, 1998)
39. M. Druchok, M. Holovko, T. Bryk, Condens. Matter Phys. **7**, 699 (2004)
40. M. Druchok, T. Bryk, M. Holovko, J. Mol. Liq. **120**, 11 (2005)
41. H. Kistenmacher, H. Popkie, E. Clementi, J. Chem. Phys. **58**, 1689 (1973), **58**, 5627 (1973), **59**, 5842 (1973)
42. K. Heinzinger, G. Palinkas, in *The Chemical Physics of Solvation*, edited by R.R. Dogonadze, E. Kalman, A.A. Kornyshev, J, Ulstrup, (Elsevier, Amsterdam, 1985), p. 313
43. P. Bopp, G. Jancso, K. Heinzinger, Chem. Phys. Lett. **98**, 129 (1983)
44. B.L. Trout, M. Parrinello, Chem. Phys. Lett. **288**, 343 (1998)
45. J.M. Hayle, *Molecular Dynamics Simulations: Elementary Methods* (Wiley, New York, 1992)
46. M.P. Allen, D.J. Tildesley, *Computer Simulation of Liquids* (Clarendon Press, Oxford, 1988)
47. T.P. Straasma, H.J. Berendsen, J. Chem. Phys. **89**, 5876 (1988)
48. J.W. Arthur, A.D.J. Haymet, J. Chem. Phys. **109**, 7991 (1998)
49. http://www.cse.clrc.ac.uk/msi/software/DL_POLY/
50. L.A. Curtis, J.W. Halley, J. Hautman, A. Rahman, J. Chem. Phys. **86**, 2319 (1987)
51. A.M. Gaspar, M. Alves-Marques, M.I. Cabaco, M.I. de Barros Marques, A.J. Kolesnikov, J. Tomkinson, J.C. Li, J. Phys.: Condens. Matter **16(36)**, 6343 (2004)
52. *The Hydrolysis of Aluminium: Conflicting Models and the Interpretation of Aluminium Geochemistry*, edited by Y.K. Kharka, A.S. Maest, (Balkema, Rotterdam, 1992)
53. E.J. Bylaska, M. Valiev, J.R. Rustad, J.H. Weare, J. Chem. Phys. **126**, 104505 (2007)
54. A. Bakker, K. Hermansson, J. Lindgren, M.M. Probst, P. Bopp, Int. J. Quant. Chem. **75**, 659 (1999)
55. A. Tongraar, K.R. Liedl, B.M. Rode, J. Phys. Chem. A **102**, 10340 (1998)
56. J.R.C. van der Maaren, J. de Bleijser Chem. Phys. Lett. **141**, 251 (1987)
57. J.F. Ahearne, Phys. Today **50**, 27 (1997)
58. S. Spencer, L. Gagliardi, N.C. Handy, A.G. Ioannou, C.-K. Skylaris, A. Willetts, A.M. Simper, J. Phys. Chem. A **103**, 1831 (1999)
59. S. Tsushima, A. Suzuki, J. Mol. Struct. **529**, 21 (2000)
60. C.X. Liu, J.M. Zachara, O. Qafoku, J.P. McKinley, S.M. Headd, Z.W. Wang, Geochim. Cosmochim. Acta **68**, 4519 (2004)
61. P. Guilbaud, G. Wipff, J. Phys. Chem. **97**, 5685 (1993)
62. P. Guilbaud, G. Wipff, J. Mol. Struct. **366**, 55 (1996)
63. J.A. Greathouse, R.J. O'Brien, G. Bemis, R.T. Pabalan, J. Phys. Chem. B **106**, 1646 (2002)
64. P. Nichols, E.J. Bylaska, G.K. Schenter, W. de Jong, J. Chem. Phys. **128**, 124507 (2008)
65. P.G. Allen, J.J. Bucher, D.K. Shuh, N.M. Edelstein, T. Reich, Inorg. Chem. **36**, 4676 (1997)
66. A.J. Dent, J.D.F. Ramsay, S.W. Swanton, J. Colloid Interface Sci. **150**, 45 (1992)
67. L. Soderholm, S. Skanthakumar, J. Neuefeind, Anal. Bioanal. Chem. **383**, 48 (2005)
68. T. Ikeda, M. Hirata, T. Kimura, J. Chem. Phys. **124**, 074503 (2006)
69. I.V. Stasyuk, O.V. Velychko, Phys. Chem. Liq. **38**, 743 (2000)
70. M.S. Wertheim, J. Stat. Phys. **42**, 459, 477 (1986)
71. M.S. Wertheim, J. Chem. Phys. **87**, 7323 (1987)
72. Yu.V. Kalyuzhnyi, G. Stell, M.F. Holovko, Chem. Phys. Lett. **235**, 335 (1995)
73. J.D.F. Ramsay, in *Water and Aqueous Solutions*, edited by G.N. Neilson, J.E. Enderby, (The Colston Research Society, Bristol, 1986), p. 207

Davydov's Solitons in DNA

Victor D. Lakhno

Abstract Charge transfer in homogeneous nucleotide chains is modeled on the basis of Holstein Hamiltonian. The path length of Davydov solitons in these chains is being studied. It is shown that in a dispersionless case, when the soliton velocity V is small, the path length grows exponentially as V decreases. In this case the state of a moving soliton is quasisteady. In the presence of dispersion determined by the dependence $\Omega^2 = \Omega_0^2 + V_0^2 \kappa^2$ the path length in the region $0 < V < V_0$ is equal to infinity. In this case the phonon environment follows the charge motion. In the region $V > V_0$ the soliton motion is accompanied by emission of phonons which leads to a finite path length of a soliton. The latter tends to infinity as $V \to V_0 + 0$ and $V \to \infty$. The presence of dissipation leads to a finite soliton path length.

An equilibrium velocity of soliton in an external electric field is calculated. It is shown that there is a maximum intensity of an electric field at which a steady motion of a soliton is possible. The soliton mobility is calculated for the stable or ohmic brunch.

Keywords Phonon emission · Landau–Pekar polaron · Quasisteady states · Dispersion

1 Introduction

An idea of modern molecular nanoelectronics is to use molecular chains as wires [1–4]. This highlights the problem of conducting properties of such molecular wires. In the pioneering works by Davydov [5–10] it was shown that the main carriers of excitation in molecular wires are solitons (or polarons). Though to date an extensive literature has been accumulated on the subject, it is a gross exaggeration to say that everything is absolutely clear now.

V.D. Lakhno
Institute of Mathematical Problems of Biology, RAS, 142290, Pushchino, Institutskaya str., 4, Russia
e-mail: lak@impb.psn.ru

N. Russo et al. (eds.), *Self-Organization of Molecular Systems: From Molecules and Clusters to Nanotubes and Proteins*, NATO Science for Peace and Security Series A: Chemistry and Biology, © Springer Science+Business Media B.V. 2009

Let us consider a molecular chain consisting of biatomic molecules in which atoms can displace from their equilibrium positions. This approximation is known as a Holstein model [11] which represents one of the simplest models of a deformable chain. It is widely used in describing charge transfer in DNA where the role of biatomic molecules is played by Watson–Crick nucleotide pairs [11–15]. Notwithstanding its simplicity the model does not have an analytical solution.

For the first time, the points to be clarified in the context of this problem were brought to notice by Landau [16]. According to Landau, even at zero temperature, an electron moving over such a chain can cause the chain's deformation which can catch the electron to form a localized state. In Pekar's works this localized state was called a polaron [17]. Polaron states have been studied for more than half a century and still remain to be understood.

One of the questions to be answered was that of whether a polaron state can move over the chain. It would seem, an unequivocal and principal answer was given in the paper by Landau and Pekar [18]. According to Landau and Pekar, the case of a deformable chain differs from the case of a rigid one in that in the former case a polaron moves as a whole while in the latter one an electron does.

This conclusion was doubted by Davydov and Enolskii [19–21]. They paid attention to the fact that in the case of optical phonons their group velocity is equal to zero and, in the absence of dispersion, the deformation induced by the electron cannot follow it. According to Davydov and Enolskii, a steady motion of solitons over a chain is possible only if the chain dispersion is taken into account.

From these considerations, the authors of [22–25] concluded that in a dispersionless case, any arbitrarily weak interaction such as, for example, an external electric field, which would force the electron to move, would lead to its destruction and transformation into Bloch state.

The conclusion of the instability of polaron states in response to weak external influences is in contradiction with the fact that a polaron state is energetically more advantageous than delocalized Bloch states.

As will be shown in this work, this contradiction results from the fact that the state of a uniformly moving polaron, considered by Landau and Pekar is, actually, not the eigen state of the initial Hamiltonian. When moving along the chain, the electron will inevitably actuate the chain's atoms, leaving behind a "tail" of oscillating molecules. This will lead not to destruction of the polaron, but to its stopping. Therefore the question arises of how suitable is the concept of moving polaron states in describing the transfer processes in polynucleotide chains.

The paper is arranged as follows.

In Section 2 we introduce a semiclassical Holstein model for a molecular chain. An expression is obtained for the Green function of the classical equation of the chain motion with regard for dissipation. In the absence of dissipation it transforms into the equation obtained by Davydov and Enolskii in [19].

In Section 3 we show that in a molecular chain with dispersionless phonons an excess electron can be in a "quasisteady" moving soliton state, the electron path length tending to infinity exponentially as the soliton velocity tends to zero. This fact justifies the idea of the possibility of "quasisteady" states of solitons in the case of small velocities. It is shown that when the soliton moves at a finite velocity, its stopping is caused by the oscillation excitations behind it.

In Section 4 we show that when dispersion is taken into account, the picture considered changes qualitatively. According to Davydov's results, in this case as $V < V_0$, the soliton does not emit phonons and represents the eigen state of the Hamiltonian. The soliton path length then goes into infinity.

In Section 5 we consider the soliton motion in the presence of dissipation. An expression is obtained for the soliton mobility in the absence of dispersion at zero temperature.

In Section 6 we investigate the steady motion of a soliton in an electric field. It is shown that there exists a maximum value of the electric field intensity at which the steady motion is possible. A general expression is obtained for the dependence of the soliton equilibrium velocity on the field intensity. The dependence contains two branches, one of which is stable ohmic, and the other is unstable nonohmic.

In Section 7 we obtain an expression for the mobility of moving soliton with regard for dispersion for $T = 0$.

In Section 8 we present a general scheme for calculation of the soliton mobility for $T \neq 0$.

In Section 9 we apply the obtained results for homogeneous polynucleotide chains.

In Section 10 we discuss the results obtained.

2 Motion Equations of Davydov's Soliton in the Continuum Approximation

Following [19], let us write down Hamiltonian H describing the electron motion in a molecular chain in the continuum approximation as:

$$H = H_e + H_{int} + H_{ph},\tag{1}$$

$$H_e = -\frac{1}{a}\int \Psi^*(x,t)\frac{\hbar^2}{2m}\frac{\partial^2}{\partial x^2}\Psi(x,t)\,dx,\tag{2}$$

$$H_{int} = \frac{\chi}{a}\int |\Psi(x,t)|^2 u(x,t)\,dx,\tag{3}$$

$$H_{ph} = \frac{M}{2a}\int \left[\left(\frac{\partial u(x,t)}{\partial t}\right)^2 + \Omega_o^2 u^2(x,t) + V_0^2\left(\frac{\partial u(x,t)}{\partial x}\right)^2\right]dx.\tag{4}$$

In Eqs. (1)–(4) the energy is reckoned from the bottom of the conductivity band, $\Psi(x,t)$ normalized electron wave function:

$$\frac{1}{a}\int |\Psi(x,t)|^2\,dx = 1\tag{5}$$

where a is the lattice constant. Unless otherwise specified, integration in Eqs. (1)–(5) and all the ensuing expressions is carried out in the infinite limit. The quantity m is the electron effective mass, χ is the constant for the electron interaction with the chain displacements $u(x,t)$, M is the reduced mass of an elementary cell, Ω_0 is the frequency of intramolecular oscillations of the chain, V_0 has the meaning of the minimum phase, or maximum group velocity of the chain oscillations. Notice that Hamiltonian $\hat{H}$ is a continuum analog of the discrete model descriptive of an excess electron inserted in a homogeneous polynucleotide chain [12–15].

Motion equations corresponding to Hamiltonian (Eqs. (1)–(4)) have the form:

$$\frac{\partial^2 u}{\partial t^2} + \Omega' \frac{\partial u}{\partial t} + \Omega_0^2 u - V_0^2 \frac{\partial^2 u}{\partial x^2} + \frac{\chi}{M} |\Psi|^2 = 0, \tag{6}$$

$$i\hbar \frac{\partial \Psi}{\partial t} + \frac{\hbar^2}{2m} \frac{\partial^2 \Psi}{\partial x^2} - \chi u \Psi = 0, \tag{7}$$

The term $\Omega' \partial u / \partial t$ in the left-hand side of Eq. (6) describes oscillations attenuation due to friction, Ω' is the frequency of the oscillations damping.

For the steady solutions of the form:

$$u(x,t) = u(\xi), \qquad \Psi(x,t) = \varphi(\xi) \exp \frac{i}{\hbar} \left[mVx - \left(W + \tfrac{1}{2} mV^2 \right) t \right],$$
$$\xi = (x - Vt)/a, \tag{8}$$

Eqs. (6) and (7) are written as:

$$-\varepsilon \Omega_0^2 \frac{d^2 u}{d\xi^2} - 2\Omega_0^2 \delta \frac{du}{d\xi} + \Omega_0^2 u + \frac{\chi}{M} |\varphi|^2 = 0, \tag{9}$$

$$-\frac{\hbar^2}{2ma^2} \frac{d^2 \varphi}{d\xi^2} + \chi u \varphi = W \varphi, \tag{10}$$

where:

$$\delta = \frac{\Omega' V}{2\Omega_0^2 a}, \qquad \varepsilon = \left(V_0^2 - V^2 \right) / a^2 \Omega_0^2 \tag{11}$$

Equation (9) is solved with the use of the Fourier transform. As a result $u(\xi)$ will take the form:

$$u(\xi) = -\frac{\chi}{M\Omega_0^2} \int d\xi' \omega(\xi' - \xi) |\varphi(\xi')|^2 \tag{12}$$

$$\omega(\xi) = \frac{1}{2\pi} \int dq \frac{e^{iq\xi}}{\varepsilon q^2 + 2i\delta q + 1} \tag{13}$$

Calculation of Eq. (13) yields:

$$\omega(\xi) = \frac{\Theta(\xi)}{\sqrt{|\varepsilon| - \delta^2}} e^{-\frac{\delta}{|\varepsilon|}\xi} \sin \frac{\xi}{|\varepsilon|} \sqrt{|\varepsilon| - \delta^2} \tag{14}$$

if $\varepsilon < -\delta^2$;

$$\omega(\xi) = \frac{2\pi\Theta(\xi)}{\sqrt{\delta^2 - |\varepsilon|}} e^{-\frac{\delta}{|\varepsilon|}\xi} \mathrm{ch}\left(\frac{\xi}{|\varepsilon|}\sqrt{\delta^2 - |\varepsilon|}\right) \tag{15}$$

if $-\delta^2 < \varepsilon < 0$;

$$\omega(\xi) = \delta(\xi) + i\Theta(\xi)e^{-\xi/2\delta} \tag{16}$$

if $\varepsilon = 0$;

$$\omega(\xi) = \frac{1}{2\sqrt{\delta^2 + |\varepsilon|}}\left\{\Theta(\xi)e^{-\frac{\xi}{|\varepsilon|}\left(\sqrt{\delta^2 + |\varepsilon|} - \delta\right)} + \Theta(-\xi)e^{\frac{\xi}{|\varepsilon|}\left(\sqrt{\delta^2 + |\varepsilon|} + \delta\right)}\right\} \tag{17}$$

if $\varepsilon > 0$, where $\delta(\xi)$ is the Dirac δ-function, $\Theta(\xi) = 1$, if $\xi > 0$, $\Theta(\xi) = 0$, if $\xi < 0$. For $\delta = 0$, i.e. in the absence of dissipation, these expressions go over into those obtained in [19]. The sign of the argument of Θ-function in Eq. (14) differs from the sign of the corresponding argument in [19] due to the fact that as $\varepsilon < 0$ the complex summand in the denominator of integrand (Eq. (13)) has a different sign.

3 Soliton Path Length in the Absence of Dispersion and Dissipation

The authors of [19–21] reasoned that in a molecular chain without dispersion a steady motion of a soliton formed by an excess electron is impossible. The arguments were based on the fact that phonons obeying the dispersion law $\Omega^2 = \Omega_0^2 + V_0^2 k^2$ have a zero group velocity as $V_0 = 0$ and therefore the phonon environment cannot follow the soliton motion.

Of interest here is to find the length of a path that a soliton having at the initial moment the velocity V travels until it finally stops. In this section we will deal with the case of the absence of dissipation ($\delta = 0$) and absence of dispersion ($V_0 = 0$).

Using Eq. (14) we express $\omega(\xi)$ as:

$$\omega(\xi) = \frac{\Theta(\xi)}{\sqrt{|\varepsilon|}}\sin\frac{\xi}{\sqrt{|\varepsilon|}}, \qquad \varepsilon < 0 \tag{18}$$

As $\varepsilon = 0$, Eq. (10) is reduced to a stationary nonlinear Schrödinger equation, the normalized solution of which has the form:

$$\varphi(\xi) = \frac{1}{\sqrt{2r}}\mathrm{ch}^{-1}(\xi/r), \qquad r = 4M(\hbar\Omega_0)^2/m\chi^2 a^2 \tag{19}$$

In the absence of dispersion and small soliton velocity, when $|\varepsilon| \ll 1$, from Eqs. (12) and (18) we get:

$$u(\xi) = c \sin\left(\xi / \sqrt{|\varepsilon|}\right), \qquad \xi < c_1 r \tag{20}$$

$$c = -\pi r \chi / 2M\Omega_0^2 a \, |\varepsilon| \, \text{sh}\left(\pi r / 2\sqrt{|\varepsilon|}\right) \tag{21}$$

where c_1 is a constant of the order of 1.

From Eq. (20) follows that for $\xi = (x - Vt)/a$ a soliton moving along a path in positive direction leaves behind a "tail" of the chain oscillations loosing on the way its kinetic energy $E_{kin} = m^{**} V^2/2$ where

$$m^{**} = \frac{1}{30} \frac{m^3 a^4}{\hbar^6} \frac{\chi^8}{\Omega_0^{10} M^4} \tag{22}$$

is the soliton effective mass, until it finally stops.

The distance L which a soliton will travel is found from the condition:

$$E_{ph} = E_{kin} \tag{23}$$

$$E_{ph} = \frac{M}{2} \int\limits_{-L/a}^{c_1 r} \left[\frac{(V^2 + V_0^2)}{a^2} \left(\frac{du}{d\xi}\right)^2 + \Omega_0^2 u^2 \right] d\xi \tag{24}$$

Substituting expression (20) into Eq. (24) and using Eq. (3) for small ε we express the soliton path length as:

$$\frac{L}{a} = \frac{1}{120\pi^2} \frac{m^5 \chi^{10} a^4 V^6}{M^5 \Omega_0^{16} \hbar^{10}} \text{sh}^2\left(\frac{\pi r a \Omega_0}{2V}\right) \tag{25}$$

The quantity L, thus defined, is the distance at which soliton velocity changes significantly.

So, in the absence of dispersion, as $V \to 0$, the soliton path length L/a tends to infinity $\sim \exp(\pi r a \Omega_0 / V)$ This fact justifies the idea of the possibility of "quasisteady" states of solitons and polarons at small V, for which all the calculations of the effective mass were made.

The physical reason why the emission of phonons by a moving soliton is exponentially small is that, according to Eq. (13), in the absence of dissipation, contribution into the emission is made by q satisfying the Cherenkov condition:

$$qV = \Omega_0 a$$

As $V \to 0$ this condition is fulfilled for $q \to \infty$. For a localized state, in this limit case, the Fourier component of the electron density is exponentially small. So, in this limit case only an exponentially small portion of the electron distribution will take part in the phonon emission.

4 Soliton Path Length in the Presence of Dispersion and Absence of Dissipation

To have the general pattern of the dependence of the soliton path length on the velocity V in the case on nonzero dispersion ($V_0 \neq 0$) let us consider some limit cases.

$\varepsilon < 0$. When $\varepsilon < 0$, the distribution of displacements over the chain is determined by Eq. (18). Since in this case the soliton velocity exceeds the maximum group velocity V_0, the soliton leaves behind a "tail" of the chain oscillations which leads to a finite path length of the soliton. For $(V - V_0)/V_0 \ll 1$, the path length will be:

$$ L/_a \sim \exp \frac{\pi\, r\, a\, \Omega_0}{\sqrt{V^2 - V_0^2}} \tag{26} $$

As $V_0 = 0$ expression (26) transforms into Eq. (25).

In another limit case of large velocities ($\pi\, r\, a\, \Omega_0/2V \ll 1$), when

$$ V \gg 2\pi \frac{M}{m} (\hbar\Omega_0)^2 \frac{\Omega_0}{\chi^2 a} > V_0 \tag{27} $$

the soliton path length is written (Eqs. (20)–(24)) as:

$$ L/_a = \frac{1}{30} \left(\frac{m}{M}\right)^3 \frac{a^2 \chi^6 V^4}{\Omega_0^{10} \hbar^6} \tag{28} $$

So, at large V the path length grows as $\sim V^4$. The physical reason is that in the case of $V \to \infty$ from Cherenkov condition it follows that $q \to 0$. As $q \to 0$ a portion of electron distribution taking part in the phonon emission is independent of V. Accordingly, the soliton path length will tend to infinity as $V \to \infty$.

$\varepsilon > 0$. When $\varepsilon > 0$, the soliton velocity is less than the maximum group velocity of the chain V_0. Using Eqs. (17) and (12) we express $u(\xi)$ as:

$$ u(\xi) = -\frac{1}{2\sqrt{|\varepsilon|}} \frac{\chi}{M\Omega_0^2} \int d\xi' e^{-|\xi'-\xi|/\sqrt{|\varepsilon|}} |\varphi(\xi')|^2 \tag{29} $$

From Eq. (29) follows that in a steady case the displacements propagate both behind the moving soliton and in front of it, dying out exponentially with distance from the soliton. In this case the quantity E_{ph} has a finite value even if the limits of integration in integral of Eq. (24) are infinite and corresponds to the energy of the phonon environment accompanying the charge motion along the chain. Then the path length will be infinite.

5 Soliton Path Length in the Presence of Dissipation—Soliton Mobility

In the presence of dissipation the soliton path length always has a finite value. As $\delta \neq 0$ the loss of energy by a soliton dH/dt due to dissipation is determined by the balance energy equation:

$$\frac{dH}{dt} = -2F, \qquad F = \frac{\gamma}{2a} \int \left(\frac{\partial u}{\partial t}\right)^2 dx, \tag{30}$$

where F is a dissipative function, $\gamma = M\Omega'$ is a friction coefficient. From Eqs. (8) and (30) follows:

$$\frac{dH}{dt} = -\frac{M\Omega'}{a^2} V^2 \int \left(\frac{du}{d\xi}\right)^2 d\xi \tag{31}$$

When the soliton moves at a small velocity, its energy can be presented as [19]:

$$H = H_0 + \frac{1}{2} m^{**} V^2, \tag{32}$$

$$m^{**} = \frac{M}{a^2} \int \left(\frac{du}{d\xi}\right)^2 d\xi, \tag{33}$$

where m^{**} is the soliton effective mass. With the use of Eqs. (32) and (33) balance Eq. (31) takes the form:

$$\frac{dV^2}{dt} = -2\Omega'V^2 \tag{34}$$

From Eq. (34) follows the expression for the soliton velocity:

$$V(t) = V(0) e^{-\Omega' t} \tag{35}$$

where $V(0)$ is the soliton initial velocity. From Eq. (35), the soliton relaxation time will be $\tau_r = 1/\Omega'$.

Using Eq. (35) we express the soliton path length as:

$$L = \int_0^\infty V(t) \, dt = V(0) / \Omega' \tag{36}$$

Now let us discuss the motion of a soliton at a small velocity in an external electric field E. Given an electric field, in the steady case the loss of energy due to dissipation ($\dot{H} = -2F$) will be counterbalanced by a gain in energy obtained by the soliton from the external field ($\dot{H} = eEV$). As a result, the balance energy will take the form:

$$eEV = \Omega' m^{**} V^2 \tag{37}$$

Whence the soliton mobility μ is:

$$\mu = e\tau_r/m^{**} \tag{38}$$

This is the ordinary expression for the electron mobility where effective mass of soliton stands in place of the effective mass of electron and the relaxation time is $\tau_r = 1/\Omega'$. We will turn back to this point in Sections 7 and 8.

Notice that according to Eq. (38), the soliton mobility is determined only by dissipation which leads to stopping of the soliton. According to the analysis carried out above, the contribution of phonon emission into the mobility vanishes in the limit considered $V \rightarrow 0$. Below we will take the case of a finite soliton velocity when the contribution of emission into its value plays a decisive role.

6 Motion in an Electric Field

To calculate the equilibrium velocity of a soliton in an external electric field of intensity E we will proceed from the energy dissipation expression (30). In the steady case the rate of energy dissipation is counterbalanced by the energy gained by the particle moving in an electric field E at the velocity V:

$$-\frac{dH}{dt} = eEV \tag{39}$$

Expanding $|\varphi(\xi)|^2$ into Fourier series:

$$|\varphi(\xi)|^2 = N^{-1} \sum_k e^{ik\xi} C_k \tag{40}$$

and using relations (12)–(14) and (39)–(40), we express the energy dissipation rate as:

$$\frac{dH}{dt} = \frac{\gamma}{2\pi a^2} \frac{\chi^2}{(M\Omega_0^2)^2} \int \frac{k^2 V^2 C_k C_{-k}}{(\varepsilon k^2 + 2i\,\delta k + 1)(\varepsilon k^2 - 2i\,\delta k + 1)} dk \tag{41}$$

From Eqs. (19) and (38) we get:

$$C_k = \frac{\pi}{2} \frac{kr}{\mathrm{sh}\,(\pi kr/2)} \tag{42}$$

Using relations (37), (41), and (42) we obtain the following general expressions for the dependence of the soliton velocity on the electric field intensity:

(1) when $\omega' < 2\omega$, $v > v_0/\sqrt{1 - (\omega'/2\omega)^2}$:

$$E = 2\omega^2\omega' \frac{\eta v}{\left(v_0^2 - v^2\right)^2} \int_0^\infty \frac{x^4 \mathrm{sh}^{-2}x}{\left(x^2 + c_1\right)^2 + c_2^2} dx \qquad (43)$$

$$c_1 = 4\pi^2 \frac{\eta^2}{\kappa^2} \frac{\omega^2}{\left(v_0^2 - v^2\right)} \left(1 + \frac{\omega'^2}{2\omega^2} \frac{v^2}{\left(v_0^2 - v^2\right)}\right),$$

$$c_2^2 = \left(\frac{2\pi\eta}{\kappa v}\right)^4 \frac{v^6 \omega'^2}{\left(v^2 - v_0^2\right)^3} \left(\omega^2 + \left(\frac{\omega'}{2}\right)^2 \frac{v^2}{\left(v_0^2 - v^2\right)}\right);$$

(2) when $\omega' < 2\omega$, $v < v_0/\sqrt{1 - (\omega'/2\omega)^2}$:

$$E = 2\omega^2\omega' \frac{\eta v}{\left(v_0^2 - v^2\right)^2} \int_0^\infty \frac{x^4 \mathrm{sh}^{-2}x}{\left(x^2 + c_1^2\right)\left(x^2 + c_2^2\right)} dx \qquad (44)$$

$$c_1^2 = \left(\frac{2\pi\eta}{\kappa}\right)^2 \frac{\omega^2}{v_0^2 - v^2} \left[1 + \frac{\omega'^2}{2\omega^2} \frac{v^2}{v_0^2 - v^2} + \frac{v\omega'/\omega^2}{\sqrt{v_0^2 - v^2}} \sqrt{\omega^2 + \left(\frac{\omega'}{2}\right)^2 \frac{v^2}{v_0^2 - v^2}}\right]$$

$$c_2^2 = \left(\frac{2\pi\eta}{\kappa}\right)^2 \frac{\omega^2}{v_0^2 - v^2} \left[1 + \frac{\omega'^2}{2\omega^2} \frac{v^2}{v_0^2 - v^2} - \frac{v\omega'/\omega^2}{\sqrt{v_0^2 - v^2}} \sqrt{\omega^2 + \left(\frac{\omega'}{2}\right)^2 \frac{v^2}{v_0^2 - v^2}}\right]$$

(3) when $\omega' > 2\omega$: for any v the dependence $E(v)$ is determined by the formulae of
  case (2).

In expressions (43) and (44) use is made of dimensionless quantities: E, η, ω, ω', κ, v, which are related to the dimensional quantities as:

$$E = \mathrm{E}\frac{ea\tau}{\hbar}, \qquad \eta = \frac{\hbar\tau}{2ma^2}, \qquad \omega = \Omega_0\tau, \qquad \omega' = \Omega'\tau,$$

$$\kappa\omega^2 = \tau^3\chi^2/M\hbar, \qquad v = V\tau/a, \qquad v_0 = V_0\tau/a, \qquad (45)$$

where τ is an arbitrary time scale.

   Recall that expressions (43) and (44) are not the exact solution of the problem (Eqs. (9) and (10)) for all the parameter values. The reason is that when deriving Eqs. (43) and (44) we used approximate expression (19) and its corresponding expression (42). Calculation experiments demonstrate that expression (19) is still a good approximation even at $r\sim1$, when the chain discreteness is essential [26]. Up to rather high velocities, the soliton shape turns out to be close to that described by Eq. (19). So, we may hope that relations (43) and (44) are valid within a wide range of the model parameters variation. At the same time it should be noted that Eqs. (43) and (44) are invalid, for example, in the limit case $\Omega_0 = 0$ which corresponds to the limit $|\varepsilon| \to \infty$. It is easy to show that any localized soliton states are lacking in this case (acoustic phonon spectrum) and the solution of Eqs. (10) and (29) will be delocalized electron states. Hence, taking account of dispersion for $\Omega_0 \neq 0$ leads to enlargement of the soliton, i.e. deviation of its shape from that given by Eq. (19). Notice that as $v = v_0$, nothing unusual peculiarity arises, in particular, the electric field intensity necessary to impart this velocity to the soliton, according to Eqs. (43) and (44), is equal to:

$$E = \frac{\pi^2}{60} \frac{\kappa^4 v_0^2 \omega'}{\omega^2 \eta^3}$$

In the limit case $v_0 = 0$, i.e. when dispersion is absent, for cases (1) and (3) we get from Eqs. (43) and (44) the following expressions:

$$E = \kappa \omega^2 \omega' v I \tag{46}$$

$$I = \frac{2\eta}{\kappa v^4} \int_0^\infty \frac{x^4/\mathrm{sh}^2 x}{\left(x^2 + c_1\right)^2 + c_2^2} dx \tag{47}$$

$$c_1 = \left(\frac{2\pi\eta}{\kappa v}\right)^2 \left(\frac{\omega'^2}{2} - \omega^2\right),$$

$$c_2^2 = \left(\frac{2\pi\eta}{\kappa v}\right)^4 \omega'^2 \left(\omega^2 - \left(\frac{\omega'}{2}\right)^2\right),$$

for $\omega > \omega'/2$. In the case of a damped motion, when friction is large, we get the following expression of I:

$$I = \frac{2\eta}{\kappa v^4} \int_0^\infty \frac{x^4/\mathrm{sh}^2 x}{\left(x^2 + c_1^2\right)\left(x^2 + c_2^2\right)} dx \tag{48}$$

$$c_1 = \frac{2\pi\eta}{\kappa v} \left(\frac{\omega'}{2} - \sqrt{\left(\frac{\omega'}{2}\right)^2 - \omega^2}\right),$$

$$c_2 = \frac{2\pi\eta}{\kappa v} \left(\frac{\omega'^2}{2} + \sqrt{\left(\frac{\omega'}{2}\right)^2 - \omega^2}\right),$$

for $\omega' > 2\omega$.

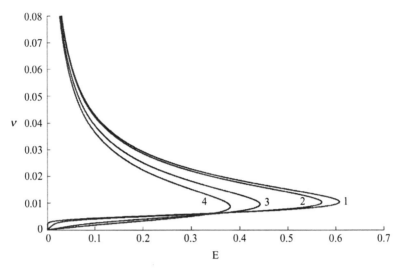

Fig. 1 The dependence of soliton velocity v on electric field E for different values of parameter ω': (1)$\omega' = 0$; (2)$\omega' = 0.001$; (3)$\omega' = 0.006$(4)$\omega' = 0.01$; ($\kappa = 4$, $\eta = 1.276$, $\omega = 0.01$)

Let us consider the limit case of the lack of friction $\omega' = 0$. From Eqs. (46) and (47) it follows:

$$E = 2\pi^2 \frac{\omega^2 \eta^2}{\kappa} \frac{1}{v^4} \frac{1}{\text{sh}^2 (2\pi\eta\omega/\kappa v)} \qquad (49)$$

So, according to Eq. (49), even in the absence of friction, for a soliton to move steadily along the chain, a nonzero field E should be applied. This result is in full agreement with the analysis carried out above. As $\omega' = 0$, the work of an electric field is necessary for a moving soliton to induce oscillations in the chain.

Figure 1 shows dependencies of the soliton velocity on the intensity of the electric field E determined by Eqs. (46)–(48) for various values of the parameter ω' in the absence of dispersion.

The limit case of $\omega' = 0$, determined by Eq. (49) corresponds to curve 1 in Fig. 1 and the limit case of $\omega' = \infty$ corresponds to the ordinate axis. As is seen from Fig. 1, a steady motion of a soliton is possible only within the interval $0 < E < E_{max}(\omega')$. On each curve, the branch for which $dv/dE = v'_E > 0$ corresponds to a steady motion of a soliton, while the branch for which $v'_E < 0$ corresponds to its unsteady motion.

From Eqs. (46)–(48) it follows that the occurrence of friction ($\omega' \neq 0$) gives rise to an ohmic branch, i.e. a linear dependence of $v(E)$ (for small E) on the stable regions of $v(E)$ curves.

7   Soliton Mobility for T = 0

At zero temperature and $v_0 \neq 0$, (Eq. (44)) yields the following expression for the soliton mobility:

$$\mu = \frac{ea^2}{\hbar} \frac{v_0^4}{2\omega^2\omega'\eta I_b} \tag{50}$$

where

$$I_b = \int\limits_0^\infty \frac{x^4/sh^2 x}{(x^2 + b^2)^2} dx, \qquad b = \frac{2\pi\eta\omega}{\kappa v_0} \tag{51}$$

From Eqs. (50) and (51) it follows that in the limit $v_0 \to \infty$:

$$\mu = 4\frac{ea^2}{\hbar} \frac{v_0^3}{\omega\omega'\kappa} \tag{52}$$

As $v_0 = 0$, from Eq. (44) we get the following expression for the soliton mobility:

$$\mu = 240\frac{ea^2}{\hbar} \frac{\eta^3\omega^2}{\kappa^4\omega'} \tag{53}$$

which coincides with earlier obtained expression (38).

Figure 2 shows in greater detail linear portions of the $v(E)$ dependencies corresponding to curves in Fig. 1.

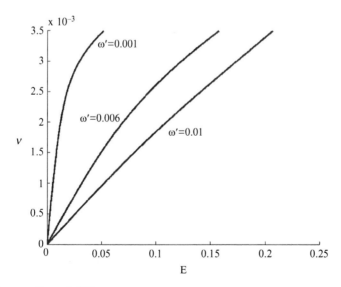

Fig. 2   Linear portions of $v(E)$ curves

Figure 2 suggests that the portions of linear dependence lengthen as ω' grows. For $\omega' \geq 0,1$ the regions of $v(E)$ linear dependence spread up to the critical value of the field intensity at which a steady motion of a soliton becomes impossible. In this case Ohm's law holds up to very high values of the field intensity.

To understand the meaning of expressions (50)–(53) for the soliton mobility let us calculate the soliton effective mass m^{**} with the use of Eq. (33). When from $v_0 \neq 0$ (Eq. (33)) we get:

$$m^{**}/m = \left(1/120\right)\left(\kappa^4/\eta^2\omega^2\right) \tag{54}$$

Substitution of Eq. (54) into Eq. (38) yields expression (53). So, the soliton mobility is described by an ordinary equation in which the effective mass is replaced by the soliton mass and the relaxation time—by the quantity $1/\Omega'$. Hence, for $V_0 = 0$, the soliton effective mass is equal to:

$$m^{**}/m = 4\eta^2\omega^2 I_b/v_0^4 \tag{55}$$

8 Soliton Mobility at Finite T

To calculate the soliton mobility at finite temperatures let us find an average length of the soliton path in the presence of dissipation with due regard for phonon emission. Let us proceed from the balance equation

$$\frac{d\mathrm{H}}{dt} = \frac{d}{dt}\left(m^{**}V^2/2\right) \tag{56}$$

Writing down $d\mathrm{H}/dt$ in the form of Eq. (41) we will express the time dependence of the soliton velocity v, providing $\omega' < 2\omega$, $v > v_0/\sqrt{1 - (\omega'/2\omega)^2}$, in the form:

$$\frac{dv^2}{dt} = -2A\omega^2\omega'\frac{\eta v^2}{(v_0 - v^2)^2}\int\limits_0^\infty \frac{x^4\mathrm{sh}^{-2}x}{\left(x^2 + c_1\right)^2 + c_2^2}dx \tag{57}$$

where $A = 2\hbar\tau/m^2a^2$, and quantities c_1, c_2 are determined by relations (43).

In the case of $\omega' < 2\omega$, $v < v_0/\sqrt{1 - (\omega'/2\omega)^2}$, the time dependence of the soliton velocity is written as:

$$\frac{dv^2}{dt} = -2A\omega^2\omega'\frac{\eta v^2}{(v_0 - v^2)^2}\int\limits_0^\infty \frac{x^4\mathrm{sh}^{-2}x}{\left(x^2 + c_1\right)\left(x^2 + c_2^2\right)}dx \tag{58}$$

where c_1 and c_2 are determined by formulae (44).

As $\omega' < 2\omega$, for any V the time dependence of the soliton velocity is determined by Eq. (56).

Equations (57) and (58) determine the dependence $v(t, v(0))$ for a given initial velocity of a soliton $v(0)$. Then for an average length of the path $\bar{L}(T)$, at finite temperature T, we get:

$$\bar{L}(T) = \int L(v(0)) f_M(v(0)) dv(0) \tag{59}$$

where

$$L(v(0)) = \int_0^\infty V(t, v(0)) dt \tag{60}$$

f_M is a Maxwell distribution of solitons over initial velocities $v(0)$.

The temperature dependence of the mobility is given by the expression

$$\mu(T) = \frac{e\bar{\tau}(T)}{m^{**}}, \qquad \bar{\tau}(T) = \frac{\bar{L}(T)}{\bar{V}(T)}, \tag{61}$$

where $\bar{V}(T)$ is the mean thermal velocity of a soliton.

By way of illustration let us consider the case of low temperatures, when $\bar{V}(T) \ll V_0$. In this case the quantities c_1 and c_2 in Eq. (58) are equal to: $c_2^2 \approx c_1^2 \approx (2\pi\eta/\kappa)^2 \omega^2/V_0^2$. Since the main contribution into integral (59) is made by $V \approx \bar{V}(T)$, then $c_1, c_2 \gg 1$.

As a result, $v(t)$, by Eq. (58), will be:

$$V(t) = V(0) \exp(-ct), \text{ where } c = \frac{1}{240} \frac{\hbar}{m^{**} a^2} \frac{\omega'}{\omega^2} \left(\kappa/\eta\right)^4.$$

Using this expression and also expressions (59)–(61) we will describe the mobility by Eq. (53) which is valid on condition that:

$$T \ll m^{**} V_0^2 \tag{62}$$

$$v_0^2 \ll (2\pi\eta/\kappa) \omega^2 \tag{63}$$

From Eq. (57)–(61) it follows that for $\omega' = 0$, $T \neq 0$, as in the case of $T = 0$, the soliton mobility becomes infinite. Of course, the above reasoning refers to the temperature which is considerably less than that of soliton decomposition [27].

9 Application to Polynucleotide Chain

Recently, modeling of the motion of a charged particle in molecular chains of various types has been the topic of considerable investigation (books and reviews [9, 28–31] and papers cited therein). A new type of conducting quasi-one-dimensional system is polynucleotide chains inter twisted into a double helix of DNA molecule.

Recall, that DNA consists of four types of nucleotides designated as A (adenine), T(thymine), C(cytosine), an G(guanine) which unite into complementary pairs in such a way that nucleotide A always pairs with T and nucleotide C always pairs with G. These nucleotide pairs are arranged in a stack to form a DNA double helix. Nowadays long sequences with a prescribed set of nucleotide pairs can be synthesized artificially. Of great interest are chains composed of uniform pairs which could serve as molecular wires in nanoelectronic devices [1]. In the majority of experiments on charge transfer in DNA the charge is carried not by electrons, but by holes. If a nucleotide in nucleotide chain be freed from electron, the hole which would arise would have a potential energy U such that $U_G < U_A < U_C < U_T$.

Overlapping of electron π-orbitals of neighboring pairs will lead to delocalization of the hole over the chain and its capture by nucleotides with lower oxidation potential. Since, according to the above inequality, guanine has the lowest oxidation potential, the hole will travel over guanines while all the other nucleotides will act as potential barriers.

In our approach Hamiltonian of a regular deformable nucleotide chain is considered in continuum approximation. To carry out numerical estimates we choose the following parameter values. For a homogeneous PolyG/PolyC chain the value of the matrix element v_{GG} is equal to 0.084 eV which corresponds to $\eta = 1{,}276$. The values of matrix elements for DNA calculated by quantum chemical methods in [32, 33], are in good agreement with experimental data on relative charge transfer rates in DNA. The values $\chi = 0.13$ eV/Å, $\Omega_0 = \sqrt{K/M} = 10^{12} \mathrm{s}^{-1}$, $\Omega' = \gamma/M = 6 \cdot 10^{11} \mathrm{s}^{-1}$ are chosen the same as in [14]. Notice, that the value of χ used by us is close to that found by quantum chemical methods in [34]: $\chi = 0{,}2349\,\mathrm{eV/Å}$. For $\tau = 10^{-14}$ s these parameter values correspond to dimensionless quantities $\kappa = 4$, $\omega = 10^{-2}$, $\omega' = 0{,}006$. The value of mobility μ calculated by formula (53) for the distance between nucleotide pairs $a = 3.4\,\text{Å}$ in a PolyG/PolyC chain is $\mu \approx 5.7 \times 10^{-2} \mathrm{cm}^2/V$ s. The characteristic size of a soliton is $r \approx 1.3$, which corresponds to the energy of the particle W_0 reckoned from the bottom of the conduction band: $W_0 = -\hbar/2m^* r^2 \approx -5 \times 10^{-2}$ eV and the soliton energy $E_0 = W_0/3 = -1.7 \times 10^{-2}$ eV. The soliton effective mass calculated by formula (54) is $m^{**} \approx 1.3 \times 10^4 m^*$ where the band mass of a hole $m^* = -\hbar/2va^2$ is equal to 3.94 m_0, m_0 is the electron mass in vacuum.

10 Discussion of Results

In Bogolubov–Tyablikov strong coupling theory [35, 36] which is presently considered the most consistent theory of a strong coupling polaron (see, for e.g., review
[37]), at $V \neq 0$ some resonance denominators appear which lead to divergence.
The results obtained here provide an explanation for this contradiction. In the case
under consideration we could have proceeded from. Froehlich Hamiltonian for the
Holstein model and get resonance denominators (see, for e.g., discussion of this
problem in [31]). So, the presence of resonance denominators stems from the fact
that a moving polaron state is nonstationary, i.e. it is not the eigen state of Froehlich
Hamiltonian. This implies an important methodological conclusion—in translation-
invariant systems, among which is Froehlich Hamiltonian, a moving polaron state
may not be its eigen state.

In a semiclassical model in a dispersionless case, a moving polaron always emits
phonons. In a quantum case this emission occurs on the condition $m^{**}v^2/2 > \hbar\,\Omega_0$.
One would think that in a quantum case, as distinct from the classical one, there
exists a minimum velocity $v_{min} = 2\hbar\,\Omega_0/m^{**}$, at which the emission is possible.
However, the very inequality from which v_{min} was obtained is fulfilled only in the
strong coupling limit, when quantum description becomes semiclassical. Since in
the strong coupling limit $m^{**} \sim \kappa^4$, where $\kappa \to \infty$, then $v_{min} \sim 1/\kappa^4 \to 0$ in complete
agreement with a semiclassical approach.

The general picture of the soliton motion along the chain at $T = 0$ in the absence
of dissipation looks as follows. For $r \gg 1$, a soliton, irrespective of the model
parameters, has a large path length which goes into infinity as $v \to 0$ and $v \to \infty$.
In the presence of dispersion when $v < v_0$ the soliton path length is $L = \infty$. When
$v < v_0$ the path length is finite and tends to infinity as $v \to \infty$.

We emphasize that despite the fact that the value of path length L at given $V > V_0$
is finite, the total soliton path for a soliton to reach the velocity V_0 is always infinite.
It is clear that as the soliton is slowing down, i.e. its velocity decreases, the local
amplitude of the phonon tail vanishing at $V \to V_0$. Accordingly the soliton path
length tends to infinity at $V \to V_0$ and an infinite time is required for a soliton to
reach the velocity V_0.

When $r \ll 1$, a soliton cannot move along the chain. The deformation produced
by it "chains" it to the site (molecule) where it is localized (when considered dis-
cretely). Formally, in the continuum model, which is invalid in this case, the soliton
path length becomes very small: $L \leq r$. The reason why the deformation "chains"
the charge to the site in the case of interest, is that the ratio of the deformation energy
E_{pot} to the oscillator energy: $E_{pot}/\hbar\,\Omega_0 = \kappa/2\,\omega$ is very large: $\kappa/2\omega \gg 1$ and
in passing to the neighboring site the charge must do some work so that to produce
an equilibrium deformation at this site. For $E_{kin} < E_{pot}$ this process is forbidden
by the energy conservation law: when $E_{pot} = \chi^2/2\,M\,\Omega_0^2 > m^{**}V^2/2$ a soliton
cannot move to the neighboring site.

In the quantum model, when $r \leq 1$, the description is carried out in terms
of a small radius polaron (SRP). Agreement with the semiclassical description is

achieved, as in the case of a large radius soliton, by passing to the limit $\kappa \to \infty$ in the equation $v_{\min} = 2\hbar\,\Omega_0/m^{**}$, where m^{**} is SRP effective mass.

So, for a SRP $v_{\min} \to 0$ as $\kappa \to \infty$. It follows that at $T = 0$ the polaron band, the concept of which was first introduced for a SRP by Tyablikov [38], does not exist. As was pointed out above, the translation invariance of the Hamiltonian, on which Tyablikov's theory relies, does not guarantee the existence of the band.

From all has been said it follows that any experimental attempts to find the band of a small radius polaron (SRP) at $T = 0$ have no prospects. Traditionally, the difficulty in finding the SRP band is associated with its narrowness when any external excitation leads to its destroy and formation of a localized state (see discussion of this problem in [39]). According to our results, even in the absence of any external effects the band will destroy by a phonon emission of a moving polaron which leads to its immediate localization.

This work was supported by RFBR, grant No 07-0700313.

References

1. Lakhno, V. D.: DNA nanobioelectronics, *Int. J. Quant. Chem.*, **108**(11), 1970–1981 (2008).
2. Dekker, C.; Ratner, M. A.: Electronic properties of DNA, *Phys. World*, **14**, 29–33 (2001).
3. Bhalla, V., Bajpai, R P., Bharadwaj, L. M.: DNA electronics. *Eur. Mol. Biol. Rep.*, **4**, 442–445 (2003).
4. Porath, D., Guniberti, G., Di Felice, R.: Long-range charge transfer in DNA. *Top. Curr. Chem.*, **237**, 183–227 (2004).
5. Davydov, A. S.: The theory of contraction of proteins under their excitation. *J. Theor Biol.*, **38**(3), 559–569 (1973).
6. Davydov, A. S.: Solitons and energy transfer along protein molecules. *J. Theor. Biol.*, **66**, 379–387 (1977).
7. Davydov, A. S.: Solitons, bioenergetics and the mechanism of muscle contraction. *Int. J. Quant. Chem.*, **16**(1), 5–17 (1979).
8. Davydov, A. S.: Solitons in molecular systems. *Phys. Scr.* **20**, 387–394 (1979).
9. Davydov, A. S.: Solitons in quasi-one-dimensional molecular structures. *Soviet Phys. Usp.*, **25**, 898–918 (1982).
10. Davydov, A. S.: *Soliton in Molecular Systems*, Reidel, Dodrecht (1985).
11. Holstein, T.: Studies of polaron motion. 1. The small polaron, *Ann. Phys. (N.Y.)*, **8**, 343–389 (1959).
12. Lakhno, V. D.: Sequence dependent hole evolution in DNA. *J Biol. Phys.*, **30**, 123–138 (2004).
13. Fialko, N. S., Lakhno, V. D.: Long-range charge transfer in DNA. *Reg. Chaotic Dyn.*, **7**(3), 299–313 (2002).
14. Fialko, N. S., Lakhno, V. D.: Nonlinear dynamics of excitations in DNA. *Phys. Lett. A*, **278**, 108–111 (2000).
15. Lakhno, V. D., Fialko, N. S.: HSSH-model of hole transfer in DNA. *Eur. Phys. J. B.* **43**, 279–281 (2005).
16. Landau, L. D.: On the motions of electrons in crystal lattice. *Phys. Zs. Sowjet.*, **3**, 664–665 (1933).
17. Pekar, S. I.: Issledovaniya po Electronnoi Teorii Kristallov (Studies in the Electronic Theory of Crystals), Gostekhizdat, Moscow, (1951).
18. Landau, L. D., Pekar, S. I.: Polaron effective mass. *ZhETF*, **18**, 419–423 (1948).
19. Davydov, A. S., Enol'skii, V. Z.: Motion of an excess electron in a molecular chain with allowance for interaction with optical phonons. *ZhETF*, **79**(11), 1888–1896 (1980).

20. Davydov, A. S., Enol'skii, V. Z.: On the effective mass of Pekar's polaron Phys. *Status. Solidi B*, **143**, 167–172 (1987).
21. Davydov, A. S., Enol'skii, V. Z.: Effective mass of Pekar polaron. *ZhETF*, **94**(2), 177–181 (1988).
22. Myasnikova, A. E., Myasnikov, E. N.: The tenzor of polaron inert mass in isotrope media. *ZhETF*, **115**(1), 180–186 (1999).
23. Myasnikova, A. E., Myasnikov, E. N.: On the conditions of Landau—Pekar polaron existence. *ZhETF*, **116**(10), 1386–1397 (1999).
24. Myasnikova, A. E.: Band structure in autolocalization and bipolaron models of high-temperature superconductivity. *Phys. Rev. B*, **52**, 10457–10467 (1995).
25. Myasnikova, A. E., Myasnikov, E. N.: Band theory of semiconductors and autolocalization of electrons. *Phys. Lett. A*, **286**, 210–216 (2001).
26. Lakhno, V. D., Korshunova, A. N.: Simulation of soliton formation in a uniform chain. *Math Model.*, **19**, 3–13 (2007).
27. Lakhno, V. D., Fialko, N. S.: Temperature destruction of soliton. In: *Mathematical Biology & Bioinformatics*, V. D. Lakhno (ed.), Proc. Int. Conf., Pushchino, 27–28 (2006).
28. Bernasconi, J., Schneider, T. (eds.), *Physics in One Dimension*, Springer-Verlag, Berlin/Heidelberg/New York, (1981).
29. Heeger, A. J., Kivelson, S., Schrieffer, J. R.: Solitons in conducting polymers. *Rev. Mod. Phys.*, **60**(3), 781–850 (1988).
30. Scott, A.C.: Davydov's soliton. *Phys. Rep.*, **217**, 1–67 (1992).
31. Lakhno, V. D.: Nonlinear models in DNA conductivity. Chapter 24 In: *Modern Methods for Theoretical Physical Chemistry of Biopolymer*, Starikov, E.B., Lewis, J.P., Tanaka, S. (eds.), Elsevier Science Ltd. 604 pp., 461–481 (2006).
32. Voityuk, A. A., Rösch, N., Bixon, M., Jortner, J.: Electronic coupling for charge transfer and transport in DNA. *J. Phys. Chem. B*, **104**, 9740–9745 (2000).
33. Jortner, J., Bixon, M., Voityuk, A. A., Rösch, N. J.: Superexchange mediated charge hopping in DNA. *J. Phys. Chem. A.*, **106**(33), 7599–7606 (2002).
34. Starikov, E. B.: Phil. Mag. Electron–phonon coupling in DNA: a systematic study. *Phil. Mag.*, **85**, 3435–3462 (2005).
35. Bogolubov, N. N.: About one new form of adiabatic perturbation theory in the problem of particle interacted with quantum field. *Ukr. Mat. Zh.*, **2**(2), 3–24 (1950).
36. Tyablikov, S. V.: Adiabatic form of perturbation theory in the problem of particle interacted with quantum field. *ZhETF*, **21**(3), 377–383 (1951).
37. Lakhno, V. D., Chuev, G. N.: Structure of a strongly coupled large polaron. *Phys.– Usp.*, **38**(3), 273–285 (1995).
38. Tyablikov, S. V.: On electron energy spectrum in polar crystal. *ZhETF*, **23**(10), 381–391 (1952).
39. Firsov, Y. A. (ed.), *Polarons*. Nauka, Moscow (1975).

Potential Energy Surfaces for Reaction Catalyzed by Metalloenzymes from Quantum Chemical Computations

Monica Leopoldini, Tiziana Marino, Nino Russo, and Marirosa Toscano

Abstract For several decades quantum mechanical (QM) computational methods have been developed and refined so that it was possible to extend their applicability field enormously. Today, they are used generally to supplement experimental techniques because the theory also affords deeper understanding of molecular processes that cannot be obtained from experiments alone. Due to their favorable scaling when compared to the *ab initio* methods, density functional theory (DFT) approach allows the treatment of very large systems such as the biomolecules. Thus, now it is possible, for instance, to study the difficult and critical reactions catalyzed by enzymes in biological systems. Here, a brief account of the studies performed on different metalloenzymes is given, focusing on methods and models used to describe their reaction mechanisms.

1 Introduction

Enzymes are biological catalysts that perform very complicated and specific reactions in mild conditions and with high efficiency and accuracy. They work enhancing considerably reaction rate, unmatched by any other type of catalyst. Metalloenzymes constitute a diverse class of enzymes that require a catalytic metal ion for activity and catalyze a wide variety of biological reactions. In some metalloenzymes the metal ion acts mainly as a Lewis acid and in these cases the cation does not suffer oxidation state change nor, generally, its protein ligands. However, many redox enzymes exist that catalyse the oxidation or reduction of a substrate or a group of substrates.

The study of the reaction pathway as it proceeds within the protein walls is the focal point of chemistry that deals with biological systems.

M. Leopoldini (✉), T. Marino, N. Russo, and M. Toscano
Dipartimento di Chimica and Centro di Calcolo ad Alte Prestazioni per Elaborazioni Parallele e Distribuite-Centro d'Eccellenza MIUR, Universita' della Calabria, I-87030 Arcavacata di Rende (CS), Italy

N. Russo et al. (eds.), *Self-Organization of Molecular Systems: From Molecules and Clusters to Nanotubes and Proteins*, NATO Science for Peace and Security Series A: Chemistry and Biology, © Springer Science+Business Media B.V. 2009

During enzymatic reactions, bonds are broken and formed, and intermediates and transition species originate. To correctly describe these events, quantum mechanical tools are expected to be utilized. In fact, enzymes mainly function by lowering transition state energies. The transition state for an enzyme-catalyzed chemical reaction represents a short-lived species having high-energy. No spectroscopic method available can detect a transition-state structure, thus a theoretical approach is often the only way to achieve molecular information that chemists seek.

Today there are computational methods for modelling molecules and reactions that can obtain results at almost any accuracy desired.

Among these tools, the density functional theory (DFT) based methods can give information about structures, properties and energetics with an accuracy comparable to the very expensive *ab initio* methods but with minor computational efforts. However, enzymes are proteins with an high molecular weight so that they cannot be entirely studied without introducing significant approximations or using chemical models to represent their usually large active region. Of course, results must be interpreted in light of the above mentioned computational limitations.

In this work, a brief summary of the computational studies performed in our laboratory on some representative metalloenzymes is presented. These studies can give insight into the elucidation of different reaction mechanisms and, in the cases these are matter of discussion or are unknown, can contribute to clarify some aspects and propose possible reaction pathways.

2 Methods and Models

2.1 Density Functional Methods

All the computations were performed with Gaussian 03, revision C02, code [1] using the DF method mainly in its B3LYP [2–5] formalism. The hybrid B3LYP functional was more frequently chosen since its performance in describing enzymatic mechanisms is widely supported by literature papers [6–9]. However, new functionals were developed during the last decade and tested towards several chemical properties, including the determination of barrier heights [10], such as MPWB1K [11], BHandHLYP [2, 3, 5], BB1K [12] and PBE1PBE [13]. Among them, the hybrid-meta MPWB1K was found to be an effective tool for calculating barriers height, thermo-chemical kinetics and non-bonded interactions, so we have chosen to use it in some case.

Solvent effects were introduced as single point computations on the optimized gas phase structures in the framework of Self Consistent Reaction Field Polarizable Continuum Model (SCRF-PCM) [14–16] in which the cavity is created *via* a series of overlapping spheres. United Atom Topological (UA0) model applied on atomic radii of the UFF force field [17] was used to build the cavity, in the gas-phase equilibrium geometry. The dielectric constant value $\varepsilon = 4$ was chosen to take into account the coupled effect of the protein itself and the water medium surrounding the protein, according to previous suggestions [6–9].

2.2 Construction of Chemical Models

The quantum mechanical (QM) studies performed on metalloenzymes use chemical models for the active sites since proteins are macromolecules that cannot be studied on the whole through theoretical methods. The developing of these models is not trivial but requires a deep analysis. First of all, X-ray structures of the proteins themselves and in complex with substrates and/or inhibitors should be available with a good resolution, so that direct information about the architecture of the active site as well as the binding mode of the substrate can be drawn. Site-specific mutagenesis of particular amino acids, together with kinetic studies, allow to identify some residues on which catalysis depends. Subsequently, a detailed analysis of the catalyzed reaction must be carried out, also making comparisons with other enzymes that catalyze similar reactions for which the work mechanism is known. This step is important not only to propose a reaction pathway but also to identify in the active sites functional groups and residues necessary to catalysis. For example, hydrolysis reactions require a nucleophilic agent that must be identified. Once we had collected all these data, the active site must be divided into two parts, the quantum mechanical cluster and the environment, that is the portion surrounding the catalytic region not involved in catalysis.

The quantum mechanical cluster is made up by the metal ion and its first coordination sphere, to which some nearby residues recognized as fundamental in catalysis are added. Ligands are represented by the functional part of the side-chains only (imidazole rings for histidines, acetates for aspartates or glutamates). One of the atoms of each amino acidic residue is usually kept frozen at its crystallographic position in order to mimic the steric effects produced by the surrounding protein and to avoid an unrealistic expansion of the cluster during the optimization procedure.

The environment not explicitly included in the quantum cluster has a double effect: steric and electrostatic. As mentioned before, the first one can be brutally reproduced by fixing some crystallographic positions. Instead, the electrostatic effect can be introduced assuming it being a homogeneous polarizable medium with a dielectric constant usually chosen to be equal to 4.

Information about the influence of the protein environment on active site structures and reactions energetics can be obtained applying the ONIOM hybrid method (QM/MM) of Morokuma and co-workers [18–24] that enables different levels of theory to be applied to different parts of a system and combined to produce a consistent energy expression. The underlying idea is that the various parts of the system each play their own role in the process under investigation, and therefore require different accuracies. According to ONIOM terminology, the full system treated at low level of theory is called "real" while the inner layer is called "model" and it is treated at both low and high level of theory. The hybrid QM/MM methods are strongly dependent on parameters and proper calibration used. In order to obtain a description that does not feel of these effects, the Orbital-Free Embedding approach of Wesolowski et al. [25, 26] can be used. Practically, using the Orbital-Free Embedding method, the system can be partitioned into two regions, one of which treated by a full density functional calculation and the second one by a frozen density, allowing us to evaluate the interactions between the active site and the environment quantum mechanically. Kohn–Sham-like equations are solved to obtain the electron density of a fragment (i.e. active site of enzyme), which is embedded in a larger frozen system (protein environment). The fact that the subsystem is embedded in a microscopic environment is represented by means of a special term in the effective potential (V_{emb}). V_{emb} is system-independent since it uses the universal functionals of electron density. It is expressed as a function of two variables ρA and ρB (the electron densities of the embedded system and its environment). It is, therefore, orbital-free. The explicit analytic form of V_{emb} is not known but it can be expressed using approximations to the non additive kinetic energy functional (the difference between the kinetic energy of the total electron densities of the two subsystems) and the exchange-correlation functional defined in KS formalism.

3   Results and Discussion

In this section we will shortly report the result obtained in our previous studies for a series of reaction catalyzed by metal containing enzymes.

3.1   Nitrate Reductase

Nitrate reductase belongs to the class of mononuclear molybdenum enzymes (molybdoenzymes), characterized by having in the active site an organic cofactor containing a dithiolene moiety, i.e. the bis-molybdopterin guanine dinucleotide (MGD) [27]. Nitrate reductase catalyses the reduction of nitrate to nitrite, thus assuming an important role in nitrogen assimilation [28].

The dissimilatory (respiratory) enzyme obtained from the microorganism *Desulfovibrio desulfuricans* consists of four domains involved in cofactor binding [29]. One of these domains is responsible for the binding with an Fe_4S_4 cluster that acts as an electrons pump in the catalytic cycle. The Mo^{VI} metal center is coordinated to four sulphur atoms coming from the two dithiolene groups, the sulphur atom of side chain of Cys140 residue and a hydroxo/water molecule. Several residues, such as Arg354, Asp155, Glu156, Asp355, Ala142, Val145, Val149, Leu359, Leu362 Gln346, Met308 and Met141 are present in the proximity of active site. For Arg354, a role in the anchor of the negative charged nitrate substrate has been proposed [29].

The nitrate reductase catalytic cycle is described in Scheme 1. The five coordinated Mo^{IV} ($Mo^{IV}S_{Cys}(SR)_4$) binds to a nitrate ion. The bond of NO_3^- with the

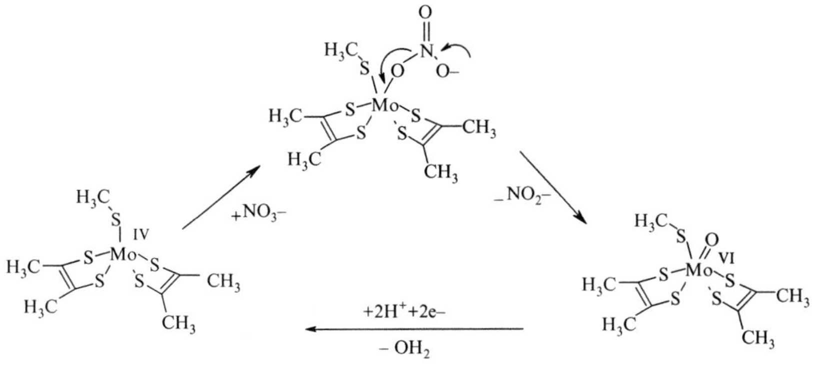

Scheme 1 Catalytic mechanism of nitrate reductase enzyme

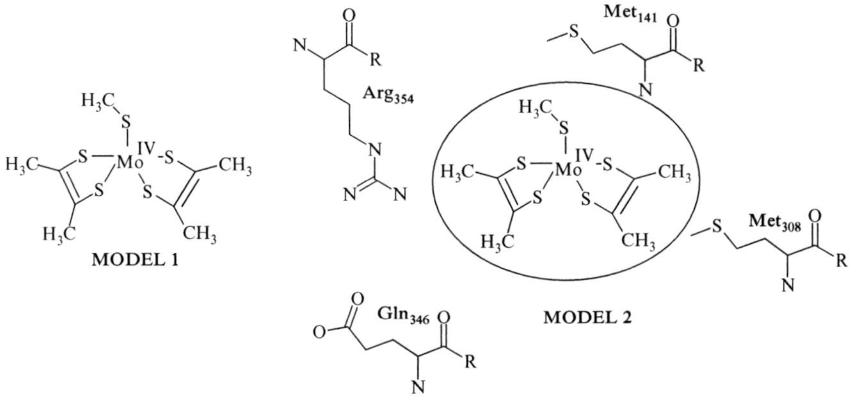

Fig. 1 Model clusters used to simulate nitrate reductase active site

metallic centre through one oxygen atom leads to a weakening of N–O bond. The Mo^{IV} is oxidized to Mo^{VI} and NO_2^- is released. Mo^{IV} is restored in another step, by two protons coming from water molecules present in the active site, and two electrons coming from the Fe_4S_4 cluster [29].

Two models were used for enzyme active site simulation. The first one (model 1 of Fig. 1) is the $[Mo^{IV}(S_2C_2CH_3)_2(SCH_3)]$, used elsewhere for the simulations of the molybdoenzymes active site [30]. The second model (model 2 of Fig. 1) was obtained adding to the model 1 the conserved residues Arg354, Gln346, Met308 and Met141, according to their crystallographic positions [29].

The B3LYP optimized geometry for the model 1 in its singlet electronic state, was found to be square pyramidal with Mo–S dithiolenes average distances of 2.36 Å, and with a Mo–S^{Cys} bond length of 2.35 Å in agreement with data reported in the experimental work of Holm and coworkers for phenoxy analogous complexes [31].

The equivalent system in the triplet electronic state presents a distorted trigonal bipyramidal geometry in which dithiolenes sulphurs occupy both equatorial and

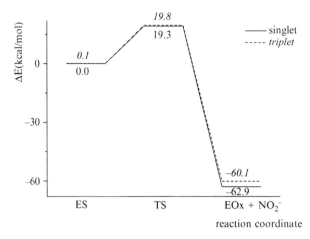

Fig. 2 Gas-phase PES for nitrate reduction performed by model 1

axial positions (equatorial Mo–S mean distance is 2.46 Å, axial Mo–S mean distance is 2.36 Å). Low- and high-spin energetic gap was found to be only 4.7 kcal/mol at B3LYP level, being the singlet the ground state.

Energetic profiles (PES) for the nitrate reduction performed for the model 1 for both spin states are reported in Fig. 2 and the geometries of stationary points in Fig. 3.

The reaction starts with the formation the enzyme-substrate complex (ES) in which the NO_3^- is linked to metallic centre through one of its oxygen atoms (Mo–O^{nitrate} distance is 2.27 and 2.26 Å in the singlet and triplet state, respectively), leading to a lengthening of the O–N bond length in the substrate. The coordination geometry around molybdenum is dependent on the electronic state since it appears to be trigonal prismatic and nearly octahedral, for singlet and triplet, respectively.

The low-high spin splitting relative to this intermediate was computed to be 0.1 kcal/mol, being also here the low spin the ground state.

In the transition state (TS) the Mo–O^{nitrate} and O^{nitrate}–N^{nitrate} distances assume the values of 1.91 (1.89) and 1.65 (1.68) Å, for singlet (triplet) spin states, respectively. The geometry in the low- and high-spin TSs appears to be very distorted with respect to the ideal trigonal prismatic and octahedral geometries of the starting ES complexes. The imaginary frequencies (589 cm^{-1} for singlet and 722 cm^{-1} for triplet) correspond to the stretching vibration mode of the Mo–O^{nitrate} and O^{nitrate}–N^{nitrate} bonds. The singlet and triplet TSs lie at 19.3 and 19.8 kcal/mol with respect to the ground state ES complex, respectively.

The final products (EOx of Fig. 3) possess a distorted geometry and lie at 62.9 kcal/mol (singlet) and at 60.1 kcal/mol (triplet) below the reference point. Mo–O bond has a length of 1.72 and 1.71 Å, for ground and excited state, respectively.

The transfer of the oxygen atom from the bound substrate to the metal centre represents the rate-limiting step.

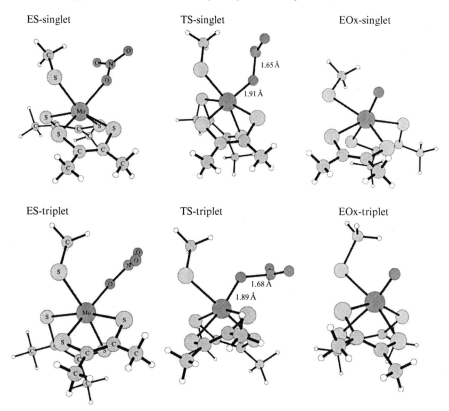

Fig. 3 Optimized geometries of the stationary points belonging to model 1 reaction profile

On model 2, ONIOM and Orbital-Free Embedding computations [31], were performed.

ONIOM procedure [18–24] was applied to the model 2 that was divided into an inner layer consisting of the Mo^{IV} thiomethyl bis dithiolenes complex, nitrate substrate and nitrite product molecules, and an outer layer made up by Arg354, Met 141, Met308 and Gln346 amino acids at their crystallographic positions.

ONIOM energetic profiles for the enzyme in the singlet and triplet electronic state were reported in Fig. 4, while equilibrium structures on the PESs were depicted in Fig. 5.

In the ES complex, the $Mo-O^{nitrate}$ and $O^{nitrate}-N^{nitrate}$ distance values (2.18 and 1.38 Å for singlet and 2.15 and 1.39 Å for triplet) suggest that the interaction between the metal and the substrate oxygen atom is stronger than that found in the analogous complex with model 1. The trigonal prismatic and octahedral arrangements around molybdenum in the low and high spin complexes appear to be much more distorted than in model 1. The ONIOM gap between the ES complexes in the singlet ground state and in the excited triplet is 2.2 kcal/mol. This value compared to data obtained for model 1 indicates a certain dependence of the gap on the presence of protein environment.

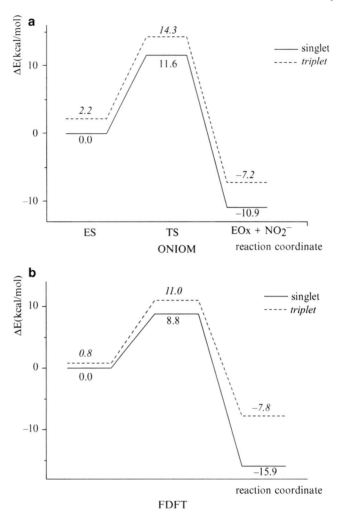

Fig. 4 PES obtained with model 2 at (**a**) ONIOM and (**b**) FDFT level

The transition states for the singlet and triplet species are characterized by Mo–$O^{nitrate}$ and $O^{nitrate}$–$N^{nitrate}$ distances that are shorter and longer with respect to those of ES complexes, respectively. Imaginary frequency values of 252 (singlet) and 444 (triplet) cm^{-1} can be attributed to the Mo–$O^{nitrate}$ and $O^{nitrate}$–$N^{nitrate}$ bonds stretching vibrational mode.

Transition states lie at 11.6 (singlet) and 14.3 (triplet) kcal/mol above the ES reference. The presence of the amino acids residues reduces, as can be noted, the barrier heights.

At 10.9 and 7.2 kcal/mol below the ES complex, for low- and high-spin case, respectively, we found the final products (EOx).

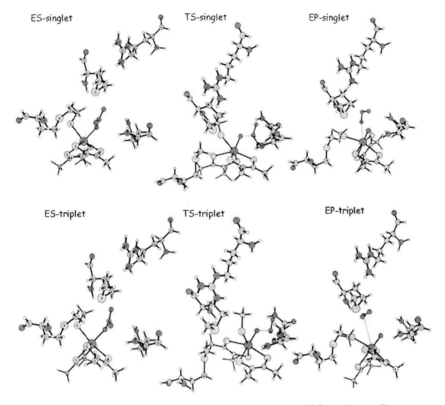

Fig. 5 Optimized geometries of stationary points belonging to model 2 reaction profile

The low- and high-spin complexes EOx exhibit a similar octahedral geometry around molybdenum. The Mo–O^{nitrate} distance is 1.73 and 1.72 Å, for singlet and triplet, respectively.

In the ONIOM treatment, the energy value of the product corresponds to a summation of the energy value of infinitely separated species (E(EOx) + E(NO$_2^-$)). Since in the computations with model 1 no geometry convergence was obtained in the presence of a long range interaction between nitrite and oxo Mo^{VI} complex, a direct comparison of the energetics obtained with the two different models cannot be done as far as products are concerned.

ONIOM computations demonstrate that the protein environment has not influence on the mechanism followed by the enzymatic reaction but affects above all the kinetics, by lowering the activation barriers. In agreement with the "entatic principle" proposed by Morokuma, this can be explained by considering that the strain to which the active site is subjected in the presence of the protein environment that acts as a device to accumulate energy, used later to overcome barriers.

The equilibrium geometries obtained at ONIOM(B3LYP:UFF) level were used to perform an orbital-free embedding analysis on model 2 [31]. This completely quantum-mechanical method uses two electronic densities for the inner and outer layers. Single point evaluation was obtained using GGA/PW91 exchange-correlation functional [32–36].

Because of the differences between the used methodology, functional and basis set, no comparison is possible with ONIOM data, thus only the qualitative aspects of these last results will be discussed.

The potential energy profiles obtained applying the orbital-free embedding procedure for both singlet and triplet states are depicted in Fig. 4b.

As in the previous cases, reaction proceeds most favorably along the singlet path.

Only 0.7 kcal/mol separate the low spin ES complex from the same species having the triplet multiplicity. The transfer of oxygen requires an activation energy of 8.8 kcal/mol for the low-spin and 11.0 kcal/mol for the high-spin. Products lie at 7.8 kcal/mol (singlet) and 15.9 kcal/mol (triplet) below the singlet ES complex.

ONIOM and orbital-free embedding results appear to be qualitatively very similar. However, it is worth to underline that the data obtained by the second treatment are characterized by the reliability peculiar to a totally quantum mechanical description.

3.2 Peptide Deformylase

During the elongation of the polypeptide chain in prokaryotes protein synthesis, the formyl group at the first formylmethionine is hydrolytically removed by the enzyme peptide deformylase (PDF) [37]. This post translational deformylation of peptides occurs only in bacterial cells but not in eukaryotes, thus peptide deformylase can be a potential target for designing new antibiotics [38].

Characterization by overexpression of deformylase gene in *Escherichia coli* [39, 40] revealed the presence of typical zinc-binding motif, HEXXH (H = histidine, E = glutamate, X = any amino acid) even if other metals than zinc were identified in the protein coming from different microorganisms [41]. So, a certain debate about the identification of the catalytic metal ion originated.

The metal ion in the active site has a tetrahedral coordination [42–44]. Ligands are the Sγ-atom of a cysteine residue (Cys 90), the Nε2-atoms of two histidines (His 132 and His 136) and the oxygen atom of a water molecule. The glutamate residue, Glu 133 of the HEXXH motif, does not bind the metal but it is required for catalysis.

In the suggested catalytic cycle [45] (see Scheme 2) the formylated peptide binds the metal through the carbonyl oxygen of the formyl group yielding to an enzyme-substrate complex. Several H-bonds with nearby residues support the nucleophilic attack of the metal bound water/hydroxide on the carbon of the formyl group, that leads to a tetrahedral intermediate. The proton of the hydroxide is then transferred to the amide at the N-terminus, with the probable aid of Glu133. The protonation of the amide group determines the C–N bond cleavage and the release of the products.

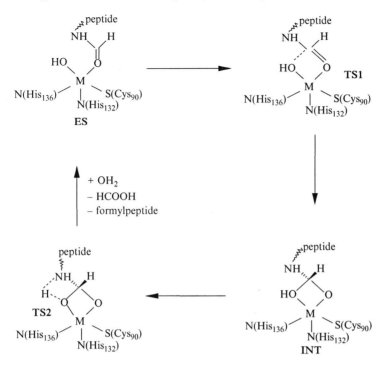

Scheme 2 Catalytic mechanism of peptide deformylase enzyme

In this work [46], a study of the hydrolysis of a formamide substrate performed by different metal forms (Zn(II), Fe(II), Ni(II) and Co(II)) of peptide deformylase active site models, is carried out with the aim to elucidate the catalytic function of metal ions.

Three models for the active site reported in the Fig. 6 were employed. The model 1 consists of a divalent metal ion (Zn, Fe, Ni, Co) coordinated to two imidazole rings and to a $-SCH_3$ group that simulate the histidine (His132 and His136) and the cysteine (Cys90) residues, respectively. In the model 2, an acetic acid molecule is added in the active site to emulate the Glu133 residue. In model 3, Gly45, Gln50 and Leu91 were further added to provide the H-bonding network involved in substrate binding. The peptide substrate was modeled by a formamide molecule. The nucleophilic attack on formamide was provided by the metal-bound hydroxide rather than the metal-bound water molecule owing the fact that the first species is generally accepted to be the nucleophile.

The first step in the catalytic reaction is the barrierless formation of a complex (ES) between the active site and the substrate. Irrespective of the metal center, model 1 ES is characterized by a bond between the carbonyl oxygen of formamide and the M^{2+} ion that becomes five-coordinated (see Fig. 7). The $-OH$ lone pair establishes a hydrogen bond with the $-NH_2$ terminal group of the substrate of 1.62, 1.67, 1.56 and 1.64 Å, for zinc, iron, nickel and cobalt, respectively.

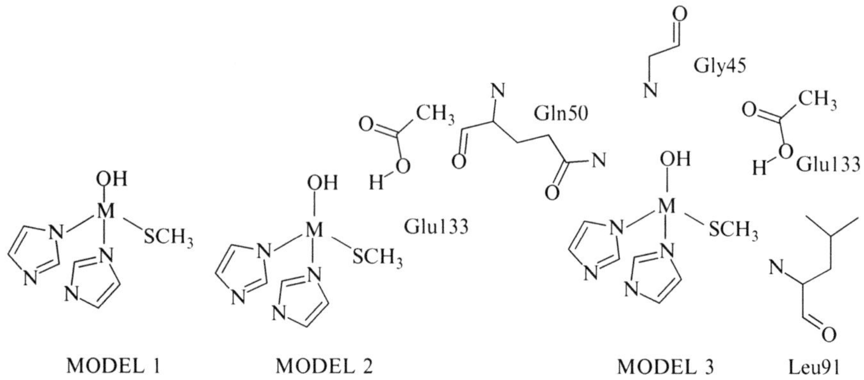

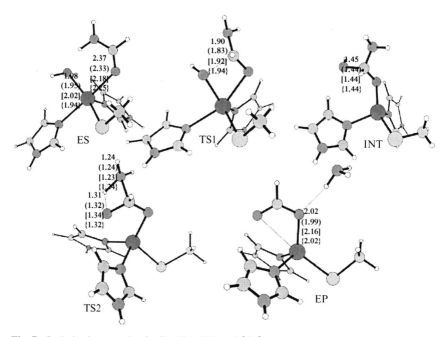

Fig. 6 Model clusters used to simulate peptide deformylase active site

Fig. 7 Optimized geometries for Zn, (Fe), [Ni] and {Co} containing model 1

The transition state TS1 for the nucleophilic attack of hydroxyl on the carbonyl carbon atom of formamide is reached when the HO–C_{sub} distance (the subscript *sub* refers to atoms in the initial substrate) assumes the values of 1.90 (Zn^{2+}), 1.83 (Fe^{2+}), 1.92 (Ni^{2+}) and 1.94 Å (Co^{2+}), while the C–N and O_{sub}–C_{sub} bonds lengthen on average (considering all the four metals) of 0.06 and 0.03 Å. Imaginary frequency values of 257 (Zn^{2+}), 306 (Fe^{2+}), 253 (Ni^{2+}) and 292 (Co^{2+}) cm^{-1} corresponds to the stretching of the HO–C_{sub} bond. In the intermediate, INT, the bond between the oxygen of the hydroxide and the carbon of the formamide is completely formed (the HO–C$_{sub}$ average distance is $\cong 1.44$ Å). The M^{2+} goes back to a four coordination after the cleavage of the bond between the –OH group and the metal. The proton transfer from the hydroxide to the amide nitrogen occurs through the transition state TS2. The O–H and OH–N$_{sub}$ average distances are $\cong 1.32$ and 1.24 Å, respectively. Imaginary vibrational mode whose frequency is 1,628 (Zn^{2+}), 1,635 (Fe^{2+}), 1,643 (Ni^{2+}) and 1,636 (Co^{2+}) cm^{-1}, respectively, sees the hydrogen atom moving between two atoms. In the final complex EP, the C–N bond is completely broken, the ammonia leaving group is still held at about 2.00 Å by a hydrogen bond with the formate that is still coordinated to the metallic centre.

The potential energy profiles obtained for zinc, iron, nickel and cobalt containing model 1 are schematically depicted in Fig. 8.

Nucleophilic addition requires an activation energy of 22.4, 21.8, 19.8 and 22.6 kcal/mol, for Zn, Fe, Ni and Co, respectively. The rate-limiting step can be recognized in the proton transfer process that converts INT into EP. It requires an amount of energy of 24.4 (Zn^{2+}), 25.7 (Fe^{2+}), 28.2 (Ni^{2+}) and 26.3 kcal/mol (Co^{2+}), when computed with respect to the intermediate INT.

Computations with model 2 were performed for the Zn- and the Fe-containing enzymes because these two metal ions have received major attention in literature as potential catalytic ions.

Also for the model 2, reaction starts with the formation of the ES complex, in which the substrate binds the metal by its carbonyl oxygen, while the metal bound –OH forms a hydrogen bond with the hydroxyl of acetic acid (Fig. 9).

The transition state (TS1) for the first ES→ INT1 interconversion step occurs for an HO–C$_{sub}$ distance value of 1.99 Å for zinc and 1.95 Å for iron. Imaginary vibrational mode (frequency at 303 (Zn^{2+}) and 229 (Fe^{2+}) cm^{-1}) clearly shows the formation of this new bond. The Glu133 residue changes significantly its orientation with respect to that in the ES complex, as to establish a further hydrogen bond (1.65 Å for zinc and 1.80 Å for iron) in which the substrate nitrogen atom acts as acceptor.

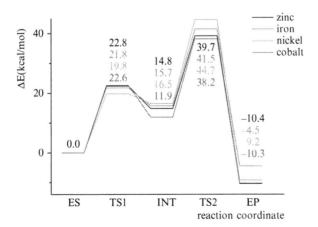

Fig. 8 Gas-phase PES obtained by peptide deformylase model 1

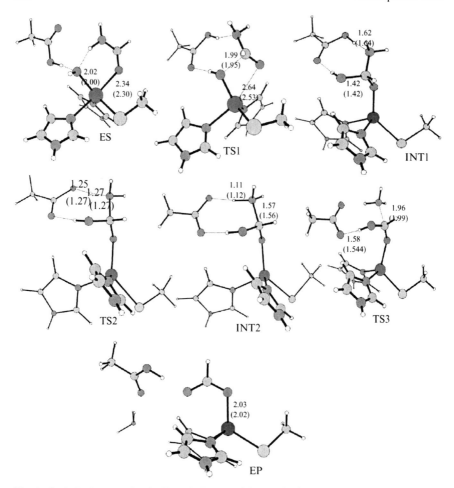

Fig. 9 Optimized geometries for Zn and (Fe) containing model 2

The nucleophilic attack leads to the intermediate INT1 characterized by an HO–C$_{sub}$ distance of 1.42 Å for both zinc and iron. Two hydrogen bonds are present, involving the –OH group of glutamate and the –NH$_2$ lone pair (1.62 Å for zinc, 1.64 Å for iron), and the –OH group of the substrate and the Glu133 carbonyl oxygen (1.86 Å for zinc and iron).

In this model 2, the protonation of –NH is mediated by the Glu133 residue, through the transition state TS2 (imaginary frequency at 773 and 691 cm^{-1}, for zinc and iron, respectively) referring to the H$^+$ shift from Glu to NH, leading to the INT2 of Fig. 9, and through the transition state TS3 (imaginary vibrational frequency of 215 (Zn^{2+}) and 223 (Fe^{2+}) cm^{-1}), in which the cleavage of the C–N bond in the substrate is coupled with the proton shift from the original –OH to Glu133.

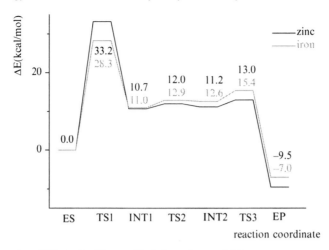

Fig. 10 Gas-phase PES obtained by peptide deformylase model 2

In the final complex EP, ammonia appears to be bound to Glu133 through a hydrogen bond (2.17 and 2.16 Å for zinc and iron). The glutamic acid residue is again produced and the formate group is strongly coordinated to the metallic centre.

As can be noted from the energetics profiles reported in the Fig. 10, the protonation of nitrogen in the substrate demands lower amounts of energy in the model II with respect to the model 1. The barrier heights for the H^+ transfers are very small when amino acids in the active site works as mediators.

The barrier heights for the H^+ transfers are very small when amino acids in the active site works as mediators. In fact, Glu133 acts lowering the activation energy required for transferring the proton on the substrate nitrogen, splitting the proton shift into two distinct processes. No influence on the relative energy of the nucleophilic addition is observed as far as the Glu133 residue is concerned.

By adding second shell amino acids to the active site as in the model 3, one should reproduce the H-bonding network able to stabilize intermediates and transition states. The Gly45, Gln50 and Leu91 residues were added to the model 2 and computations were performed for zinc and iron. The enzyme–substrate complex (ES) when compared with the model 2 one shows that three additional stabilizing hydrogen bonds are present (see Fig. 11). No significant differences are encountered in going from zinc to iron ES complexes.

The main difference encountered in going from model 2 to model 3 is that now the nucleophilic addition occurs in a concerted way with the shift from the neutral Glu133 to the amide nitrogen in the substrate. The critical distance HO–C_{sub} assumes the value of 1.82 for Zn^{2+} and 1.83 Å for Fe^{2+}. Vibrational analysis gave a low imaginary frequency at 297 cm^{-1}, for zinc, and at 291 cm^{-1} for iron, whose corresponding vibration mode indicates the simultaneous occurrence of the stretching of the heavy atoms in the HO–C_{sub} bond, and the approaching of the hydroxyl of Glu133 residue to the nitrogen atom of the –NH_2 group of the substrate. The next point after the TS1 along the model 3 reaction path is an intermediate that shows a

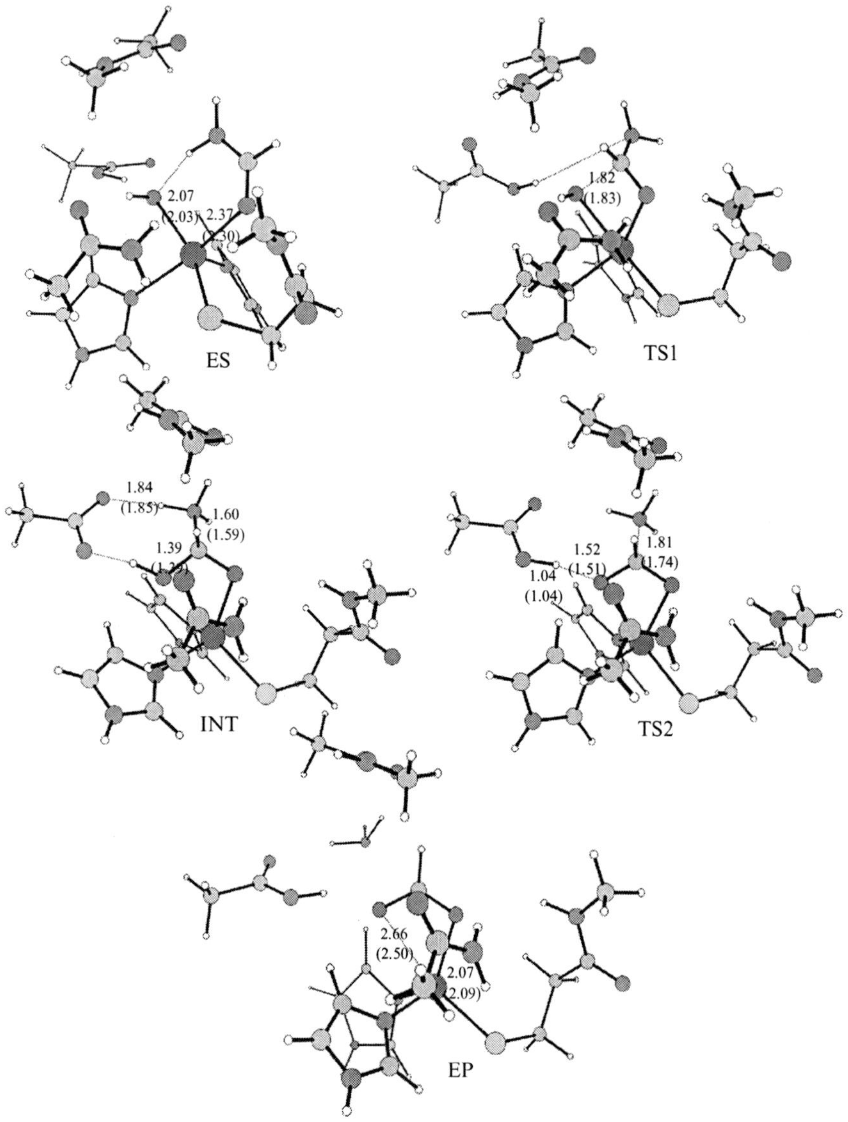

Fig. 11 Optimized geometries for Zn and (Fe) containing model 3

protonated nitrogen atom and a stable hydroxyl–carbon bond in the substrate. The carbon-nitrogen bond in the substrate is quite long (1.60 and 1.59 Å, for zinc and iron, respectively) as compared to a covalent σ C–N bond. Because of the established hydrogen bonds with the substrate (OH–O_{glu} and O_{glu}–H_{NH2} distances are 1.57 and 1.84 Å in the case of Zn^{2+}, and 1.54 and 1.85 Å in the case of Fe^{2+}), the unprotonated Glu133 residue is fixed at a good position to accept the hydroxyl proton.

The conversion of the INT into the product complex EP occurs *via* the transition state TS2, in which the unprotonated Glu133 group easily takes the proton from the hydroxide (O_{glu}–HO distance is 1.04 Å, for both zinc and iron cation). Consequently, the C_{sub}–N_{sub} bond lengthens up to 1.81 Å (Zn^{2+}) and 1.74 Å (Fe^{2+}).

In the EP structure, the formate is coordinated to the metal cation, with a protonated glutamate residue and ammonia molecule interacting with the Gly45 and Glu133 residues *via* two hydrogen bonds.

Within model 3 computations, the nucleophilic addition on the substrate carbon is the rate-limiting step, requiring an activation energy of 15.6 and 17.0 kcal/mol, for Zn^{2+} and Fe^{2+}, respectively (see Fig. 12). The INT species lies at 13.7 and 15.6 kcal/mol above the ES minimum. The transition state TS2 is responsible for both the heterolytical dissociation of C_{sub}–N_{sub} bond and for the H^+ transfer from the –OH nucleophile to the oxygen atom of Glu133 residue. A feature of model 3 energetic profile is that the Zn and Fe TS2 lie more or less at the same energy of INT species. This can be translated into a *single step* mechanism for peptide deformylase, as it occurs for other metallopeptidases.

Upon comparison between the PESs obtained for models 2 and 3, it is worth to note that the presence of additional amino acids surrounding the active site decreases the energy of the rate-determining transition state TS1, passing from 33.2 to 15.6 kcal/mol, and from 28.3 to 17.0 kcal/mol, for zinc and iron, respectively. Furthermore, the H^+ transfer process from Glu133 –OH to –NH_2 group in the substrate becomes concerted with the nucleophilic attack. The C–N bond cleavage, mediated by H^+ movement from the –OH to the Glu133 carboxyl group in the intermediate, becomes a barrierless process, implying that once the TS1 is overtaken, the tetrahedral complex directly collapses into products.

The protonation of the substrate may be recognized as the rate-limiting step when no assistance by amino acidic residues is present, demanding an higher amount of energy with respect to the nucleophilic attack of the substrate carbonyl carbon

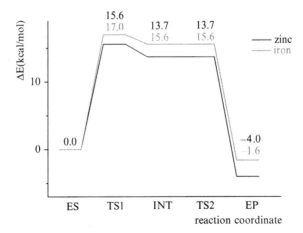

Fig. 12 Gas-phase PES obtained by peptide deformylase model 2

atom. If it is assisted by Glu133, the barrier decreases so it becomes lower than that required for the tetrahedral intermediate formation.

Adding the nearby residues Gly45, Gln50 and Leu91 the rate-limiting hydroxide attack activation energy decreases of about $\approx 10\,kcal/mol$, reproducing a biologically reasonable reaction rate.

3.3 Methanol Dehydrogenase

The pyrroloquinoline quinone (PQQ, see Fig. 13) containing enzymes belong to the quinoproteins dehydrogenase class, involved in the conversion of alcohols and amines to the corresponding aldehydes and lactones [47].

Methanol dehydrogenase is found in the periplasm of methylotrophic and autotrophic bacteria where it catalyzes the oxidation of methanol (or other primary alcohols) to the corresponding aldehydes, with the release of two protons and two electrons. Apart from the PQQ cofactor, MDH requires a divalent calcium cation for its catalytic activity [48–50].

The X-ray structure of methanol dehydrogenase from the *Methylophilus methylotrophus* W3A1 (*M*. W3A1) [51] revealed in the active site a Ca^{2+} cation and a PQQ cofactor not covalently bound to the protein. The oxygen atoms of the PQQ are involved in several hydrogen bonds with the residues Glu55, Arg109, Thr153, Ser168, Arg324, Asn387. The calcium ion is coordinated to the O_5, N_6 and $O_7\alpha$ atoms of PQQ, the $O_1\varepsilon$ and $O_2\varepsilon$ of Glu171, $O_1\delta$ of Asn255 and the $O_1\varepsilon$ of Asp297.

For MDH, two mechanisms were proposed (see Scheme 3): the addition-elimination (A) [52–54] and the hydride transfer (B) [54, 55].

Mechanism (A) involves the nucleophilic addition of the methanol oxygen to the PQQ carbonyl C_5, followed by the protonation of the cofactor and the H^+ abstraction from the substrate CH_3 by the PQQ oxygen attached to C_4, and the concomitant formation of a C_4–C_5 double bond in the PQQ, reduced to $PQQH_2$. In the mechanism (B), a direct transfer of the H^- to the C_5 in the PQQ entails the direct formation of the product. The reduction of the cofactor to $PQQH_2$ is achieved through an internal enolization step.

In both mechanisms, Asp297 residue acts as acid/base catalyst, i.e. as a base by abstracting the H^+ from the substrate hydroxyl, and as an acid by donating the same proton to the cofactor.

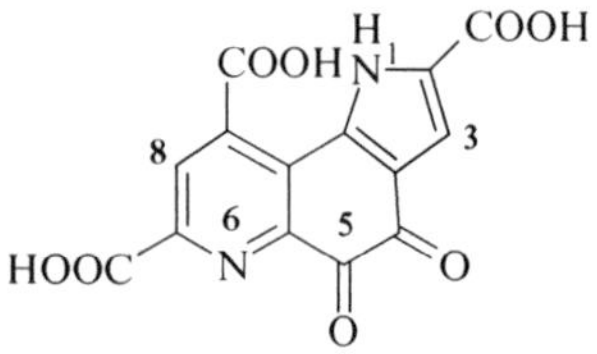

Fig. 13 PQQ cofactor   PQQ

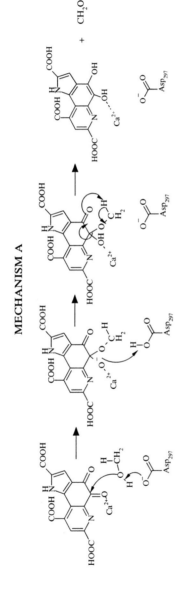

Scheme 3 Addition–elimination (A) and hydride transfer (B) mechanisms proposed for methanol dehydrogenase enzyme

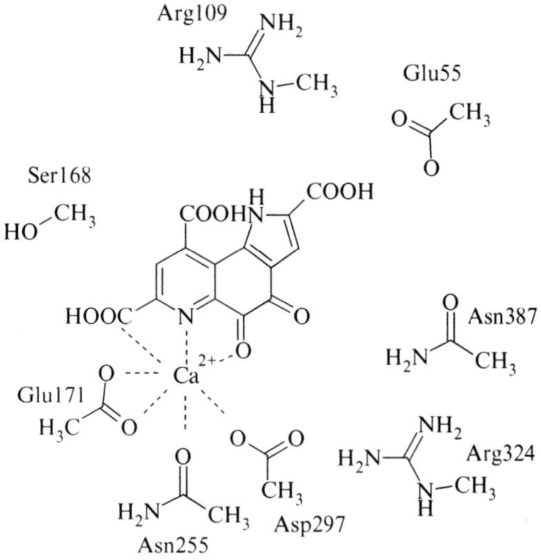

Fig. 14 Model cluster used
to simulate methanol dehydrogenase active site

Even if several experimental studies have been devoted to the identification of the real mechanism, literature data are still not sufficient to clearly identify the catalytic mechanism followed by PQQ MDH.

With the purpose to highlight the favoured catalytic mechanism of MDH among the previously proposed ones, a model system including the PQQ cofactor, Ca^{2+} coordinating groups and some nearby residues, was used (see Fig. 14) to build the Potential Energy Surfaces (PESs) for both mechanisms [56].

The B3LYP gas-phase PESs for both mechanisms (A) and (B) are is reported in Fig. 15.

The Michaelis–Menten complex (ES) between the substrate and the catalytic centre shows a substrate oxygen interacting with the calcium cation, establishing a coordination bond of 2.51 Å. The substrate proton is involved in a H-bond with the negatively charged oxygen of the Asp297 (1.90 Å) [56].

As far as the addition–elimination mechanism is concerned, a saddle point for the nucleophilic addition of the substrate oxygen on the PQQ C_5 carbon is found (TS1a) at 1.4 kcal/mol below the reference (E+S). The formation of a O_{met}–C_5 bond (critical distance is 1.80 Å) occurs in a concerted way to the shift of a proton from the alcoholic –OH group to the O1ε atom of Asp297 coordinating to the cation. The normal vibration mode corresponds to the coupled stretchings of the incoming O_{met}–C_5 and H–O1ε bonds, even if the low value of the imaginary frequency of $160\,cm^{-1}$ seems to indicate that the TS1a is mainly characterized by the motion of heavy atoms while the potential energy surface for the transfer of the H^+ is very flat. The activation energy is computed (with respect to the ES species) to be 11.5 kcal/mol [56].

The tetrahedral intermediate INT1a is characterized by a O_{met}–C_5 σ bond is of 1.48 Å and suggests a hexa-coordinated calcium cation because of a consistent

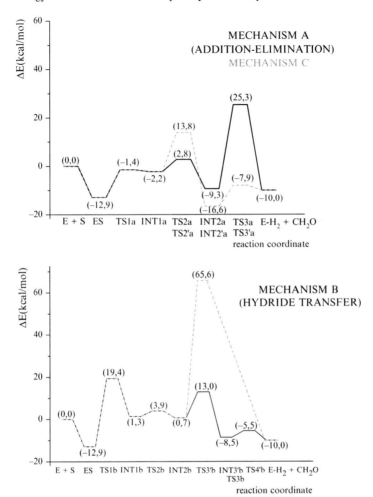

Fig. 15 Gas-phase PES for methanol oxidation according to mechanism A (top) and B (bottom)

weakening of $Ca^{2+}-O_{Asp}$ bond (2.82 Å in the INT1a versus 2.38 Å in the ES adduct) [56].

The next point along the potential energy profile is the proton transfer from HO–Asp to the PQQ C_5–O carbonyl oxygen that occurs through the transition state TS2a [56]. The proton is shared between the oxygen atoms of the PQQ and the Asp297 (1.21 and 1.19 Å, respectively), while the Asp297 oxygen atom moves again closer to the metallic centre ($Ca^{2+}-O_{Asp}$ distance is 2.53 Å). The imaginary frequency at $1,416\,cm^{-1}$ refers to the stretching vibrational mode of the $O_{Asp}-$ and $H-O_{PQQ}$ bonds. The energy required to overcome this TS2a is 5.0 kcal/mol as computed with respect to the INT1a intermediate (solid line). From the TS2a, another intermediate along the reaction profile, INT2a, is found in which the $-CH_3$ group of methanol approaches the PQQ oxygen atom attached to C_4 so that the latter is in good position to favour the breaking of the H–CH$_2$ bond in the substrate (the distance H_{met}–O_{PQQ} is 2.72 Å) [56].

The transition state TS3a that finally leads to the PQQH$_2$ reduced species, is found for C–H and H–O distance values of 1.31 and 1.32 Å, respectively [56]. The bond between O_{met} and C_5 in the cofactor, lengthens up to 2.13 Å underling that it is going to breaking. The imaginary frequency at 1,402 cm^{-1} corresponds to the movement of the H$^+$ between the carbon and the oxygen atoms. The amount of energy of 34.6 kcal/mol required to overcome this barrier indicates that this step, responsible for the cleavage of the σ C–H bond in the substrate, is the rate-determining one.

The reaction products, that is the reduced EH$_2$ species and CH$_2$O, are found 10.0 kcal/mol below the Ref. [56].

In the hydride transfer mechanism (B) (see Scheme 3), the formation of the Michaelis–Menten complex ES is followed by a direct transfer of an hydride ion from the alcohol to the C_5 in the PQQ system. The H$^-$ transfer should be concerted with the proton abstraction by the O1ε atom of Asp297 residue, so that formaldehyde is immediately produced.

The first transition state TS1b that refers to the hydride addition, shows an imaginary frequency at 1,413 cm^{-1} corresponding to the stretching of the bond between a substrate methyl hydrogen and the C_5 atom of PQQ coupled with the methanol deprotonation by Asp297 [56]. As in the TS1 on path A, also in this case the potential energy surface for the methanol deprotonation is very flat. Activation energy was computed to be 32.3 kcal/mol (with respect to the ES). The cleavage of the covalent C–H bond in the methanol is energetically very expensive. The height of the barrier underlines that the hydride transfer step is quite slow.

TS1b evolves into the INT1b intermediate lying at 1.3 kcal/mol where the formaldehyde product already formed is leaving the active site. The C_5 carbon establishes a covalent bond with the substrate hydrogen of 1.12 Å. The C_5O bond lengthens up to 1.37 Å [56].

The next step along the reaction profile, that is the H$^+$ transfer from the Asp297 residue to the oxygen attached to C_5, occurs through the transition state TS2b that lies at 3.9 kcal/mol. The normal vibration mode at 1,410 cm^{-1} corresponds to the shift of the proton from the Asp297 oxygen to the PQQ C_5O one.

The INT2b intermediate formed after the proton transfer is characterized by a tetrahedral C_5 carbon atom and by a network of hydrogen bonds that involve the PQQ oxygen atoms [56].

The enolization process leading to the final reduced PQQH$_2$ cofactor occurs by transferring the hydrogen bound to C_5 to the C_4O carbonyl oxygen. The transition state TS3b for the direct hydrogen transfer on C_4O oxygen atom occurs when the CH–O and C–HO critical distances assume the values of 1.34 and 1.47 Å, respectively [56]. The imaginary frequency at 2,260 cm^{-1} is assigned to the stretching of these bonds. TS3b is a highly stressed species, so that its formation, requiring 65.6 kcal/mol, is energetically prohibitive.

Alternately, the enolization process may be mediated by Asp297 residue (red line on the bottom of Fig. 15), emphasizing the role of this amino acid as catalytic base. Two steps are necessary in this case: one that transfers the H^+ from the quinone to the O2ε of the aspartate, and another in which the protonation of the C_4O oxygen atom by the same residue occurs.

The transition state TS3'b, in which we can observe the incipient breaking and formation of C_5–H and O2ε–H (in the Asp297) bonds, respectively, is characterized by an imaginary frequency of $1,851 \, cm^{-1}$. The proton appears to be shared between the two C_5 and O2ε atoms (C_5–H and O2ε–H distances are 1.44 and 1.25 Å, respectively) [56]. The energy cost for this process is 12.3 kcal/mol. From TS3'b, a stable intermediate INT3'b lying 8.5 kcal/mol below the reference, originates. The Asp297 residue is now involved into two hydrogen bonds with the PQQ oxygen atoms. Through the transition state TS4'b, the second proton is transferred from the O2ε of Asp297 to the negatively charged oxygen atom attached to C_4 atom of PQQ, finally leading to the $PQQH_2$ reduced species. The imaginary frequency at $1,474 \, cm^{-1}$ refers to the stretching of the two O2ε–H (1.20 Å) and C_4O–H (1.24 Å) bonds [56]. The computed energetic expense to overcome this transition state is of 3.0 kcal/mol.

A look to the paths in Fig. 15 shows that the enolization may occur only if mediated by the aspartate residue.

As we have seen from the PESs reported in the Fig. 15, both the proposed mechanisms (A) and (B) lead to two energetic profiles kinetically very slow and in particular out of rates characterizing the enzymatic catalysis. This fact puts us in doubt that the suggested mechanisms may not represent the real ones. Thus, we have revised the literature proposal, addressing particular to the addition–elimination process in which we have foreseen some possibility of improvement.

The new reaction sequence is depicted in the Scheme 4. The optimized structures of the stationary points belonging to the new reaction profile (traced in Fig. 15 with a red dashed line on the top) are depicted in Fig. 16.

In this mechanism, named (C), we propose that the proton shift from the aspartate residue to the PQQ oxygen atom belonging to the C_5O carbonyl group (TS2a), can occur after the substrate C–H bond cleavage. On the other hand, no experimental evidences can thwart this hypothesis.

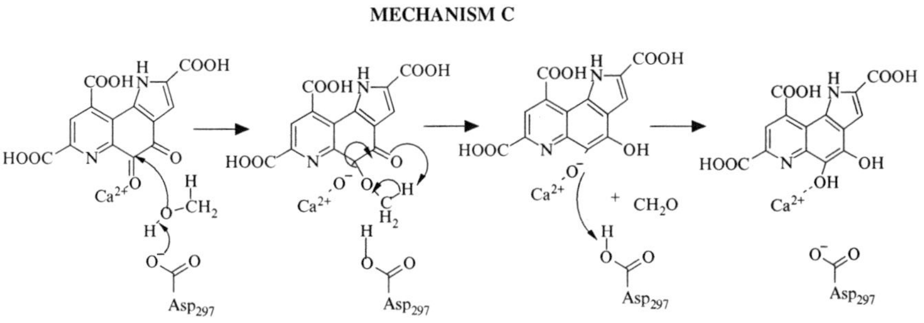

MECHANISM C

Scheme 4 Modified addition–elimination mechanism C proposed in this study

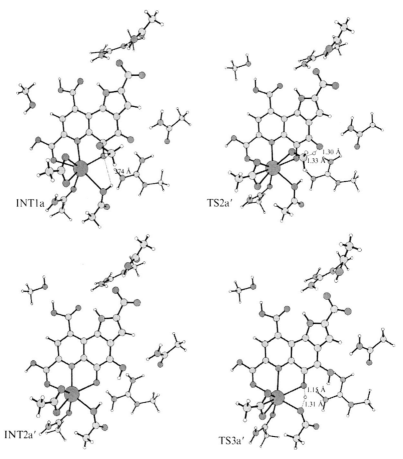

Fig. 16 Optimized geometries of the stationary points belonging to the modified addition elimination mechanism C

If we assume the protonation of the cofactor as the last step, the obtained PES becomes quite different with respect to before (red dashed line of the top of Fig. 15). Starting from the INT1a intermediate, the transition state (TS2'a) describing the C–H bond breaking by the oxygen atom linked to C_4 (imaginary frequency at 1,389 cm^{-1}), lies at 16.0 kcal/mol (dashed line). In this species the C–H and a H–O distances are of 1.33 and 1.30 Å, while the O_{met}–C_{5PQQ} bond is 2.37 Å. The lower energy of TS2'a with respect to that of TS3a, corresponding to the breaking of C–H in the mechanism (A), can be explained by considering that the negative charge of the cofactor makes easier the abstraction of the H$^+$ from the substrate CH by the C_4O carbonyl oxygen and that the instability of INT1a with respect to INT2a in the previous reaction path, due to a weaker bond between the calcium cation and a protonated aspartate, reduces the energetic cost to break the substrate CH bond. The barrier height (16.0 kcal/mol) may be now compatible with an enzymatic catalysis.

From the TS2'a, an intermediate (INT2'a), in which the C_5O is negatively charged and the C_4O is protonated, is found 16.6 kcal/mol below the reference point, with CH_2O product leaving the metallic centre.

Finally, through the transition state TS3'a lying at 8.7 kcal/mol above INT2'a, responsible for the protonation of the PQQ C_5O oxygen atom, the reduction of the cofactor to the $PQQH_2$ is complete.

Both mechanisms (A) and (B) were proven to be not compatible with the kinetics requirements of the enzymatic processes. In fact, too high energy barriers (34.6 kcal/mol for addition–elimination, and 32.3 kcal/mol for hydride transfer) that exceed considerably the usual limits of 15–20 kcal/mol, were found in correspondence of the rate-determining step that is, in both cases, the cleavage of the covalent C–H bond in the substrate.

The third mechanism that we propose [47] as an "addition–elimination–protonation" process (mechanism C), in which the sequence of the steps sees the cleavage of the C–H bond in the substrate to occur before the cofactor protonation by the Asp297 amino acid residue, was found to be more reliable being the activation energy in the rate-determining step 16.0 kcal/mol in the gas-phase, and 11.1 kcal/mol in the protein environment simulations.

The mechanism (C) proposed [56] can be regarded with confidence as the preferred reaction path followed by the PQQ containing methanol dehydrogenase enzyme.

3.4 Formate Dehydrogenase

Formate dehydrogenase is a molybdoenzyme belonging to the anaerobic formate hydrogen lyase complex of *Escherichia coli* microrganism, that catalyzes the oxidation of formate to carbon dioxide [57]. The first crystal structure of both oxidized and reduced forms revealed a molybdenum, a Fe_4S_4 cluster and a selenocysteine residue central to the catalytic activity [57]. In the reduced Mo(IV) form, the molybdenum in the active site is coordinated by four sulphur atoms coming from the cofactors bis-molybdopterin guanine dinucleotide (MGD), as encountered for nitrate reductase, and by the selenium atom of the SeCys140 residue. Recently, another formate-reduced *E. Coli* FDH crystal structure was solved [58] and has shown a selenium atom of the SeCys140 away from the molybdenum, implying that the selenium is not a metal ligand in the reduced form of the enzyme.

From these experimental evidences, two reaction mechanisms for the oxidation of the formate were proposed (see Scheme 5).

Mechanism A [59] implies the coordination of the formate substrate by its oxygen atom to the molybdenum, probably by replacing an –SH ligand. The selenium atom abstracts the proton by cleaving the C–H bond in the substrate while two electrons are transferred to the Mo center. The Mo (VI) form is restored in another step, with the two electrons traveling to the Fe_4S_4 cluster through the MGD moiety.

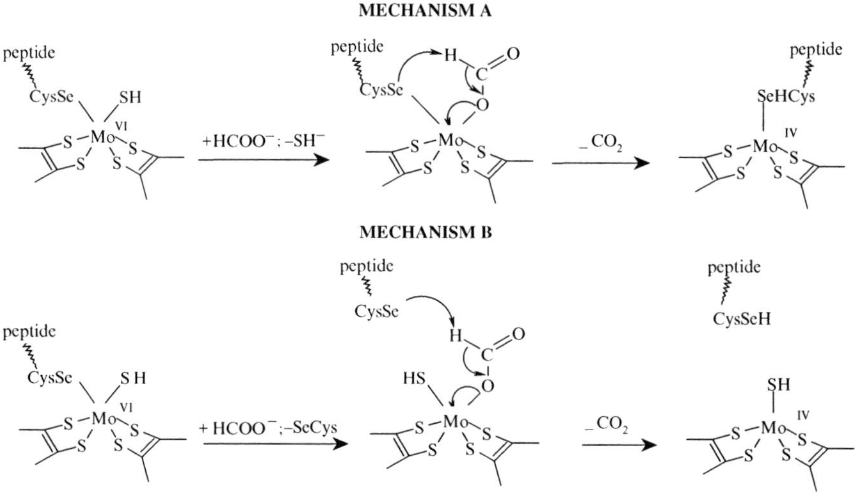

MECHANISM A

MECHANISM B

Scheme 5 Catalytic mechanisms A and B proposed for formate dehydrogenase enzyme

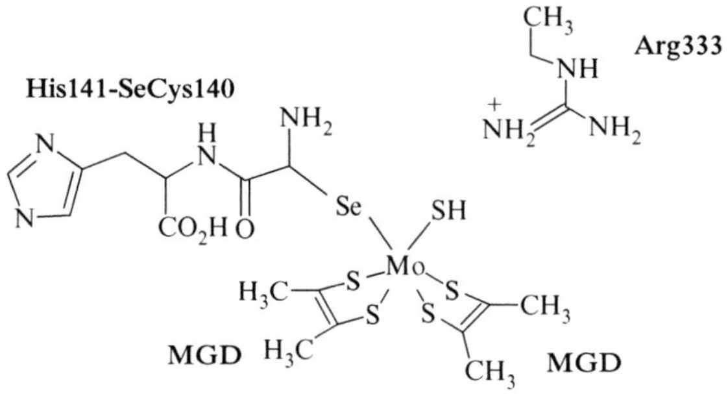

Fig. 17 Model cluster used to simulate formate dehydrogenase active site

In the mechanism B [60], the formate binding displaces the selenocysteine ligand that away from the metallic centre abstracts the proton from the substrate leading to the CO_2 product, that is released from the active site, and to the Mo(IV) form of the enzyme. The two mechanisms, A and B, differ in the role played by the SeCys140 residue as a Mo ligand or as unbound form during catalysis.

With the aim to determine which one may represent the preferred reaction path followed by FDH enzyme among the two proposals, we performed a density functional based study [61] on formate oxidation by a Mo containing cluster as simplified model of the active site (Fig. 17).

For both mechanisms, two spin multiplicities, singlet and triplet, for the active site were found close in energy.

The Potential Energy Surfaces (PESs) for both oxidation mechanisms A and B of the formate substrate by the Mo containing cluster are reported in the Fig. 18.

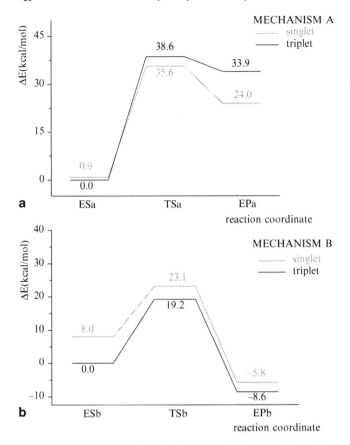

Fig. 18 Gas-phase potential energy surfaces for formate oxidation according to (**a**) mechanism A and (**b**) B

In both cases, the oxidation reaction starts with the formation of a complex between the substrate and the active site occurring by the coordination of formate oxygen atom to the metal [61]. The next abstraction of the proton from the formate carbon by the selenium atom of the SeCys140 residue, leads finally to the products [61].

For both reaction paths, the rate-determining step is the H abstraction from the C atom of the substrate performed by the selenium belonging to the SeCys residue.

As far as the mechanism A involving the SeCys140 residue as Mo ligand is concerned, the two ESa complexes in the different multiplicities are practically degenerate (the energetic gap is only 0.9 kcal/mol), so they may coexist. The TSa is found at 35.6 (singlet) and 38.6 (triplet) kcal/mol, above the initial minimum. Products complexes lie at 24.0 and 33.9, for singlet and triplet, respectively.

Although the energetic profile seems to be more favorable for the reaction in the ground state, the activation energies required to oxidize the substrate to carbon dioxide are in both cases very high (34.7 and 38.6 kcal/mol, for singlet and triplet, respectively). In addition, the complex between the reduced active site and the CO_2, in which the latter is still bound to Mo, is decidedly less stable than ESa, so that the reaction results to be very endothermic.

In the mechanism in which the SeCys140 is in the unbound form (mechanism B), the energy gap between the singlet and triplet ESb species is 8 kcal/mol, being the latter the minimum energy structure. Transition states are found at 23.1 and 19.2 kcal/mol, for singlet and triplet, respectively. Finally, products are found 5.8 and 8.6 kcal/mol below the minimum. The triplet potential energy profile lies below that obtained for the low-spin state, thus representing the minimum energy path.

In the mechanism B the height of the barriers for the formate oxidation are lower than those computed for mechanism A (19.2 for triplet, and 23.1 kcal/mol, for singlet). In particular, that computed in the high-spin path, is now in the range normally accepted for the enzymatic catalysis. The formation of the EPb product is exothermic for both electronic states.

A simple explanation for the reliability of the mechanism B with respect to the mechanism A can be found in the different values of net charge exhibited by selenium. When Se is coordinated to the molybdenum (mechanism A), the lone pair necessary to abstract the proton from the substrate is not fully available, but involved in the bonding with the cluster (Se net charge is 0.149 and 0.062|e|, in the ESa singlet and triplet species, respectively). In the case of mechanism B, the net charge on Se assumes values of −0.242 for singlet ESb and −0.319|e| for the triplet ESb. This different charge value accounts for the better character of catalytic base exhibited by the SeCys140 residue in the mechanism B.

Thus, we can conclude that the selenium atom is a better base in performing the proton abstraction from the substrate when it is not bound to the metallic centre since in this case the lone pair present on the negative selenocysteine is more available.

In the Table 1, gas-phase activation energies computed for the mechanism B at different levels of theory are reported. They are computed as far as the high spin electronic state is concerned. H-GGA BH and HLYP and PBE1PBE provide an activation energy of 17.9 and 19.9 kcal/mol, respectively. These values are very similar to those obtained with B3LYP (19.2 kcal/mol), being the PBE1PBE the one marking mostly. This is quite expected since the behaviour of B3LYP functional is closer to the one of PBE1PBE rather than to the BH and HLYP. The values obtained with the HM-GGA functionals are 23.6 kcal/mol, for MPWB1K, and 22.8 kcal/mol, for the BB1K, with a difference of 0.8 kcal/mol between them. The resemblance of these two methods in determining barrier heights is in agreement with the available literature data.

H-GGA methods are found to provide the smallest activation energies with respect to the HM-GGA ones, that seem to overestimate barrier heights. Taking into account the most recent benchmarking studies, also for this investigation the B3LYP functional seems to provide lower activation energies.

Based on these findings, the "unbound SeCys" mechanism [61] can be regarded with confidence as the preferred reaction path followed by the Mo containing formate dehydrogenase enzyme.

Table 1 Activation energy (in kcal/mol) for the formate oxidation by formate dehydrogenase at different DF levels

	Activation energy (kcal/mol)
B3LYP	19.2
BHandHLYP	17.9
MPWB1K	23.6
BB1K	22.8
PBE1PBE	19.9

3.5 Cadmium Carbonic Anhydrase

Carbon dioxide (CO_2) represents a key metabolite in all living organisms. It exists in equilibrium with bicarbonate (HCO_3^-), which unlike CO_2 is poorly soluble in lipid membranes. Carbon dioxide can freely spreads in and out of the cell, while bicarbonate must be transported. The interconversion of carbon dioxide and bicarbonate proceeds slowly at physiological pH, so organisms must produce enzymes to accelerate the process [62]. Carbonic anhydrases are enzymes that catalyse the reversible hydration of CO_2.

$$H_2O + CO_2 \longleftrightarrow H^+ + HCO_3^-$$

Carbonic anhydrases were found in all kingdoms of life [62]. Their essential roles consists in facilitating the transport of carbon dioxide and protons in the intracellular space, but are also involved in many other processes, from respiration and photosynthesis in eukaryotes to cyanate degradation in prokaryotes. Carbonic anhydrase (CA) isozymes are metalloenzymes consisting of a single polypeptide chain (Mr~29,000) complexed to an atom of zinc. They are incredibly active catalysts, with a turnover rate (kcat) of about 10^6 reactions per second! Three distinct classes of CA are known (called α, β and γ) that show very little sequence or structural similarity, yet they all have the same function and require a zinc ion at the active site in their catalytic activity. CA from mammals belong to the α class, the plant enzymes belong to the β class, while the enzyme from methane-producing bacteria that grow in hot springs forms the γ class.

The Zn^{2+} ion is the second most abundant transition element in biology (after iron) and although it is an integral component of more than 400 enzymes involved in different vital processes, in CA it can be replaced with other divalent transition metal ions such as Co^{2+}, Mn^{2+}, Ni^{2+}, Hg^{2+}, Cd^{2+} and Cu^{2+} that conserve in part the catalytic activity of enzyme [61, 62]. The substitution in CA of the native zinc with other metal ions as those above mentioned was used to gain structural details about the Zn–CA [61, 62].

Evidence of in vivo utilization of Cd^{2+} in CA in some marine organisms was published by Price and Morel [63] and Morel et al. [64]. Only in the 2000, Lane and Morel [65] arrived at the surprising result of the identification of the first special cadmium carbonic anhydrase (Cd–CA) on behalf of the same marine diatom under conditions where the zinc concentration is low [66–68].

The mechanism of action of the mammalian zinc ion containing carbonic anhydrase was studied in depth at both experimental [69–76] and theoretical [77–86] levels. The enzymatic cycle consists of a two-step mechanism: the first step is a nucleophilic attack of a zinc-bound hydroxide ion on carbon dioxide; the second step consists in the ionisation of the zinc-bound water molecule and the removal of a proton from the active site for its regeneration.

Because of the presence in literature of a previous theoretical investigation on the Zn–CA system [86], we have undertaken a comparative study of the catalytic mechanism of the cadmium carbonic anhydrase performed at same level of theory of Zn–CA and devoted to establish the main structural and energetical differences present in the catalytic reactions governed by the two different transition metal ions [87].

To simulate the Cd–CA active site we have used the same model (Fig. 19) employed in the previous investigation of the Zn^{2+}–carbonic anhydrase mechanism [86]. It includes a Cd^{2+} cation linked to an –OH group and to three imidazole rings belonging to the three histidine residues His94, His96 and His119 present in the inner coordination shell of the ion. In addition, the active site interacts with the protein via the side chains of Glu106, Thr199. These residues, as suggested by Thoms [75], could act as acceptors of the proton released by the zinc-bound water molecule instead of the imidazole ring of the His64, as previously indicated [77, 86]. In addition, another water molecule named 'deep water' is present. In order to reduce the computational efforts, the residue Glu106 was replaced with an acetate fragment.

The B3LYP/DZVP reaction profile is reported in Fig. 20 together with that obtained by Bottoni et al. in their study on the human Zn–CA [86].

The nucleophilic attack of an –OH lone pair of the M0 (naked enzyme) on the carbon atom of CO_2 gives rise to the intermediate M1 (see Fig. 21) without activation barrier. Contrarily to what happens for Zn–CA, in which M1 is located at –6.03 kcal/mol, in the case of Cd–CA, this species results to be strongly stabilized (it lies at –34.33 kcal/mol below M0 and CO_2 separated reactants energy taken as reference).

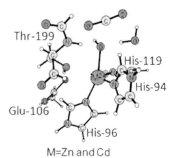

Fig. 19 Model used to simulate carbonic anhydrase active site

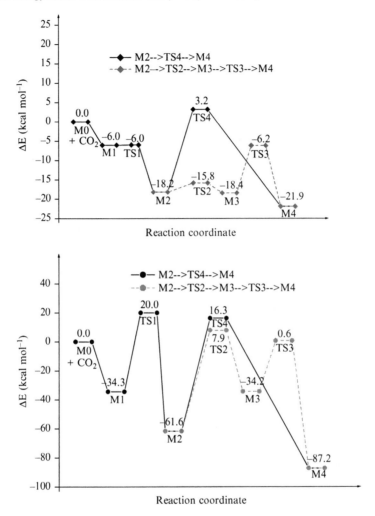

Fig. 20 B3LYP potential energy profile obtained for Cd–CA (*top*) in comparison with the one of Zn–CA [86] (*bottom*)

The comparison of Zn– and Cd–M1 geometrical features (Fig. 21) shows clearly the different coordination chemistry of the two metal ions. The cadmium ion appears to be tetracoordinated. The distance between the O and C atoms is, in the case of cadmium containing enzyme, much shorter than in the case of Zn–CA [86] (1.481 versus 2.630 Å, respectively) and the Cd^{2+}–O and Cd^{2+}–O_a bonds are 2.437 and 3.161 Å long, respectively.

A charge transfer of about 0.41e from the –OH group to CO_2 moiety occurs as evidenced by the NBO analysis. A strongly polarized σ bond is formed between O and C, with 72.67% contribution from oxygen), in which the carbon and oxygen atoms use an orbital sp and p, respectively. The oxygen atoms of the CO_2 fragment

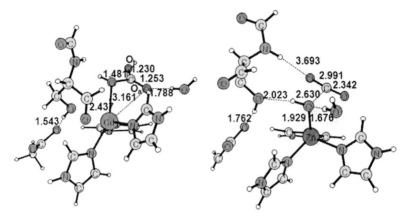

Fig. 21 Optimized geometry for M1 Cd–CA (*left*) and Zn–CA [95] (*right*)

are differently charged (q_{Oa} = −0.873e and q_{Ob} = −0.792e) corroborating the already happened nucleophilic attack. On the contrary, in the case of the Zn^{2+} ion the Mulliken population analysis indicated the predominantly electrostatic nature of the interaction between the –OH and CO_2 groups [86]. Consequently, the activation energy required for the evolution of the process, is in the case of Cd–CA considerably higher (54.34 kcal/mol) than that required by the Zn–CA. In TS1, the distance between –OH group and carbon atom of CO_2 is 1.476 Å. The hydrogen carbonate fragment, lying out from the molecular plane, is practically monocoordinated to the Cd^{2+} ion (Cd^{2+}–O = 2.239 Å) since the Cd–O_a distance is still quite long (2.809 Å). For Zn–CA, TS1 is rashly defined as a barrier because of the small energy difference with respect to M1 (only 0.04 kcal/mol). Bottoni et al. [86] explained this low barrier with the presence of a stabilizing H-bond network, that is absent in Cd–CA. It is worth noting that Cd–TS1 is substantially different from Zn–TS1. In fact, in Cd–TS1 the imaginary vibrational frequency is associated to the stretching mode of the incoming Cd^{2+}–O_a bond coupled to the out of plane of the HCO_3^- fragment that assumes a planar disposition with respect to the metal center. The Cd^{2+} ion is pentacoordinated in Cd–TS1 and the Cd^{2+}–O_a bond is shorter than in M1. Instead, Zn–TS1 represents the transition state for the C–O bond formation [87].

The tendency to the pentacoordination of Cd^{2+} is still evident in M2, the next minimum encountered on the potential energy surface [87].

M2 can be connected to the M4 intermediate, through two distinct reaction channels. The first, M2 → TS4 → M4, is a one-step process, whereas the second, M2 → TS2 → M3 and M3 → TS3 → M4, proceeds in two steps.

In the former channel, the M2 → M4 interconversion occurs through an internal rotation (TS4) of the bicarbonate moiety around the C–O_a bond. The rotation requires a very big amount of energy (about 77.94 kcal/mol) because it is accompanied by several geometrical modifications. In TS4, the cadmium ion becomes tetracoordinated since the Cd^{2+}–O bond is broken and the hydrogen carbonate group remains anchored to Cd^{2+} through the O_a atom (Cd^{2+}–O_a = 2.344 Å).

Imaginary vibrational frequency corresponds to the torsion of bicarbonate around the C–O_a bond. In M4, the Cd^{2+} ion is still linked to the O_a atom with a distance shorter (2.248 Å) than in M2. The water molecule, bond to the O_b atom in TS4, is now attached to O_a atom (2.008 Å) likewise than in M2. It is found in a suitable orientation to attack the cadmium ion, that now presents a free coordination position, in order to restore the catalyst [87].

In the second reaction channel, the transition state TS2 connects the M2 and M3 intermediates. The conversion, occurring through two simultaneous proton transfers involving the two threonine and glutamate residues and bicarbonate, requires 69.52 kcal/mol. Vibrational imaginary mode confirms these motions. The geometrical features of TS2 show a Cd^{2+} ion linked to the O_a and O atoms with distances of 2.445 Å and 2.302 Å, respectively. The water molecule forms a H-bond with O_b oxygen of the bicarbonate fragment (2.018 Å) that presents a negative charge due to the proton loss. The M3 intermediate lies at 27.40 kcal/mol above M2. In this species, the metal center appears to be still pentacoordinated (the Cd^{2+}–O and Cd^{2+}–O_a distances are of 2.325 and 2.390 Å, respectively). The deep water molecule is, like in the transition state TS2, involved in a H-bond with the O_b atom (1.834 Å). Starting from M3, a barrier of about 35 kcal/mol should be overcome to reach the final M4 complex. TS3 appears as a four-coordinated species where the Cd^{2+}–O_a distance is shortening (2.215 Å) and the Cd^{2+}–O bond is completely broken (4.109 Å). The product of the bicarbonate rearrangement (M4) is 42.97 kcal/mol more stable than M3 [87].

Although also for Zn–CA, two different reaction channels leading from M2 to M4 were hypothesized [86], the results are significantly different from that obtained in the case of the Cd–CA [87]. The differences concern both the mutual stability of intermediates and the height of barriers that should allows to pass from a minimum to another (path b of Fig. 20). Contrarily to the reaction path of Zn–CA [86], the one of Cd–CA is characterized by deep holes and high barriers thus, despite the major exothermicity of the reaction with respect to the most common Zn–CA, the Cd–CA enzyme is prevent from carrying out its activity easily. This fact was underlined by the experimental studies on cadmium CA present in literature [88, 89] that attribute the different activity of cadmium and zinc enzymes to the different coordination chemistry of metal ions that in turn depends on their size. Furthermore, they point their attention to the minor acidic character of cadmium with respect to zinc.

The fact that the Cd^{2+} is a Lewis acid weaker than Zn^{2+} and that it has a minor polarizing effect on its environment can decide if the fourth metal ion ligand, in the initial form of enzyme, can be present as an –OH group or a water molecule. However, although we knew that the cadmium–hydroxide intermediate is generally present at more basic pH values, we have chosen to consider the same starting geometry for both enzymes. This means that in our case, the minor acidity of the cadmium ion can have significance only as far as the strength with which the metal cation binds the CO_2 is concerned. In fact, in the case of cadmium ion, the oxygen atom of the –OH group in M0 has a charge more negative than that on the same atom in the zinc M0 ($-1.312e$ versus $-1.010e$, respectively), and thus can establish with the substrate a more strong bond. This is confirmed by NBO analysis and by the greater stability of all cadmium intermediates with respect to those of zinc along the whole reaction path [87].

Our findings are in agreement with the experimental observations that indicate how the Zn–CA be decisively more active than Cd–CA (the activity of Cd–CA was estimated to be only the 2% of that of Zn–CA [90]).

Really, is sufficient to stop to the first intermediate M1 and to the following transition state TS1 to understand that, too large modifications in the geometry and hence a very big amount of energy are necessary to continue the next transformations until M4. TS1 represents the rate determining step of Cd–CA mechanism for all the two possible channels, while, in the case of Zn–CA, the crucial steps are TS1 or TS4 depending if the transformation of M2 occurs through the first or the second channel.

As evident, the coordination chemistry of the two cations plays an important role in determining the catalytic activity of the CA enzyme [87]. The more flexible coordination geometry, the major rapidity in ligand exchange, the lack of redox activity and the more marked role as Lewis acid are just a few examples to explain the better performance of zinc with respect to cadmium ion. This can be really observed in the comparison of two catalytic paths where the limits of cadmium ion are evidenced by the presence of high interconversion barriers between intermediates having different coordination.

4   Conclusions

In this summary, computational approaches and chemical models applied to the quantum chemical studies of reactions catalyzed by metalloenzymes were presented and discussed. The studied enzymes cover a wide range of biological processes, such as oxidation of small molecules involved in nitrogen fixation, hydrogen production and energy accumulation, hydrolysis of peptides, transport of carbon dioxide and protons.

Even if accurate quantum chemical treatment of transition metals containing enzymes is a relatively new area, it is becoming a powerful tool for description of intermediates and transition states originating during chemical reactions, that integrates classical experimental methodologies.

The protocols used here were proven to give results that are usually in very good agreement with experimental indication.

Acknowledgement Financial support from the Università della Calabria and Regione Calabria (POR Calabria 2000/2006, misura 3.16, Progetto PROSICA) is gratefully acknowledged.

References

1. M. J. Frisch, G. W. Trucks, H. B. Schlegel, G. E. Scuseria, M. A. Robb, J. R. Cheeseman, J. A. Jr. Montgomery, T. Vreven, K. N. Kudin, J. C. Burant, J. M. Millam, S. S. Iyengar, J. Tomasi, V. Barone, B. Mennucci, M. Cossi, G. Scalmani, N. Rega, G. A. Petersson, H. Nakatsuji, M. Hada, M. Ehara, K. Toyota, R. Fukuda, J. Hasegawa, M. Ishida, T. Nakajima, Y. Honda, O. Kitao, H. Nakai, M. Klene, X. Li, J. E. Knox, H. P. Hratchian, J. B. Cross, C. Adamo, J. Jaramillo, R. Gomperts, R. E. Stratmann, O. Yazyev, A. J. Austin, R. Cammi, C. Pomelli, J. W. Ochterski, P. Y. Ayala, K. Morokuma, G. A. Voth, P. Salvador, J. J. Dannenberg, V. G. Zakrzewski, S. Dapprich, A. D. Daniels, M. C. Strain, O. Farkas, D. K. Malick, A. D. Rabuck, K. Raghavachari, J. B. Foresman, J. V. Ortiz, Q. Cui, A. G. Baboul, S. Clifford, J. Cioslowski, B. B. Stefanov, G. Liu, A. Liashenko, P. Piskorz, I. Komaromi, R. L. Martin, D. J. Fox, T. Keith, M. A. Al-Laham, C. Y. Peng, A. Nanayakkara, M. Challacombe, P. M. W. Gill, B. Johnson, W. Chen, M. W. Wong, C. Gonzalez, J. A. Pople, Gaussian, Inc., Pittsburgh, PA, 2003.
2. A. D. Becke, Density-functional thermochemistry.3. The role of exact exchange. *J. Chem. Phys.* **98**, 5648–5652 (1993).
3. C. Lee, W. Yang, R. G. Parr, Development of the Colle–Salvetti correlation-energy formula into a functional of the electron density. *Phys. Rev. B* **37**, 785–789 (1988).
4. A. D. Becke, A new mixing of Hartree–Fock and local density-functional theories *J. Chem. Phys.* **98**, 1372 (1993).
5. A. D. Becke, Density-functional exchange-energy approximation with correct asymptotic behavior *Phys. Rev. A* **38**, 3098 (1988).
6. P. E. M. Siegbahn, M. R. A. Blomberg, Transition metal systems in biological studied by high-accuracy quantum chemical methods *Chem. Rev.* **100**, 421–437 (2000).
7. L. Noodleman, T. Lovell, W. G. Han, J. Li, F. Himo, Quantum chemical studies of intermediates and reaction pathways in selected enzymes and catalytic synthetic systems *Chem. Rev.* **104**, 459–508 (2004).
8. P. E. M. Siegbahn, Mechanisms of metalloenzymes studied by quantum chemical methods *Quart. Rev. Biophys.* **36**, 91–145 (2003).
9. F. Himo, Quantum chemical modelling of enzyme active sites and reaction mechanisms *Theor. Chem. Acc.* **116**, 232–240 (2006).
10. S. F. Sousa, P. A. Fernandes, M. J. Ramos, General performance of density functionals *J. Phys. Chem. A*, **111**, 10439–10452 (2007), and references there in.
11. Y. Zhao, D. G. Truhlar, Hybrid meta density functional theory methods for thermochemistry, thermochemical kinetics, and noncovalent interactions: The MPW1B95 and MPWB1K models and comparative assessments for hydrogen bonding and van der Waals interactions *J. Phys. Chem.* **108**, 6908–6918 (2004).
12. Y. Zhao, B. J. Lynch, D. G. Truhlar, Development and assessment of a new hybrid density functional model for thermochemical kinetics *J. Phys. Chem. A* **108**, 2715–2719 (2004).
13. J. P. Perdew;, K. Burke, M. Ernzerhof, Generalized gradient approximation made simple *Phys. Rev. Lett* **77**, 3865–3868 (1996).
14. S. Miertus, E. Scrocco, J. Tomasi, Electrostatic interaction of a solute with a continuum. A direct utilization of ab initio molecular potentials for the prevision of solvent effects *Chem. Phys.* **55**, 117–129 (1981).
15. S. Miertus, E. Scrocco, J. Tomasi, Approximate evaluations of the electrostatic free energy and internal energy changes in solution processes *Chem. Phys.* **65**, 239–245 (1982).
16. M. Cossi, V. Barone, R. Commi, J. Tomasi, Ab initio study of solvated molecules: a new implementation of the polarizable continuum model *Chem. Phys. Lett.* **255**, 327 (1996).
17. V. Barone, M. Cossi, B. Menucci, J. Tomasi A new definition of cavities for the computation of solvation free energies by the polarizable continuum model *J. Chem. Phys.* **107**, 3210–3221 (1997).
18. F. Maseras, K. Morokuma, IMOMM: A new integrated *ab initio* + molecular mechanics geometry optimization scheme of equilibrium structures and transition states *J. Comput. Chem.* **16**, 1170–1179 (1995).

19. S. Humbel, S. Sieber, K. Morokuma, The IMOMO method: Integration of different levels of molecular orbital approximations for geometry optimization of large systems: Test for n-butane conformation and $S_N 2$ reaction: RCl + Cl$^-$ *J. Chem. Phys.* **105**, 1959–1967 (1996).

20. T. Matsubara, S. Sieber, K. Morokuma, A test of the new "integrated MO + MM " (IMOMM) method for the conformational energy of ethane and n-butane *Int. J. Quantum Chem.* **60**, 1101–1109 (1996).

21. M. Svensson, S. Humbel, R. D. J. Froese, T. Matsubara, S. Sieber, K. Morokuma, ONIOM: A multilayered integrated MO + MM method for geometry optimizations and single point energy predictions. A test for Diels–Alder reactions and Pt(P(t − Bu)$_3$)$_2$ + H$_2$ oxidative addition *J. Phys. Chem.* **100**, 19357–19363 (1996).

22. M. Svensson, S. Humbel, K. Morokuma, Energetics using the single point IMOMO (integrated molecular orbital + molecular orbital) calculations: Choices of computational levels and model system *J. Chem. Phys.* **105**, 3654–3661 (1996).

23. S. Dapprich, I. Komuromi, K. S. Byun, K. Morokuma, M. J. Frisch, A new ONIOM implementation in Gaussian98. Part I. The calculation of energies, gradients, vibrational frequencies and electric field derivatives *J. Mol. Struct.(THEOCHEM)* **461–462**, 1–21 (1999).

24. T. Vreven, K. Morokuma, On the application of the IMOMO (integrated molecular orbital + molecular orbital) method *J. Comput. Chem.* **21**, 1419–1432 (2000).

25. T. A. Wesolowski, A. Warshel, Frozen density functional approach for ab-initio calculations of solvated molecules *J. Phys. Chem.* **97**, 8050–8053 (1993).

26. G. Hong, M. Strajbl, T. A. Wesolowski, A. Warshel, Constraining the electron densities in DFT method as an effective way for ab initio studies of metal-catalyzed reactions *J. Comp. Chem.* **21**, 1554–1561 (2000).

27. R. Hille, The mononuclear molybdenum enzymes *Chem. Rev.* 96, 2757–2816 (1996).

28. J. A. Craig, R. H. Holm, Reduction of nitrate to nitrite by molybdenum-mediated atom transfer: a nitrate reductase analog reaction system *J. Am. Chem. Soc.* **111**, 2111–2115 (1989).

29. M. J. Dias, M. E. Than, A. Humm, R. Huber, G. P. Bourenkov, H. D. Bartunik, S. Bursakov, J. Calvete, J. Caldeira, C. Carneiro, J. J. G. Moura, I. Moura, M. J. Romao, Crystal structure of the first dissimilatory nitrate reductase at 1.9 solved by MAD methods *Structure* **7**, 65–79 (1999).

30. C. E. Webster, M. B. Hall, The theoretical transition state structure of a model complex bears a striking resemblance to the active site of DMSO reductase *J. Am. Chem. Soc.* **123**, 5820–5821 (2001).

31. M. Leopoldini; N. Russo; M. Toscano; M. Dulak; A. T. Wesoloski, Mechanism of nitrate reduction by *Desulfovibrio desulfuricans* nitrate reductase. A theoretical investigation *Chem. A Eur. J.* **12**, 2532–2541 (2006).

32. K. Burke, J. P. Perdew, Y. Wang in *Electronic Density Functional Theory: Recent Progress and New Directions*, Eds. J. F. Dobson, G. Vignale, and M. P. Das (Plenum, New York, 1998).

33. J. P. Perdew in *Electronic Structure of Solids* '91, Eds. P. Ziesche and H. Eschrig (Akademie Verlag, Berlin, 1991) 11.

34. J. P. Perdew, J. A. Chevary, S. H. Vosko, K. A. Jackson, M. R. Pederson, D. J. Singh, C. Fiolhais, Atoms, molecules, solids, and surfaces: Applications of the generalized gradient approximation for exchange and correlation *Phys. Rev. B* **46**, 6671–6687 (1992).

35. J. P. Perdew, J. A. Chevary, S. H. Vosko, K. A. Jackson, M. R. Pederson, D. J. Singh, C. Fiolhais, Erratum: Atoms, molecules, solids, and surfaces: Applications of the generalized gradient approximation for exchange and correlation *Phys. Rev. B* **48**, 4978 (1993).

36. J. P. Perdew, K. Burke, Y. Wang, Generalized gradient approximation for the exchange-correlation hole of a many-electron system *Phys. Rev. B* **54**, 16533–16539 (1996).

37. M. Kozak, Comparison of initiation of protein synthesis in procaryotes, eucaryotes, and organelles. *Microbiol. Rev.* **47**, 1–45 (1983).

38. H. C. Neu, The crisis in antibiotic resistance. *Science* **257**, 1064–1073 (1992).

39. T. Meinnel, S. Blanquet, Evidence that peptide deformylase and methionyl-tRNA(fMet) formyltransferase are encoded within the same operon in *Escherichia coli. J. Bacteriol.* **175**, 7737–7740 (1993).

40. T. Meinnel, S. Blanquet, Enzymatic properties of *Escherichia coli* peptide deformylase *J. Bacteriol.* **177**, 1883–1887 (1995).
41. P. T. R. Rajagopalan, A. Datta, D. Pei, Purification, characterization, and inhibition of peptide deformylase from *Escherichia coli Biochemistry* **36**, 13910–13918 (1997).
42. T. Meinnel, C. Lazennec, S. Blanquet, Mapping of the active site zinc ligands of peptide deformylase *J. Mol. Biol.* **254**, 175–183 (1995).
43. M. K. Chan, W. Gong, P. T. R. Rajagopalan, B. Hao, C. M. Tsai, D. Pei, Crystal structure of the *Escherichia coli* peptide deformylase *Biochemistry* **36**, 13904–13909 (1997).
44. T. Meinnel, S. Blanquet, F. Dardel, A new subclass of the zinc metalloproteases superfamily revealed by the solution structure of peptide deformylase *J. Mol. Biol.* **262**, 375–386 (1996).
45. D. Groche, A. Becker, I. Schlichting, W. Kabsch, S. Schultz, A. F. V. Wagner, Iron center substrate recognition and mechanism of peptide deformylase *Nat. Struc. Biol.* **5**, 1053–1058 (1998).
46. M. Leopoldini; N. Russo; M. Toscano, The role of the metal ion in formyl-peptide bond hydrolysis by a peptide deformylase active site model *J. Phys. Chem. B.* **110**, 1063–1072 (2006).
47. J. A. Duine; J. Frank, The prosthetic group of methanol dehydrogenase. Purification and some of its properties *Biochem. J.* **187**, 221–226 (1980).
48. S. White, G. Boyd, F. S. Mathews, Z. X. Xia, W. W. Dai, Y. F. Zhang, V. L. Davidson, The active site structure of the calcium-containing quinoprotein methanol dehydrogenase *Biochemistry* **32**, 12955–12958 (1993).
49. K. Matsushita, K. Takahashi, O. Adachi, A novel quinoprotein methanol dehydrogenase containing an additional 32-kilodalton peptide purified from Acetobacter methanolicus: Identification of the peptide as a MoxJ product *Biochemistry* **32**, 5576–5582 (1993).
50. J. J. Meulenberg, E. Sellink, N. H. Riegman, P. W. Postma, Nucleotide sequence and structure of the Klebsiella pneumoniae PQQ operon *Mol. Gen. Genet.* **232**, 284–294 (1992).
51. Z. Xia, W. Dai, Y. Zhang, S. A. White, G. D. Boyd, F. S. Mathews, Determination of the gene sequence and the three-dimensional structure at 2.4 angstroms resolution of methanol dehydrogenase from Methylophilus W3A1 *J. Mol. Biol.* **259**, 480–501 (1996).
52. C. Anthony, Quinoprotein-catalysed reactions *Biochem. J.* **320**, 697–711 (1996).
53. J. J. Frank, M. Dijkstra, J. A. Duine, C. Balny, Kinetic and spectral studies on the redox forms of methanol dehydrogenase from Hyphomicrobium X. *Eur. J. Biochem.* **174**, 331–338 (1988).
54. A. J. Olisthoorn, J. A. Duine, On the mechanism and specificity of soluble, quinoprotein glucose dehydrogenase in the oxidation of aldose sugars *Biochemistry* **37**, 13854–13861 (1998).
55. Y. J. Zheng, T. C. Bruice, Conformation of coenzyme pyrroloquinoline quinone and role of Ca^{2+} in the catalytic mechanism quinoprotein methanol? dehydrogenase *Proc. Natl. Acad. Sci USA* **94**, 11881–11886 (1997).
56. M. Leopoldini, N. Russo, M. Toscano, The preferred reaction path for the oxidation of methanol by PQQ-containing methanol dehydrogenase. Addition–elimination versus hydride transfer mechanism *Chem. A Eur. J.* **13**, 2109–2117 (2007).
57. J. C. Boyington, V. N. Gladyshev, S. V. Khangulov, T. C. Stadtman, P. D. Sun, crystal structure of formate dehydrogenase H: Catalysis involving Mo, molybdopterin, selenocysteine, and an Fe_4S_4 cluster *Science* **275**, 1305–1308 (1997).
58. H. C. A. Raaijmakers, M. J. Romao, Formate-reduced *E. coli* formate dehydrogenase H: the reinterpretation of the crystal structure suggests a new reaction mechanism *J. Biol. Inorg. Chem.* **11**, 849–854 (2006).
59. M. Leopoldini, S. G. Chiodo, M. Toscano, N. Russo, Reaction mechanism of molybdoenzyme formate dehydrogenase *Chem. A Eur. J.* **14**, 8674–8681 (2008).
60. P. Woolley, Models for metal ion function in carbonic anhydrase *Nature* **258**, 677–682 (1975).
61. D. R. Garmer, M. Krauss, Metal substitution and the active site of carbonic anhydrase *J. Am. Chem. Soc.* **114**, 6487–6493 (1992).
62. J. Solà, J. Mestres, M. Duran, R.Carbo, Ab initio quantum molecular similarity measures on metal-substituted carbonic anhydrase (M(II)CA, M = Be,Mg,Mn,Co,Ni,Cu,Zn, and Cd) *J. Chem. Inf. Comput. Sci.* **34**, 1047–1053 (1994).
63. N. M. Price, F. M. M. Morel, Cadmium and cobalt substitution for zinc in a marine diatom *Nature* **344**, 658–666 (1990).

64. F. M. M. Morel, J. R. Reinfelder, S. B. Roberts, C. P. Chamberlain, J. G. Lee, D. Yee, Zinc and carbon co-limitation of marine phytoplankton *Nature* **369**, 740–742 (1994).
65. T. W. Lane, F. M. M. Morel, A biological function for cadmium in marine diatoms *Proc. Natl. Acad. Sci. USA* **97**, 4627–4631 (2000).
66. H. Strasdeit, The first cadmium-specific enzyme *Angew. Chem. Int. Ed.* **40**, 707–709 (2001).
67. J. T. Cullen, T. W. Lane, F. M. M. Morel, R. M. Sherrel, Modulation of cadmium uptake in phytoplankton by seawater CO_2 concentration *Nature* **402**, 165–167 (1999).
68. J. G. Lee, F. M. M. Morel, Replacement of zinc by cadmium in marine phytoplankton. *Mar. Ecol. Prog. Ser.* **127**, 305–309 (1995).
69. I. Bertini, C. Luchinat, Cobalt(II) as a probe of the structure and function of carbonic anhydrase *Acc. Chem. Res.* **16**, 272–279 (1983).
70. D. N. Silverman, S. Lindskog, The catalytic mechanism of carbonic anhydrase: implications of a rate-limiting protolysis of water *Acc. Chem. Res.* **21**, 30–36 (1988).
71. D. W. Christianson, C. Fierke, Carbonic anhydrase: Evolution of the zinc binding site by nature and by design *Acc. Chem. Res.* **29**, 331–339 (1996).
72. E. Kimura, Model studies for molecular recognition of carbonic anhydrase and carboxypeptidase *Acc. Chem. Res.* **34**, 171–179 (2001).
73. X. Zhang, C. D. Hubbard, R. van Eldik, Carbonic anhydrase catalysis: A volume profile analysis *J. Phys. Chem.* **100**, 9161–9171 (1996).
74. C. Tu, B. C. Tripp, J. G. Ferry, D. N. Silverman, Bicarbonate as a proton donor in catalysis by Zn(II)- and Co(II)-containing carbonic anhydrases *J. Am. Chem. Soc.* **123**, 5861–5866 (2001).
75. S. Thoms, Hydrogen bonds and the catalytic mechanism of human carbonic anhydrase II *J. Theor. Biol.* **215**, 399–404 (2002).
76. D. Schroder, H. Schwartz, S. Schenk, E. Anders, A gas-phase reaction as a functional model for the activation of carbon dioxide by carbonic anhydrase *Angew. Chem. Int. Ed.* **42**, 5087–5090 (2003).
77. K. M. Merz, R. Hoffmann, M. J. S. Dewar, The mode of action of carbonic anhydrase *J. Am. Chem. Soc.* **111**, 5636–5649 (1989).
78. J-Y. Liang, W. N. Lipscomb, study of carbonic anhydrase catalysed hydration of CO_2. A brief review *Int. J. Quant. Chem.* **36**, 299–312 (1989).
79. Y-J. Zheng, K. M. Merz, Mechanism of the human carbonic anhydrase II-catalyzed hydration of carbon dioxide *J. Am. Chem. Soc.* **114**, 10498–10507 (1992).
80. M. Sola, A. Lledos, M. Duran, J. Bertran, Ab initio study of the hydration of carbon dioxide by carbonic anhydrase. A comparison between the Lipscomb and Lindskog mechanisms *J. Am. Chem. Soc.* **114**, 869–877 (1992).
81. K.M. Merz, L. Banci, Binding of bicarbonate to human carbonic anhydrase II: A continuum of binding states *J. Am. Chem. Soc.* **119**, 863–871 (1997).
82. M. Mauksch, M. Brauer, J. Weston, E. Anders, New insights into the mechanistic details of the carbonic anhydrase cycle as derived from the model system $[(NH_3)_3Zn(OH)]^+/CO_2$: How does the H_2O/HCO_3^- replacement step occur? *ChemBioChem* **2**, 190–198 (2001).
83. M. Brauer, J. L. Perez-Lustres, J. Weston, E. Anders, Quantitative reactivity model for the hydration of carbon dioxide by biomimetic zinc complexes *Inorg. Chem.* **41**, 1454–1463 (2002).
84. Z. Smedarchina, W. Siebrand, A. Fernandez-Ramos, Q. Cui, Kinetic isotope effects for concerted multiple proton transfer: A direct dynamics study of an active-site model of carbonic anhydrase II *J. Am. Chem. Soc.* **125**, 243–251 (2003).
85. Q. Cui, M. Karplus, Is a "proton wire" concerted or stepwise? A model study of proton transfer in carbonic anhydrase *J. Phys. Chem. B* **107**, 1071–1078 (2003).
86. A. Bottoni, C. Z. Lanza, G. P. Miscione, D. Spinelli, New model for a theoretical density functional theory investigation of the mechanism of the carbonic anhydrase: How does the internal bicarbonate rearrangement occur? *J. Am. Chem. Soc.* **126**, 1542–1550 (2004).
87. T. Marino, N. Russo, M. Toscano, A comparative study of the catalytic mechanisms of the zinc and cadmium containing carbonic anhydrase *J. Am. Chem. Soc.* **127**, 4242–4253 (2005).
88. C. Kimblin, V. J. Murphy, T. Hascall, B. M. Bridgewater, J. B. Bonanno, G. Parkin, Structural studies of the [tris(imidazolyl)phosphine]metal nitrate complexes $\{[Pim^{Pri,But}]M(NO_3)\}^+$ (M = Co, Cu, Zn, Cd, Hg): Comparison of nitrate-binding modes in synthetic analogues of carbonic anhydrase *Inorg. Chem.* **39**, 967–974 (2000).

89. J. Ejnik, A. Munoz, T. Gan, C. F. Shaw III, D. H. J. Petering, Interprotein metal ion exchange between cadmium carbonic anhydrase and apo or zinc-metallothionein *J. Biol. Inorg. Chem.* **4**, 784–790 (1999).
90. I. Bertini, C. Luchinat, M. S. Viezzoli, *Zinc Enzymes*; in *Progress in Inorganic Biochemistry and Biophysics*, eds. I., Bertini, C., Luchinat, W., Maret, and M. Zeppezauer, Vol. 1; Birkhauser: Boston, MA, 1986; Chapter 3.

Gold in Hydrogen Bonding Motif—Fragments of Essay

Demonstration of Nonconventional Hydrogen Bonding Patterns Between Gold and Clusters of Conventional Proton Donors

Eugene S. Kryachko

Abstract These are the fragments of essay that highlight the proneness of gold to form nonconventional hydrogen bonds with the conventional proton donor molecules and the fair relationship of the computational model with the experiments on anion photoelectron spectroscopy for the gold–water complexes $[Au(H_2O)_{1 \leq n \leq 2}]^-$.

1 Introduction

The hydrogen bonding interaction is the well-recognized and widely studied phenomenon that manifests in the formation of a so-called conventional hydrogen $(:= H-)$ bond. According to Pimentel and McClellan [1] (see also [2–17]), "a hydrogen bond is said to exist when (1) there is evidence of a bond, and (2) there is evidence that this bond sterically involves a hydrogen atom already bonded to another atom". This definition assumes that a conventional hydrogen bond is at least a three-party interaction. One party is a proton donor atom or molecule X. It donates the hydrogen atom H, a second party, or speaking precisely, the hydron $H^{+\delta}$ $(0 < +\delta \leq 1)$ bonded to X at the bond length R(X–H), to the third party which is a proton acceptor group Y. The latter, while interacting with X–H, yields the complex $X-H \cdots Y$ hydrogen bond. The above word "complex" means that Y is bound to X–H and therefore, this hydrogen bonding interaction is attractive. Energetically, it implies that the binding energy $E_b[X-H \cdots Y]$ of the complex $X-H \cup Y := X-H \cdots Y$ is defined in a standard way as the energy difference $\Delta E[X-H \cdots Y] := E[X-H \cdots Y] - (E[X-H] + E[Y])$. The ZPE-corrected binding energy $E_b^{ZPE}[X-H \cdots Y] := E[X-H \cdots Y] - (E[X-H] + E[Y]) + ZPE[X-H \cdots Y] - (ZPE[X-H] + ZPE[Y])$.

E.S. Kryachko
Bogolyubov Institute for Theoretical Physics, Kiev-143, 03680 Ukraine
e-mail: e_kryachko@yahoo.com; eugene.kryachko@ulg.ac.be

N. Russo et al. (eds.), *Self-Organization of Molecular Systems: From Molecules and Clusters to Nanotubes and Proteins*, NATO Science for Peace and Security Series A: Chemistry and Biology, © Springer Science+Business Media B.V. 2009

Geometrically, the X–H $\cdots$ Y bond is characterized by the bond length R(X–H), the H-bond separation r(H $\cdots$ Y), and the bond angle $\angle$XHY. By definition, the H-bond X–H $\cdots$ Y is formed if the following conditions are satisfied [1–17]:

(i) There exists a clear evidence of the bond formation—this might be, e.g., the appearance of the H-bond stretching mode v_σ(X $\cdots$ Y);

(ii) There exists a clear evidence that this bond specifically involves a hydrogen atom (hydron) bonded or bridged to Y predominantly along the bond direction X–H (see particularly [13, 15]);

(iii) The X–H bond elongates relative to that in the monomer, i.e.

$$\Delta R(X–H) := R_{complex}(X–H) - R_{monomer}(X–H) > 0;$$

(iv) The H-bond separation r(H $\cdots$ Y) defined as the distance between the bridging proton and the proton acceptor Y is shorter than the sum of van der Waals radii of H and Y, that is, shorter than the so-called van der Waals cutoff (see particularly Refs. [9, 10, 14] and also Ref. [21] $\in$ [15]):

$$r(H \cdots Y) < w_H + w_Y$$

where w_Z is the van der Waals radius of Z (Z = H, Y). Note that w_H varies and is usually taken the value either equal to 1.20 Å (see e. g. [18a]) or to 1.10 Å [18b]. $w_{Au} = 1.66$ Å. The distance r(X $\cdots$ Y) between the proton donor X and the proton acceptor Y is often referred to as the H-bond length;

(v) The stretching vibrational mode v(X–H) undergoes a red shift with respect to that of the isolated X–H group, that is,

$$\Delta v(X–H) := v_{complex}(X–H) - v_{monomer}(X–H) < 0,$$

and its IR intensity significantly increases;

(vi) The proton nuclear magnetic resonance (^{1}H NMR) chemical shift in the X–H $\cdots$ Y hydrogen bond is shifted downfield compared to the monomer.

It is worth mentioning that the conditions (iii)–(vi) can also be treated as some indirect justification of validity of (ii).

Throughout a hydrogen bridge, a hydrogen bond connects together X and Y whose electronegativities must be larger than the hydrogen one. Hence, X and Y can be particularly chosen as the following atoms: F (3.98), N (3.04), O (3.44), C (2.55), P (2.19), S (2.58), Cl (3.16), Se (2.55), Br (2.96), and I (2.66) where the Pauling electronegativity is given in parentheses. The latter possess a lone pair of electrons and that is why they cast as typical conventional proton acceptors Y under the formation of conventional hydrogen bonds [19].

The above definition of the attractive hydrogen bond interaction is rather general and allows to unify many types of interaction under the "hydrogen bonding" category, thus considerably extending its conventional manifold (see in particular [20–40] and references therein), either its X- or Y-submanifolds, or both. Few directions are particularly undertaken [20–24]. One is to include the X–H⋯H–Y dihydrogen bond which is formed between a conventional hydrogen bond donor such as an N–H or O–H bond as the weak acid component and the hydride bond as the weak base component where X can be a transition metal or boron [20, 21, 23–25]. The other is to deal with the π-hydrogen bond as formed, for instance, between water and aromatic moieties [26].

Since the 1990s, the conventional Y-submanifold is largely extended by including transition-metal "nonconventional" proton acceptors [21–24, 27–39] such as Co, Rh, Ir, Ni, Pd, Pt, Ru, and Os, which are prone to cast as the proton acceptors and are thus capable to form nonconventional hydrogen bonds. Though, they do not possess free pairs of electrons, in contrast to the conventional ones. The specific criteria which additionally characterize such nonconventional hydrogen bonds are formulated by Brammer and coworkers [21, 24]: (α) the bridging hydrogen is bonded to a rather electronegative element; (β) the acceptor metal atom is electron-rich (e. g. late transition metals) with filled d shells (see also [5]); (γ) the bonding arrangement is approximately linear (see also [39]).

What about gold — the "cornerstone" of nanoscience due to the discovery of more than two decades ago that gold nanoparticles supported on metal oxides reveal the exclusively high catalytic activity for CO oxidation [41]? Whether gold can belong to the Y-submanifold? Or put in the other words: Whether the gold atom or clusters of gold are prone to play, interacting with conventional proton donors such as the O–H and N–H groups, a role of a proton acceptor and hence to participate in the formation of nonconventional hydrogen bonds?

2 Gold as Nonconventional Proton Acceptor

2.1 Introductory Background

The Pauling electronegativity of the noble, coinage metal atom Au is equal to 2.54, that is, it is greater in comparison with H. It obviously obeys the foregoing condition (β). We may therefore raise a question of whether the gold atom or clusters of gold are prone to play, interacting with conventional proton donors such as the O–H and N–H groups, a role of a proton acceptor and hence to participate in the formation of nonconventional hydrogen bonds?

The story of answering this question goes back to the end of 2004 when the work [42] which reports a strong computational evidence of the propensity of a triangular gold cluster to behave as a proton acceptor with the O–H group of formic acid and the N–H one of formamide, was submitted. What precisely was demonstrated in this work was that the triangular Au_3 cluster forms the cyclic and planar complexes with formamide and formic acid by means of two bonding ingredients: the anchoring bond that anchors the gold atom to the carbonyl oxygen and the N–H⋯Au or O–H⋯Au contact between Au and the amino group of formamide or the hydroxyl group of formic acid. It was argued therein that the latter contacts share all the common features (i)–(vi) of the conventional hydrogen bonds and they can therefore be treated as their nonconventional analogs.

Since this work [42], the existence of the X–H $\cdots$ Au$_n$ nonconventional hydrogen bond was computationally demonstrated for a wide variety of molecules in different charge states Z = 0, ±1 [43–52], ranging from the Au$_n$–DNA bases and Au$_n$–DNA duplexes [43–45, 53, 54], to Au$_n$–(HF)$_m$ [46], [Au$_n$–(H$_2$O)$_m$]^Z [47, 51], and [Au$_n$–(NH$_3$)$_m$]^Z [48–50] complexes. The latter family also includes the smallest nano-sized tetrahedral gold cluster Au$_{20}^Z(T_d)$ [55] (see also [56] for current review and references therein). The charge-state specificity of the bonding ingredients of the [Au$_n$–(NH$_3$)$_m$]^Z complexes unveiled in [48, 49] has recently been explored to formulate the bonding encoding approach for molecular logic [50].

2.2   O–H $\cdots$ Au and N–H $\cdots$ Au Neutral Hydrogen Bonds

Recalling the work [42], in Table 1 we reproduce the key features of the bonding patterns of the original Au$_3$–formamide (FO) and Au$_3$–formic acid (FA) complexes on the background of those that exist in the corresponding lower-gold complexes of Au$_{1 \leq n \leq 2}$–FO and Au$_{1 \leq n \leq 2}$–FA. These bonding patterns are composed of two concomitant bonding ingredients. One is the Au–O anchoring that dominates and governs the major part of the charge transfer between the interacting species. Notice that the Au–O anchoring bond of the Au$_3$–FO complex is shorter by 0.026 Å compared to that of the the Au$_3$–FA one, and this mainly determines a higher stability of the former complex. Another bonding ingredient is the H-bonding that involves the proton donor X–H group and the gold atom of the Au$_3$ cluster different from that anchors O of FO or FA. True, its existence is evidently predetermined by the anchoring, though it, as acting cooperatively through the Au–Au bonds within the clusters Au$_2$ and Au$_3$, apparently reinforces the former. Table 1 also illustrates the maintenance of the above criteria (α) and (γ) in the nonconventional hydrogen bonds with gold as a proton acceptor and the dependence of their strength on the size of a given gold cluster.

To conclude this section and to make a closer link to the next section, we present in Table 2 the complexes Au$_3$–FO and Au$_3$–FA in the Z = -1-charge state which, as appears, are exclusively bound by the nonconventional X–H $\cdots$ Au hydrogen bond. This however is expected. The reason is rather simple: the X–H $\cdots$ Au hydrogen bond is strengthened in the -1-charge state since the gold cluster acquires the negative charge that reinforces its proton acceptor capability compared to the neutral case. This is precisely the case with [Au$_3$–FA]$^-$ which $-\Delta v$(O–H) as the most significant hallmark (v) of the H-bond formation increases from 520 to 616 cm^{-1}. It is however not the case with [Au$_3$–FO]$^-$ which $-\Delta v$(N–H) decreases, on the contrary, from 221 to 212 cm^{-1}. This implies that the anionic charge state is the necessary but not the sufficient precondition of strengthening of the nonconventional hydrogen bond—in the other words, it endorses this nonconventional hydrogen bonding interaction. The anchoring bond is a different one that however does not exist in [Au$_3$–FO]$^-$.

Table 1 The representative features of the complexes $Au_{1 \leq n \leq 2}$–FO and $Au_{1 \leq n \leq 2}$–FA [51] calculated within the following computational methodology. The Kohn–Sham self-consistent field formalism with the hybrid density functional B3LYP potential is used in conjunction with the basis set comprised of the standard Pople basis set 6–$31^{2+}G(d,p)$ for non-gold atoms and the energy-consistent $19 - (5s^2 5p^6 5d^{10} 6s^1)$ valence electron relativistic effective core potential (RECP) developed by Ermler, Christiansen and coworkers with the primitive basis set $(5s5p4d)$ [57] for the gold (see also [58]). The harmonic vibrational frequencies and unscaled zero-point energies (ZPE) are also calculated. The ZPE-corrected binding energies E_b^{ZPE} are given in kcal·mol^{-1}, $\Delta R(X–H)$, $r(H \cdots Au)$ and $R(Au–O)$ in Å, $\angle XHAu$ in degrees, and $\Delta \nu(X–H)$ in cm^{-1} ($X = N$ for FO and O for FA)

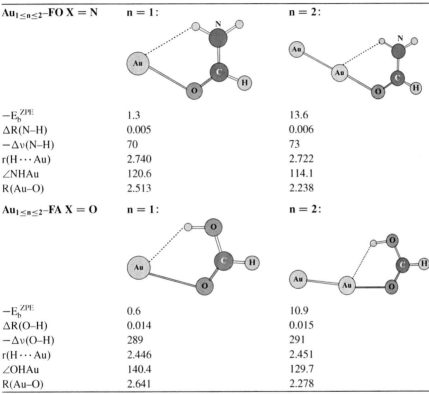

$Au_{1 \leq n \leq 2}$–FO $X = N$	$n = 1$:	$n = 2$:
$-E_b^{ZPE}$	1.3	13.6
$\Delta R(N–H)$	0.005	0.006
$-\Delta \nu(N–H)$	70	73
$r(H \cdots Au)$	2.740	2.722
$\angle NHAu$	120.6	114.1
$R(Au–O)$	2.513	2.238
$Au_{1 \leq n \leq 2}$–FA $X = O$	$n = 1$:	$n = 2$:
$-E_b^{ZPE}$	0.6	10.9
$\Delta R(O–H)$	0.014	0.015
$-\Delta \nu(O–H)$	289	291
$r(H \cdots Au)$	2.446	2.451
$\angle OHAu$	140.4	129.7
$R(Au–O)$	2.641	2.278

2.3 Experimental and Computational State of Art

On the experimental side, the hydrogen acceptor propensity of the gold atom and some its clusters has been experimentally detected for the complexes $[Au(H_2O)]^-$–Ar_n [59], $[Au(H_2O)_{n=1,2}]^-$ [60], and $[Au(H_2O)]^-$ [61], the crown compound $[Rb([18]crown$-$6)(NH_3)_3]Au$–NH_3 [62–64], for the complexes of small gold clusters with acetone [65] and with amino acids [66, 67], and for the gold(III) antitumor complex [68]. The latter work reports the synthesis and properties of the Au(III) compound of tridentate ligand 1,4,7-triazacyclononane (TACN)

Table 2 The complexes Au_3–FO and Au_3–FA. For the notations see Table 1

Au_3–FO X = N	n = 3: Z = 0	Z = −1
$-E_b^{ZPE}$	17.2	9.5
$\Delta R(N\text{–}H)$	0.014	0.013
$-\Delta v(N\text{–}H)$	221	149[a], 212[b], 148[c]
$r(H\cdots Au)$	2.714	2.708
$\angle NHAu$	172.3	167.4
$R(Au\text{–}O)$	2.214	
Au_3–FA X = O	**n = 3: Z = 0**	**Z = −1**
$-E_b^{ZPE}$	14.4	8.1
$\Delta R(O\text{–}H)$	0.026	0.031
$-\Delta v(O\text{–}H)$	520	616
$r(H\cdots Au)$	2.493	2.288
$\angle OHAu$	173.8	176.0
$R(Au\text{–}O)$	2.240	

[Au(TACN)Cl$_2$][AuCl$_4$]. In its cationic state, the Au(III) atom is bound by two N atoms of TACN and two atoms of Cl. The unbound amine group of TACN forms with Au(III) the Au(I)–H(3C) (in the notations of [68]) bond length of 1.91 Å. As concluded therein, "but as far as we know, no Au(III)–H short contact has ever been discussed before except for some relevant calculations on Au(O)–H or Au(I)–H″ performed in [44].

The experimental works [59–64] are focused on the nonconventional X–H $\cdots$ Au$^-$ hydrogen bonds that involve the auride anion Au$^-$. The latter behaves very similar to the heavier halides Br$^-$ and I$^-$ which are known as rather good proton acceptors in hydrogen-bonded systems [14]. The bonding patterns of the compound [Rb([18]crown-6)(NH$_3$)$_3$]Au–NH$_3$ [62, 63] comprise of the hydrogen bonds which are formed between neutral ammonia molecules playing as proton donors, on the one hand, and auride anions as proton acceptors on the other. Four NH$_3$ molecules are coordinated to Au$^-$ and simultaneously three of them, characterized by the distance $r(N\cdots Au^-) = 3.73$ Å, are coordinated to Rb centers.

The fourth is uncoordinated to the latter and separated from Au^- by $r(N \cdots Au^-) =$ 3.63 Å. The corresponding bridged hydrogen atom is distanced from Au^- by $r(H \cdots Au^-) = 2.581$ Å and forms the bonding angle $\angle NHAu^- = 158.1°$. Thus, the geometric criteria (i)–(iv) are obeyed for the $X–H \cdots Au^-$ contact that is then definitely the hydrogen bond. Moreover, the auride anion has the $5d$ shell filled with the $5d^{10}6s^2$ valence electrons (see e.g. [47])—that is, the condition (β) is also satisfied. Besides, the proton donor N atom is highly electronegative [condition (α)], and the angle $\angle NHAu^-$ is within the range determined by the condition (γ) [62]. Nuss and Jansen [62] then concluded: "To our knowledge, only theoretical investigations on gold as an acceptor in hydrogen bonds exist up to now. Kryachko and Remacle have reported DFT studies on $N–H \cdots Au$ interactions between formamide or DNA bases as donors and gold atoms from small Au_n ($n = 3, 4$) clusters as acceptors. ... the experimental values agree very well with the calculations."

To pursue this theme and to particularly explain the above experimental observations, let consider a quite large class of the nonconventional $X–H \cdots Au^-$ hydrogen bonds formed in the complexes $D_{1,2}–Au^-$ with $D = HF$, H_2O, and NH_3, displayed in Figs. 1 and 2, and compare the proton acceptor propensity of the auride anion with that of Ag^-. Before summarizing in Table 3 the computational data for these nonconventional $X–H \cdots Au^-$ and $X–H \cdots Ag^-$ H-bonds, let us make few preliminary comments. First: in general, the bonding patterns between the molecules such as HF, $(HF)_2$, H_2O, $(H_2O)_2$, NH_3 and $(NH_3)_2$ and the gold cluster Au_n^Z where $n \geq 1$ and $Z = 0, \pm1$ are governed by the Au–X anchoring and by the nonconventional $X–H \cdots Au$ hydrogen bonding interactions and exhibit distinct characteristics as the Z-charge state of the gold cluster varies [47–49]. Second: if $n = 1$ and $Z = -1$, the nonconventional hydrogen bonding likely plays the leading role. And third: as demonstrated in Table 3 and in Fig. 2, the potential energy surfaces, or PESs for short, of the interaction of the coinage anions Au^- and Ag^- with the dimers $(HF)_2$ and $(NH_3)_2$ comprise of two nearly iso-energetic conformers which can be considered either as involving a whole dimer (type I) or two separated monomers (type II), whereas the PESs involving $(H_2O)_2$ consist of a single conformer of the type I.

As particularly follows from the computational data reported in [51], the bond length $r(H \cdots Au^-) = 2.581$ Å and the bonding angle $\angle NHAu^- = 158.1°$ which together characterize the nonconventional $N–H \cdots Au^-$ H-bond in the compound $[Rb([18]crown-6)(NH_3)_3]Au–NH_3$ [62, 63] fairly agree with $r(H \cdots Au^-) = 2.690$ and 2.565 Å and $\angle NHAu^- = 157.4°$ and 177.6° of the complexes $Au^-–NH_3$ and $Au^-–(NH_3)_2{}^I$, respectively.

The full agreement between experiment and theory or computation, if speaking precisely and can say so, has also been recently confirmed by the anion photoelectron spectroscopy measurements of $[Au(H_2O)]^-$ and $[Au(H_2O)_2]^-$ conducted by Bowen and coworkers [60]. That is why the next section provides a number of computational mise-en-scènes behind these experiments on small gold–water complexes $[Au(H_2O)_{1 \leq n \leq 2}]^-$ and extend the latter to the larger ones, $[Au(H_2O)_{3 \leq n \leq 5}]^-$. These complexes exist only due to the nonconventional $X–H \cdots Au^-$ hydrogen

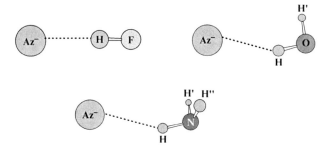

Fig. 1 Schematic representation of the complexes D–Az⁻ with the nonconventional hydrogen bond involving the Az atom (z = u, g and D = HF, H_2O, and NH_3)

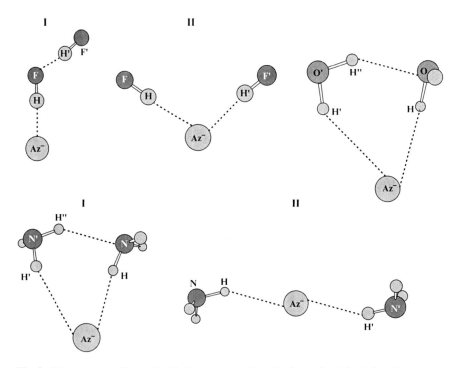

Fig. 2 The complexes D_2–Az⁻ with the nonconventional hydrogen bond involving the Az atom as a proton acceptor (z = u, g and D = HF, H_2O, and NH_3). If D = HF and NH_3, D_2–Az⁻ admits two conformations, whereas if D = H_2O, D_2–Az⁻ exists in a single conformation

bonds that glue together the auride anion and water clusters. It is worth mentioning in this regard that, as recently shown in [69], the anionic dimer Pt_2^- does not form any hydrogen bond with NH_3.

Table 3 The selected characteristics of the nonconventional hydrogen bonds formed in the complexes $Y_{1,2}$–Au^- and $Y_{1,2}$–Ag^- with $Y = HF$, H_2O, and NH_3, and $Y_2 = (HF)_2$, $(H_2O)_2$, and $(NH_3)_2$. The ZPE-corrected binding energies E_b^{ZPE} are given in $kcal \cdot mol^{-1}$, $\Delta R(X–H)$ and $r(H \cdots Az^-)$ in Å, $\angle AHAz$ in degrees ($z = u$ or g), and $\Delta v(X–H)$ in cm^{-1}. $\Delta E_{I–II}$ is the difference in energy (in $kcal \cdot mol^{-1}$) between the conformers **I** and **II**, if any. The asymptotic dissociation limit is chosen as $Az^- + Y_2$ for the conformer **I** and Y–$Az^- + Y$ for the conformer **II**

	$-E_b^{ZPE}$	$\Delta R(X–H)$	$-\Delta v(X–H)$	$r(H \cdots Az^-)$	$\angle XHAz$
Az^-HF (X = F)					
Au^-	20.3	0.052	1,076	2.156	178.2
Ag^-	18.6	0.042	901	2.291	179.7
Az^-–$(HF)_2$					
Au^-					
I	31.2	0.097	1,806; 499	1.985	178.5
II	14.9	0.044, 0.044	943, 858	2.190, 2.190	176.6, 176.6
$\Delta E_{I–II} = -0.7$					
Ag^-					
I	28.6	0.075	1,498; 432	2.116	178.4
II	13.9	0.037, 0.037	812, 774	2.305, 2.305	175.6, 175.6
$\Delta E_{I–II} = -0.6$					
Az^-–H_2O (X = O)					
Au^-	11.9	0.024, 0.001	394, 82	2.444, 3.406	156.3, 79.8
Ag^-	12.4	0.016, 0.004	243, 135	2.637, 3.215	138.2, 94.6
Az^-–$(H_2O)_2$					
Au^- I	18.9	0.029, 0.012	525, 43, 196, 50	2.348, 2.758	160.1, 148.1
Ag^- I	19.0	0.023, 0.010	281, 17, 197, 81	2.488, 2.955	149.8, 144.8
Az^-–NH_3 (X = N)					
Au^-	6.0	0.015	160	2.690	157.4
Ag^-	7.0	0.012	119	2.835	145.6
Az^-–$(NH_3)_2$					
Au^-					
I	10.7	0.018, 0.007	169, 155	2.565, 3.070	177.6, 146.2
II	5.2	0.013, 0.013	140, 134	2.705, 2.705	158.9, 158.9
$\Delta E_{I–II} = -0.3$					
Ag^-					
I	10.1	0.015, 0.006	174, 56	2.689, 3.254	168.6, 141.5
II	5.3	0.009, 0.011	107, 100	2.877, 2.856	149.5, 149.2
$\Delta E_{I–II} = -0.3$					

3 Hydrogen Bonding Patterns Between Auride Anion and Clusters of Water

The key computational facts about the most stable anionic complexes $[Au(H_2O)_{1 \leq n \leq 5}]^-$ are collected in the left column of Table 4. It is worth mentioning that the computational electron affinity of the gold atom is high, namely, $EA^{theor}(Au) = 2.129\,eV$:

Table 4 The computational mise-en-scènes of bonding between the auride anion (left column) or the gold atom (right column) and the selected clusters of water molecules $(H_2O)_{1 \leq n \leq 5}$ [52]. The vertical detachment energies, VDE, and adiabatic detachment energy, ADE, are given in eV. The ZPE-corrected binding energies E_b^{ZPE} and energy differences are given in kcal·mol^{-1}, R(O–H) and r(H···Y) in Å, ∠XHY in °, and ν(X–H) in cm^{-1}. The reference asymptote for the complex $[Au(H_2O)_n]^Z$ is the infinitely separated Au^Z and $(H_2O)_n$ where $Z = -1, 0$ and $(H_2O)_n$ designates the ground-state cluster of n molecules of water. The exception is $n = 4$, as indicated below by asterisk, when the R-sided conformer $[Au(H_2O)_4]_R^Z$ is treated with respect to Au^Z and the corresponding water cluster $(H_2O)_4^{3D}$ which is displayed in Fig. 3. The quantities chosen to characterize the nonconventional hydrogen bonds O–H···Au$^-$ are underlined. The MP2 values are presented in curly brackets

$\mathbf{Z = -1}$	$\mathbf{Z = 0}$
$\mathbf{n = 1}$	

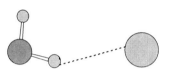 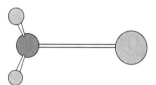

VDE = 2.71 (VDE^{expt} = 2.76 [60])

ADE^{ZPE} = 2.61

q_M(Au) = −0.902 q_M(H$_2$O) = −0.098

R(O–H) = 0.9865 {0.9796}; 0.9631 {0.9615}

ΔR(O–H$_{bonded}$) = 0.0244 r(H$_{bonded}$···Au) = 2.437

{2.464} ∠OH$_{bonded}$Au = 157.3

ν(O–H) = 3,403; {3,556}; 3,849 {3,849}

Δν(O–H$_{bonded}$) = −414

−E$_b^{ZPE}$ = 12.30 {12.92} (14.32 [61a]; 12.43 [59])

q_M(Au) = −0.053 q_M(H$_2$O) = 0.053

R(Au–O) = 2.670 {2.634}

R(O–H) = 0.9640 {0.9618}

ν$_{sym,asym}$(O–H) = 3790, 3897 {3852, 3973}

−E$_b^{ZPE}$ = 1.21 {1.30} (2.61 [61a]; 1.11 [71])

$\mathbf{n = 2}$

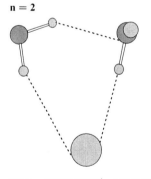 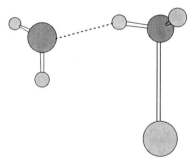

VDE = 3.17 (VDE^{expt} = 3.20 [60])

ADE^{ZPE} = 2.84

q_M(Au) = −0.839 q_M(H$_2$O) = −0.117, −0.044

R(O–H) = 0.9925 {0.9878}, 0.9738

{0.9698}; 0.9703

{0.9668}; 0.9635 {0.9616}

q_M(Au) = −0.097 q_M(H$_2$O) = 0.108, 0.011

R(Au–O) = 2.520 {2.411}

R(O–H) = 0.9778 {0.9733}; 0.9663

{0.9622}; 0.9637 {0.9616}, 0.9630 {0.9609}

r(H···O) = 1.840 {1.828}

∠OHO = 160.6 {164.9}

(continued)

Table 4 (continued)

Z = −1	Z = 0
ΔR(O–H) = 0.0296, 0.0127; −0.0004	ν(O–H) = 3,563 {3665}; 3774 {3862}; 3855
r(H_bonded ··· Au) = 2.346 {2.322}, 2.745 {2.762}	{3924}, 3890 {3,977}
r(H···O) = 2.095 {2.096}	Δν(O–H) = −143
∠OH_bonded Au = 161.1 {161.3}, 148.8 {149.8}	−E_b^ZPE = 3.24 {3.61}
∠OHO = 152.3 {152.4}	
ν(O–H) = 3279 {3369}, 3643 {3742};	
3727 {3,830}; 3,846 {3,916}	
Δν(O–H) = −535, −171, +21	
−E_b^ZPE = 19.72 {20.74}	

n = 3

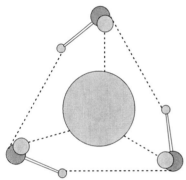 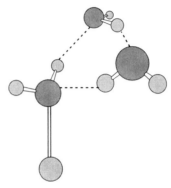

VDE = 3.56	R(Au–O) = 2.602
ADE^ZPE = 3.03	R(O–H) = 0.9845, 0.9756, 0.9717
R(O–H) = 0.9789; 0.9716	ΔR(O–H) = 0.0044, 0, −0.0039
ΔR(O–H) = 0.0179; −0.0040	r(H···O) = 1.792, 1.918, 2.018
r(H_bonded ··· Au) = 2.577 r(H···O) = 2.120	∠OHO = 155.1, 145.2, 142.5
∠OH_bonded Au = 149.8	ν(O–H) = 3439, 3620, 3690;
ν(O–H) = 3527, 3530, 3555; 3681, 3707, 3,709	3857, 3888, 3895
Δν(O–H) = −361, − 363, − 339; +121, +89, +80	Δν(O–H) = −121, +2, +61
−E_b^ZPE = 22.70	−E_b^ZPE = 1.93

n = 4

L-sided isomer:

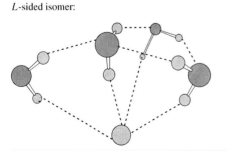 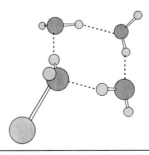

(continued)

Table 4 (continued)

Z = −1	Z = 0
VDE = 3.98	R(Au–O) = 2.604
ADE^{ZPE} = 2.93	R(O–H) = 0.9937, 0.9851, 0.9824, 0.9778
R(O–H) = <u>0.9848, 0.9805, 0.9761, 0.9688;</u>	ΔR(O–H) = 0.0100, 0.0014,
0.9729, 0.9720, 0.9703, 0.9679	−0.0013, −0.0059
r(H$_{bonded}$ ⋯ Au) = 2.363, 2.414, 2.547, 2.988	r(H ⋯ O) = 1.696, 1.758, 1.787, 1.851
r(H ⋯ O) = 2.091, 2.103, 2.103, 2.363	∠OHO = 168.4, 165.5, 165.2, 163.3
∠OH$_{bonded}$Au = 146.0, 160.7, 152.7, 147.4	ν(O–H) = 3252, 3424, 3491, 3576;
ν(O–H) =	3853, 3880, 3881, 3886
<u>3407, 3510, 3595, 3679</u>; 3699, 3705, 3763, 3777	Δν(O–H) = −123, −45, +22, +59
−E$_b$^{ZPE} = 20.45	−E$_b$^{ZPE} = 1.94

R-sided isomer:

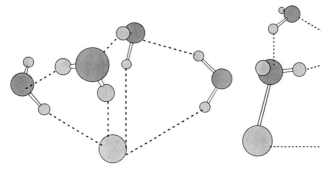

VDE = 3.96	R(Au–O) = 2.473
ADE^{ZPE} = 3.11	R(O–H$_{bonded}$) = 0.9736
R(O–H) =	r(H$_{bonded}$ ⋯ Au) = 2.548
<u>0.9881, 0.9794, 0.9743, 0.9679</u>; 0.9736, 0.9732,	∠OH$_{bonded}$Au = 155.5 ν(O–H$_{bonded}$) = 3631
0.9727; 0.9639	R(O–H) = 1.0026, 0.9759, 0.9709, 0.9692;
r(H$_{bonded}$ ⋯ Au) = 2.393, 2.523,	0.9639, 0.9628, 0.9614
2.726, 3.014	r(H ⋯ O) = 1.651, 1.867, 2.048, 2.111
r(H ⋯ O) = 2.025, 2.026, 2.045	∠OHO = 164.8, 160.5, 138.0, 140.0
∠OH$_{bonded}$Au = 160.8, 155.3, 141.9, 145.8	ΔE^{ZPE}(R–L) = 4.08 −E$_b$^{ZPE} = 4.44*
ν(O–H) = <u>3354, 3522, 3621</u>;	
3665, 3677, 3685; 3769, 3843	
ΔE(R–L) ≈ 0.09 ΔE^{ZPE}(R–L) ≈ 0.12	
−E$_b$^{ZPE} = 27.15*	

(continued)

Table 4 (continued)

Z = −1	Z = 0
n = 5	

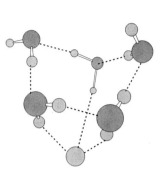

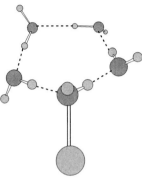

Z = −1	Z = 0
VDE = 3.71	R(Au–O) = 2.624
ADE^{ZPE} = 3.07	R(O–H) = 0.9936, 0.9878, 0.9858, 0.9841,
R(O–H) =	0.9796
0.9848, 0.9842, 0.9780; 0.9926, 0.9903, 0.9792,	r(H···O) = 1.685, 1.720, 1.737, 1.751, 1.807
0.9768, 0.9760;	∠OHO = 173.3, 175.5, 174.3, 174.2, 174.8
r(H_{bonded}···Au) = 2.451, 2.462, 2.606	ν(O–H) = 3250, 3362, 3418, 3465, 3542;
r(H···O) = 1.732, 1.755, 1.910, 1.930, 1.933	3856, 3878, 3881, 3886, 3893
∠OH_{bonded}Au = 156.2, 166.1, 155.2	−E_b^{ZPE} = 1.82
ν(O–H) =	
3269, 3316; 3431, 3446, 3528; 3581, 3599, 3624;	
3874, 3877	
−E_b^{ZPE} = 23.63	

The experimental value of $EA^{expt}(Au) = 2.30 \pm 0.10\,eV$ according to [70a]; $2.308664 \pm 0.000044\,eV$ according to [70b]; and $2.927 \pm 0.050\,eV$ according to [70c]. $EA^{theor}(Au) = 2.33\,eV$ [70d] and $2.166\,eV$ [70e]. With the used basis set, MP2 yields $1.536\,eV$. The $EA^{theor}(Au) = 1.86\,eV$ was calculated at the MCPF computational level in [70f]. The PW91PW91 density functional potential in conjunction with the basis set [57] yields 2.25 and 2.31 eV with the LANL2DZ basis set, as reported in [70g]. Therefore, the gold atom of $[Au(H_2O)_{1 \le n \le 5}]^-$ where the most excess electron charge is located on. This is witnessed by the Mulliken charges of gold which are e.g. equal to $q_M^{n=1}(Au) = -0.902$ and $q_M^{n=2}(Au) = -0.839$, and therefore, as anticipated, the gold atom mainly exists in $[Au(H_2O)_{1 \le n \le 5}]^-$ as the auride anion. The latter casts as the strong proton acceptor, even stronger in some cases than the oxygens of the studied water clusters: this can readily be seen by juxtaposing the stretching frequency $\nu(O\text{–}H(\cdots Au)) = 3,279\,cm^{-1}$ of the hydron H_{bonded}, which belongs to the proton–donor water molecule of the water dimer and H-bonded to Au^- in $[Au(H_2O)_2]^-$, and the $\nu(O\text{–}H(\cdots O)) = 3,727\,cm^{-1}$ of the H (comparing with the ground-state water dimer, this stretching mode $\nu(O\text{–}H)$ is red-shifted by only $21 \, \text{cm}^{-1}$.), that bridges the water dimer within $[Au(H_2O)_2]^-$. It is worth mentioning that the aforementioned stretching vibrational mode $\nu(O–H(\cdots Au))$ of $[Au(H_2O)_2]^-$ is the lowest one among the considered series of complexes. The following two straightforward and rather important conclusions can be drawn from the fact that the auride anion functions as the strong proton acceptor in the complexes $[Au(H_2O)_{1 \leq n \leq 5}]^-$:

1. This predetermines rather large, by the absolute value, binding energies E_b^{ZPE}. As demonstrated in Table 4, the latter range from $12.3 \, \text{kcal} \cdot \text{mol}^{-1}$ for $n = 1$ to ~ 19.7–22.7 for $n = 2$–4, and apparently approaches the saturation threshold of ca. $23.6 \, \text{kcal} \cdot \text{mol}^{-1}$ for $n = 5$. Note that the presented computational E_b^{ZPE} are consistent with the experimental $E_b^{ZPE-expt}$ [60], for instance, with $E_b^{ZPE-expt}([AuH_2O]^-) = 10.4 \, \text{kcal} \cdot \text{mol}^{-1}$ and with the stabilization energy of $[Au(H_2O)_2]^-$ taken with respect to the asymptote $[AuH_2O]^- + H_2O$. The latter, as estimated in [60] as equal to ca. $10.2 \, \text{kcal} \cdot \text{mol}^{-1}$, fairly agrees with our value of $10.9 \, \text{kcal} \cdot \text{mol}^{-1}$.

Actually, the strength of the nonconventional hydrogen bonds of the studied complexes is underestimated because they mainly involve so-called "free" O–H groups of water clusters that do not participate in the water–water hydrogen bonds. Consider for instance the complex $[Au(H_2O)_3]^-$. Its water trimer is considerably enlarged in comparison to the equilibrium one in the neutral charge state in order to accommodate the auride anion. The difference in energy between these two forms of water trimer, equal to $10.78 \, \text{kcal} \cdot \text{mol}^{-1}$, is the additional contribution that, together with $E_b^{ZPE}([Au(H_2O)_3]^-)$, determines the strength of the three nonconventional O–H $\cdots$ Au$^-$ hydrogen bonds of $[Au(H_2O)_3]^-$.

2. The auride anion is a strong proton acceptor that, while interacting with a water cluster, significantly perturbs it. This perturbation manifests in a number of ways.

One of them is spectroscopic—it is the formation of a wide infrared window $\Delta\nu_w(O–H)$ of ca. $450 \, \text{cm}^{-1}$ in $[Au(H_2O)]^-$ between the most red-shifted O–H stretching mode(s) and the next one(s). This window narrows to $\sim 80 \, \text{cm}^{-1}$ in $[Au(H_2O)_2]^-$ and to $\sim 130 \, \text{cm}^{-1}$ in $[Au(H_2O)_3]^-$ and is superimposed with the stretches of the conventional O–H $\cdots$ O hydrogen bonds in larger complexes.

The other is that the auride anion can also be a "breaker" of the water–water hydrogen bonds as e.g. occurs under the formation of the complexes $[Au(H_2O)_{4,5}]^-$.

4 Mapping the Experiment: Electron Detachment in Auride–Water Clusters

Let us now turn to the computational mise-en-scènes which are gathered in the right column of Table 4 and which are served for deeper understanding of the experiments on anion photoelectron spectroscopy of gold–water complexes [60]. Since gold is the key carrier of the excess electron charge of the complexes $[Au(H_2O)_{1 \leq n \leq 5}]^-$, a removal of this charge, formally implying the alternation $Z = -1 \Rightarrow Z = 0$ of the charge states, converts the auride anion into the neutral gold atom. Despite the high electron affinity of the latter, Au may only induce a small charge transfer from the adjacent oxygen atom and, as a result, forms with the latter the so-called Au–O anchoring bond (see [45] for more detailed discussion and for the related references). This anchoring bond appears to be very weak that is reflected in the corresponding binding energies. For n = 1–5, the shortest Au–O anchoring bond of 2.520 Å is formed in $[Au(H_2O)_2]$ – naturally, its formation is characterized by the relatively large binding energy $E_b^{ZPE}([Au(H_2O)_2]) = 3.2\,kcal \cdot mol^{-1}$.

Nevertheless, despite its weakness, the anchoring bond enables to re-polarize the adjacent O–H bond of the neighboring water molecule and substantially activates it within the $O–H \cdots O$ hydrogen bond if n ≥ 2. For n = 2, $\Delta R(O–H) = 0.008$ Å, $\Delta r(H \cdots O) = -0.093$ Å, and $\Delta v(O–H) = -143\,cm^{-1}$, relative to the gas-phase water dimer. Since the anchoring interaction is weak, the relaxation of the water cluster within $[Au(H_2O)_{1 \leq n \leq 5}]$ is not significant, in contrast to the anionic charge state. For example, the energy difference between the equilibrium water trimer and that of $[Au(H_2O)_3]$ amounts only to $0.4\,kcal \cdot mol^{-1}$.

The experiments on anion photoelectron spectroscopy of $[Au(H_2O)]^-$ and $[Au(H_2O)_2]^-$ that are conducted in [60] measure the vertical detachment energies, $VDE_1^{expt} = 2.76\,eV$ and $VDE_2^{expt} = 3.20\,eV$ which correspondingly agree with the computational ones, $VDE_1^{theory} = 2.708\,eV$ and $VDE_1^{theory} = 3.187\,eV$. From the viewpoint of the chemical bonding patterns that are formed in the studied gold–water complexes, the charge state alternation $Z = -1 \Rightarrow Z = 0$, which can be achieved either by using different metallic supporters or/and applied voltage, the NeNePo ("A Negative ion–to Neutral–to Positive ion") experimental technique (see [72] and references therein), the resonant photoionization [73], or by varying pH in different solvents [74], executes a simple switch-type operation. In the most studied cases, the latter transforms the nonconventional $O–H \cdots Au$ hydrogen bonding interaction to the Au–X anchoring one, except the cyclic R-sided conformer $[Au(H_2O)_4]_R^{Z=0}$ which is stabilized by both the anchoring and nonconventional hydrogen bonding interactions.

The total relaxation of the studied systems under the charge state alternation $Z = -1 \Rightarrow Z = 0$ is rationalized in terms of the adiabatic detachment energy or shortly ADE. As follows from Table 4, the VDE-ADE difference amounts to 0.15 eV for n = 1, 0.36 eV for n = 2, 0.53 eV for n = 3, then rises to 1.05 eV for n = 4 and falls to 0.62 eV for n = 5. On the one hand, this difference can be interpreted as the effect of solvent on the electron detachment that causes the significant relaxation of the solvent molecules. On the other, it implies that the concept of VDE is not generally true, as far as larger solvent clusters are on the stage. The latter solvation, as is remarkably seen from the case of n = 4, results in that the bottom of the solvent potential energy surface is quite dense, accommodating many conformers with nearly equal energies. Nevertheless, their VDEs and ADEs can be essentially different as in the present case of n = 4 when the $\Delta VDE(L–R)$ is only 0.02 eV, whereas $\Delta ADE(L–R) = -0.18\,eV$.

As was noticed above, the key source of that VDE drastically distinguishes from the ADE is the considerable solvent relaxation that is actually a breakage of the hydrogen bonding patterns of water clusters by the auride anion. This is

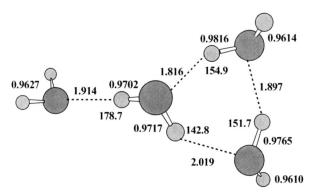

Fig. 3 The lower-energy 3D water tetrameric cluster $(H_2O)_4^{3D}$. The selected bond distances are given in Å and bond angles in degrees. Its stretching frequencies $\nu(O-H)$ are the following: 3,483; 3,595; 3,671; 3,744; 3,816; 3,885; 3,893; and 3,918 cm^{-1}. For comparison, the corresponding frequencies of $(H_2O)_4^{ring}$: 3,375, 3,469 (doublet), 3,507, 3,884 (quartet). Interestingly, the latter spectrum possesses a large window of $\sim$380 cm^{-1} between 3,507 and 3,884 cm^{-1}. In the gas phase, the water cluster $(H_2O)_4^{3D}$ lies naturally higher the $(H_2O)_4^{ring}$ by $\Delta E^{ZPE} = 6.58$ kcal·mol^{-1}. Their enthalpy difference $\Delta H = 7.45$ kcal · mol^{-1}. Due to the large entropy difference, $\Delta S = 9.88$ cal · T^{-1}·mol^{-1}, the Gibbs free energy difference is lowered to $\Delta G = 4.51$ kcal·mol^{-1}. Furthermore: the total dipole moment of $(H_2O)_4^{3D}$ is rather large, i.e., 3.63 D, compared to that $(H_2O)_4^{ring}$ is non-polar. This implies that the 3D cluster $(H_2O)_4^{3D}$ can be energetically favorable in polar environment

transparently observed for the L-sided isomer of $[Au(H_2O)_4]_L^-$ when the water tetrameric ring is broken to accommodate the auride anion. The number of nonconventional hydrogen bonds is equal to 4. And remarkably, all hydrogen are involved in the hydrogen bonds of $[Au(H_2O)_4]_L^-$ and $[Au(H_2O)_4]_R^-$, either conventional or nonconventional. It is worth to mention that herein we exploit two isoenergetic isomers, $[Au(H_2O)_4]_L^-$ and $[Au(H_2O)_4]_R^-$ which however take different pathways under the electron detachment. If the former proceeds to the conformer $[Au(H_2O)_4]_L$ where water molecules arrange in the well-known tetrameric ring structure $(H_2O)_4^{ring}$ with the planar oxygen frame [75], the latter adapts the 3D shape $(H_2O)_4^{3D}$, as demonstrated in Fig. 3.

Summarizing: in these essay's fragments, we have demonstrated what matters and relevant to the experiments on anion photoelectron spectroscopy for the gold–water complexes and, especially, the computational evidence of interesting synergetic effect between the nonconventional O–H$\cdots$Au hydrogen bonding interaction and the Au–X anchoring one that both govern the gold–water complexes in the charge states $Z = -1$ and $Z = 0$. What about $Z = +1$? For example, the recent work [75] reveals a rather surprising dumbbell-type structure around Au$^+$ exhibiting two Au–O anchoring bonds, each ended by rigid tetrameric rings of water. Then, let us pose the question of how many nonconventional hydrogen bonds with water the auride anion enables to form simultaneously? Comparing with maximum two anchoring Au–O bonds that the gold atom can form at once [76], Fig. 4

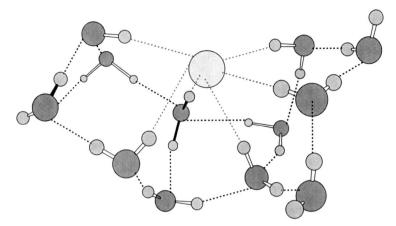

Fig. 4 The cluster $[Au(H_2O)_{12}]^-$ with six blue nonconventional O–H $\cdots$ Au$^-$ hydrogen bonds. Its electronic energy is equal to $-1,053.751779$ hartrees and ZPE $= 190.04\,kcal \cdot mol^{-1}$ [52]. Yellow ball designates the gold atom, the oxygen atoms are indicated by red and the hydrogens by light blue balls. The nonconventional hydrogen bonding patterns of this cluster are summarized in the following table where the bond distances are given in Å, bond angles in degrees, and concomitant stretching frequencies in cm^{-1}

R(O–H)	r(H$_{bonded}\cdots$Au)	$\angle$OH$_{bonded}$Au	ν(O–H)
0.9819	2.469	170.2	3,465; 3,476
0.9752	2.572	162.7	3,604
0.9740	2.610	163.1	3,500; 3,582; 3,635; 3,656
0.9739	2.644	164.1	3,635; 3,637; 3,656
0.9738	2.632	165.1	3,635; 3,637; 3,656
0.9732	2.539	174.1	3,476; 3;550; 3,519; 3,550

demonstrates that at least six nonconventional O–H $\cdots$ Au$^-$ hydrogen bonds the auride anion can form with 12 molecules of water to hold the cluster $[Au(H_2O)_{12}]^-$ bound. It is a belief that this fascinating feature of gold is awaiting its experimental proof.

Acknowledgements The author gratefully thanks Francoise Remacle, Kit Bowen, Alfred Karpfen, Camille Sandorfy†, Pekka Pyykkö, Lucjan Sobczyk, George V. Yukhnevich, and Georg Zundel† for encouraging discussions, useful suggestions, and valuable comments.

References and Notes

1. C. G. Pimentel, A. L. McClellan, The Hydrogen Bond, Freeman, San Francisco, 1960.
2. D. Hadži, H. W. Thompson (Eds.), Hydrogen Bonding, Pergamon Press, London, 1959.
3. W. C. Hamilton, J. A. Ibers, Hydrogen Bonding in Solids, Benjamin, New York, 1968.
4. S. N. Vinogradov, R. H. Linell, Hydrogen Bonding, Van Nostrand-Reinhold, New York, 1971.
5. M. D. Joesten, L. J. Schaad, Hydrogen Bonding, Dekker, New York, 1974.

6. P. Schuster, G. Zundel, C. Sandorfy (Eds.), The Hydrogen Bond. Recent Developments in Theory and Experiments, North-Holland, Amsterdam, 1976.

7. P. Schuster, in: B. Pullman (Ed.), Intermolecular Interactions: From Diatomics to Biopolymers, Wiley, Chichester, 1978, p. 363.

8. P. Schuster (Guest Ed.), Top. Curr. Chem. 120 (1984).

9. G. A. Jeffrey, W. Saenger, Hydrogen Bonding in Biological Structures, Springer, Berlin, 1991.

10. G. A. Jeffrey, An Introduction to Hydrogen Bonding, Oxford University Press, Oxford, 1997.

11. S. Scheiner, Hydrogen Bonding. A Theoretical Perspective, Oxford University Press, Oxford, 1997.

12. D. Hadži (Ed.), Theoretical Treatment of Hydrogen Bonding, Wiley, New York, 1997.

13. T. Steiner, G. R. Desiraju, Chem. Commun. 891 (1998).

14. G. R. Desiraju, T. Steiner, The Weak Hydrogen Bond in Structural Chemistry and Biology, Oxford University Press, Oxford, 1999.

15. T. Steiner, Angew. Chem. Int. Ed. 41 (2002) 48.

16. A. Karpfen, Adv. Chem. Phys. 123 (2002) 469.

17. (a) S. Grabowski (Ed.), Hydrogen Bonding— New Insights, Vol. 3 of Challenges and Advances in Computational Chemistry and Physics, in: J. Leszczynski (Ed.), Springer, Dordrecht, 2006; (b) For the recent IUPAC categorization of the hydrogen bond see http://www.iupac.org/projects/2004/2004-026-2-100.html

18. (a) A. Bondi, J. Phys. Chem. 68 (1964) 441; (b) R. S. Rowland, T. Taylor, Ibid. 100 (1996) 7384.

19. The existence of the lone-pair electrons in the proton acceptor is not apparently mandatory, as has recently been argued by Olovsson: I. Olovsson, Z. Phys. Chem. 220 (2006) 963.

20. R. H. Crabtree, P. E. M. Siegbahn, O. Wisenstein, A. L. Rheingold, T. F. Koetzle, Acc. Chem. Res. 29 (1996) 348.

21. L. Brammer, D. Zhao, F. T. Ladipo, J. Braddock-Wilking, Acta Crystallogr. Sect. B 51 (1995) 632.

22. A. Martin, J. Chem. Ed. 76 (1999) 578.

23. L. M. Epstein, E. S. Shubina, Coord. Chem. Rev. 231 (2002) 165 and references therein.

24. L. Brammer, Dalton Trans. (2003) 3145 and references therein.

25. E. S. Shubina, N. V. Belkova, L. M. Epstein, J. Organomet. Chem. 536 (1997) 17.

26. J. L. Atwood, F. Hamada, K. D. Robinson, G. W. Orr, R. L. Vincent, Nature 349 (1991) 683.

27. L. Brammer, M. C. McCann, R. M. Bullock, R. K. McMullan, P. Sherwood, Organometallics 11 (1992) 2339.

28. S. G. Kazarian, P. A. Hanley, M. Poliakoff, J. Am. Chem. Soc. 115 (1993) 9069.

29. F. Cecconi, C.A. Ghilardi, P. Innocenti, C. Mealli, S. Midollini, A. Orlandini, Inorg. Chem. 23 (1984) 922.

30. A. Albinati, C. G. Anklin, F. Ganazzoli, H. Ruegg, P. S. Pregosin, Inorg. Chem. 26 (1987) 503.

31. A. Albinati, P.S. Pregosin, F. Wombacher, Inorg. Chem. 29 (1990) 1812.

32. J. M. Casas, L.R. Falvello, J. Fornies, A. Martin, A. J. Welch, Inorg. Chem. 35 (1996) 6009.

33. A. Albinati, F. Lianza, P.S. Pregosin, B. Müller, Inorg. Chem. 33 (1994) 2522.

34. T. W. Hambley, Inorg. Chem. 37 (1998) 3767.

35. M. S. Davies, R. R. Fenton, F. Hug, E. C. H. Ling, T. W. Hambley, Aust. J. Chem. 53 (2000) 451.

36. L. Brammer, J. M. Charnock, P. L. Goggin, R. J. Goodfellow, T. F. Koetzle, A. G. Orpen, J. Chem. Soc. Dalton Trans. (1991) 1789.

37. K. Wieghardt, H.J. Kupper, E. Raabe, C. Kruger, Angew. Chem., Int. Ed. 25 (1986) 1101.

38. G. Orlova, S. Scheiner, Organometallics 17 (1998) 4362.

39. D. Braga, F. Grepioni, E. Tedesco, K. Biradha, G. R. Desiraju, Organometallics 16 (1997) 1846.

40. H. Jacobsen, Chem. Phys. 345 (2008) 95.

41. (a) M. Haruta, T. Kobayashi, H. Sano, N. Yamada, Chem. Lett. (1987) 405; (b) M. Haruta, N. Yamada, T. Kobayashi, S. Iijima, J. Catal. 115 (1989) 301.

42. E. S. Kryachko, F. Remacle, Chem. Phys. Lett. 404 (2005) 142.

43. E. S. Kryachko, F. Remacle, in: J.-P. Julien, J. Maruani, D. Mayou, S. Wilson, G. Delgado-Barrio (Eds.), Recent Advances in the Theory of Chemical and Physical Systems, Vol. 15, Springer, Dordrecht, 2006, p. 433.
44. E. S. Kryachko, F. Remacle, Nano Lett. 5 (2005) 735.
45. E. S. Kryachko, F. Remacle, J. Phys. Chem. B 109 (2005) 22746.
46. E. S. Kryachko, A. Karpfen, F. Remacle, J. Phys. Chem. A 109 (2005) 7309.
47. E. S. Kryachko, F. Remacle, in: A. Torro-Labbe (Ed.), Theoretical Aspects of Chemical Reactivity, Vol. 16 of Theoretical and Computational Chemistry, P. Politzer (Ed.), Elsevier, Amsterdam, 2006, p. 219.
48. E. S. Kryachko, F. Remacle, in: S. Lahmar, J. Maruani, S. Wilson, G. Delgado-Barrio (Eds.), Topics in the Theory of Chemical and Physical Systems, Vol. 16 of Progress in Theoretical Chemistry and Physics, Springer, Dordrecht, 2007, p. 161.
49. E. S. Kryachko, F. Remacle, J. Chem. Phys. 127 (2007) 194305.
50. E. S. Kryachko, F. Remacle, Mol. Phys. 106 (2008) 521.
51. E. S. Kryachko, J. Mol. Struct. 880 (2008) 23.
52. E. S. Kryachko, Coll. Czech. Chem. Comm. R. Zahradnik Festschrift 73 (2008) 1457.
53. (a) M.-V. Vázquez, A. Martínez, J. Phys. Chem. A 112 (2008) 1033; (b) J. Valdespino-Saenz, A. Martínez, Ibid. A 112 (2008) 2408.
54. A. Martínez, O. Dolgounitcheva, V. G. Zakrzewski, J. V. Ortiz, J. Phys. Chem. A 112 (2008) 10399.
55. J. Li, X. Li, H.-J. Zhai, L.-S. Wang, Science 299 (2003) 864.
56. E. S. Kryachko, F. Remacle, Int. J. Quantum Chem. 107 (2007) 2922.
57. R. B. Ross, J. M. Powers, T. Atashroo, W. C. Ermler, L. A. LaJohn, P. A. Christiansen, J. Chem. Phys. 93 (1990) 6654.
58. (a) F. Remacle, E. S. Kryachko, Adv. Quantum Chem. 47 (2004) 423; (b) F. Remacle, E. S. Kryachko, J. Chem. Phys. 122 (2005) 044304.
59. H. Schneider, A. D. Boese, J. M. Weber, J. Chem. Phys. 123 (2005) 084307.
60. W. Zheng, X. Li, S. Eustis, A. Grubisic, O. Thomas, H. de Clercq, K. Bowen, Chem. Phys. Lett. 444 (2007) 232.
61. (a) D.-Y. Wu, S. Duan, X.-M. Liu, Y.-C. Xu, Y.-X. Jiang, B. Ren, X. Xu, S. H. Lin, Z.-Q. Tian, J. Phys. Chem. A 112 (2008) 1313; (b) Y. X. Chen, S. Z. Zou, K. Q. Huang, Z. Q. Tian, J. Raman Spectrosc. 29 (1998) 749.
62. H. Nuss, M. Jansen, Angew. Chem. Int. Ed. 45 (2006) 4369.
63. H. Nuss, M. Jansen, Z. Naturforsch. Sect. B, J. Chem. Sci. 61 (2006) 1205.
64. Notice that in the [(NMe$_4$)Au] compound which was synthesized in: P. D. C. Dietzel, M. Jansen, Chem. Commun. (2001) 2208, the hydrogen atoms of the methyl groups and the auride anion are separated by 2.921(0) Å, with $r_{min}(C \cdots Au^-) = 3.663(1)$ Å. Such contact separation obeys the van der Waals cutoff condition (iv). However, as the authors of this work pointed out, "whether this contact is a true hydrogen bond or not has not yet been proven", since a short separation N^+–C–H $\cdots Y^-$ can be originated from the attractive $N^+ \cdots Y^-$ Coulomb interaction where, in addition, the H $\cdots Y^-$ contact prevents a further approach of Y^- simply due to steric effects [14].
65. G. S. Shafai, S. Shetty, S. Krishnamurty, D. G. Kanhere, J. Chem. Phys. 126 (2007) 014704.
66. A. H. Pakiari, Z. Jamshidi, J. Phys. Chem. A 111 (2007) 4391.
67. (a) A. Aqil, H. Qiu, J.-F. Greisch, R. Jérôme, E. De Pauw, C. Jérôme, Polymer 49 (2008) 1145; (b) A bit of pedantry: the reference "Remacle et al." on p. 1151 of [67] to the works [25, 29b–c] therein is grammatically inconsistent since: (i) "et al." (precisely "et alii" or "et alia") means "and others" (the plural number) and is used for a list of names of more than two authors (e.g., the American Psychological Association style), more than three authors (e.g., the Modern Language Association of America style), or four and more authors (e.g., the American Physical Society style), or even 16 and more authors according to the American Chemical Society style, and (ii) the order of the authors of the cited works [25, 29b–c] is actually opposite.
68. P. Shi, Q. Jiang, J. Lin, Y. Zhao, L. Lin, Z. Guo, J. Inorg. Biochem. 100 (2006) 939.
69. M. Onák, Y. Cao, M. K. Beyer, R. Zahradnik, H. Schwarz, Chem. Phys. Lett. 450 (2008) 268.

70. (a) G. Ganteför, S. Krauss, W. Eberhardt, J. Electron Spectrosc. Relat. Phenom. 88 (1998) 35; (b) H. Jotop, W. C. Lineberger, J. Phys. Chem. Ref. Data 14 (1985) 731; (c) K. J. Taylor, C. L. Pettiettehall, O. Cheshnovsky, R. E. Smalley, J. Chem. Phys. 96 (1992) 3319; (d) S. Buckart, G. Ganteför, Y. D. Kim, P. Jena, J. Am. Chem. Soc. 125 (2003) 14205; (e) A. M. Joshi, W. N. Delgass, K. T. Thomson, J. Phys. Chem. B 109 (2005) 22392; (f) C. W. Bauschlicher, Jr., S. R. Langhoff, H. J. Partridge, J. Chem. Phys. 93 (1990) 8133; (g) A. V. Walker, J. Chem. Phys. 122 (2005) 094310.

71. A. Antušek, M. Urban, A. J. Sadlej, J. Chem. Phys. 119 (2003) 7247.

72. (a) S. Wolf, G. Sommerer, S. Rutz, E. Schreiber, T. Leisner, L. Wöste, R. S. Berry, Phys. Rev. Lett. 74 (1995) 4177; (b) L. D. Socaciu-Siebert, J. Hagen, J. Le Roux, D. Popolan, M. Vaida, S. Vajda, T. M. Bernhardt, L. Wöste, Phys. Chem. Chem. Phys. 7 (2005) 2706; (c) R. Mitri, M. Hartmann, B. Stanca, V. Bonai-Koutecký, P. Fantucci, J. Phys. Chem. A 105 (2001) 8892; (d) T. M. Bernhardt, J. Hagen, L. D. Socaciu-Siebert, R. Mitri, A. Heidenreich, J. Le Roux, D. Popolan, M. Vaida, L. Wöste, V. Bonai-Koutecký, J. Jortner, Chem. Phys. Chem. 6 (2005) 243.

73. S.-I. Ishiuchi, M. Sakai, Y. Tsuchida, A. Takeda, Y. Kawashima, M. Fujii, O. Dopfer, K. Müller-Dethlefs, Angew. Chem. Int. Ed. 44 (2005) 6149.

74. D. Margulies, G. Melman, A. Shanzer, J. Am. Chem. Soc. 128 (2006) 4865.

75. K. S. Kim, M. Dupuis, G. C. Lie, E. Clementi, Chem. Phys. Lett. 131 (1986) 451.

76. J. U. Reveles, P. Calaminici, M. R. Beltrán, A. M. Köster, S. N. Khanna, J. Am. Chem. Soc. 129 (2007) 15565.

77. (a) J. Hrušák, D. Scröder, H. Schwarz, Chem. Phys. Lett. 225 (1994) 416; (b) H. Hertwig, J. Hrušák, D. Scröder, W. Koch., H. Schwarz, Ibid. 236 (1995) 194; (c) D. Feller, E. D. Glendening, W. A. de Jong, J. Chem. Phys. 110 (1999) 1475; (d) H. M. Lee, S. K. Min, E. C. Lee, J.-H. Min, S. Odde, K. S. Kim, Ibid. 122 (2005) 064314; (e) H. M. Lee, M. Diefenbach, S. B. Suh, P. Tarakeshwar, K. S. Kim, Ibid. 123 (2005) 074328; (f) L. Poisson, F. Lepetit, J.-M. Mestdagh, J.-P. Visticot, J. Phys. Chem. A 106 (2002) 5455; (g) A. J. Karttunen, T. A. Pakkanen, Ibid. B 110 (2006) 25926.

Interatomic Potential for Platinum and Self-Diffusion on Pt(111) Surface by Molecular-Dynamics Simulation

N.I. Papanicolaou and N. Panagiotides

Abstract We present a many-body interatomic potential for Pt within the second-moment approximation of the tight-binding model by fitting to the volume dependence of the total energy of the metal, computed by first-principles augmented-plane-wave calculations. This was used, in conjuction with molecular-dynamics simulations, to study the diffusion of Pt adatoms and dimers on Pt(111) surface. The diffusion coefficient of the adatoms and dimers was computed and was found to present Arrhenius behavior. The migration energies and pre-exponential factors for hopping diffusion mechanism were determined as well and compared with experimental data obtained by scanning tunnelling microscopy, field ion microscopy methods and previous calculations. Both quantities were found to be in good agreement with measurements. At high temperatures we have also investigated a concerted exchange adatom diffusion mechanism, where there is a participation of two surface atoms belonging to nearest-neighbour rows.

Keywords Interatomic potentials · Molecular dynamics simulation · Surface diffusion · Adatoms · Dimers · Vibrations · Platinum

1 Introduction

The diffusion of adatoms on metal surfaces plays an essential role in epitaxial crystal growth, heterogeneous catalysis, surface reconstruction, reactivity and other surface processes [1–5]. Experimental techniques such as field ion microscopy (FIM) and scanning tunnelling microscopy (STM) are important methods for investigating surface diffusion [1–3]. In parallel, atomistic simulations have played an important role in this area by providing information on different elementary diffusion mechanisms, migration energy barriers and pre-exponential factors associated with each diffusion process [6, 7]. The description of interactions between atoms is very

N.I. Papanicolaou (✉) and N. Panagiotides
Department of Physics, University of Ioannina, P.O. Box 1186, GR-45110 Ioannina, Greece
e-mail: nikpap@uoi.gr

N. Russo et al. (eds.), *Self-Organization of Molecular Systems: From Molecules and Clusters to Nanotubes and Proteins*, NATO Science for Peace and Security Series A: Chemistry and Biology, © Springer Science+Business Media B.V. 2009

crucial in any atomistic simulation, ranging from first-principles electronic structure calculations to semi-empirical methods. *Ab-initio* techniques are superior, accurate and can supply information on diffusion barriers, but they are limited by high computational costs. They are therefore used for relatively short time scales and small systems. In order to simulate longer time scales and larger systems parameter-based empirical methods such as the many-body potentials of the embedded-atom method (EAM) [8, 9], Finnis–Sinclair [10] potentials and the second-moment approximation (SMA) of the tight-binding (TB) method [11–15] are found to be very efficient. The semi-empirical methods, although less accurate, are very fast and able to simulate the dynamical character of diffusion and reveal different elementary processes (e.g., hopping or adatom-surface exchange). In particular, the TB-SMA expression [13–15] of the total energy of a metallic system is based on a small set of adjustable parameters, which can be determined by adjusting to experimental data [15, 16] or *ab-initio* results [6, 17, 18].

The aim of the present work is firstly to construct a reliable TB-SMA atomistic potential for Pt, with parameters determined from *ab-initio* calculations rather, than from experimental quantities; secondly, to use this interatomic potential in tandem with molecular-dynamics (MD) simulations to study self-diffusion on Pt(111) surface at various temperatures. Our goal is to investigate different diffusion mechanisms, to determine the relevant migration energies and pre-exponential factors for each process and compare with FIM [19], STM experiments [20] and previous computations. In order to test the transferability of the new potential, we computed the bulk modulus, elastic constants, linear thermal expansion coefficient, zone boundary phonon frequencies, as well as the surface relaxations of the metal. The simulated data are compared with available experimental results.

The paper is structured as follows: in Section 2 we present the computational scheme. The obtained results are presented and discussed in Section 3. Finally our conclusions are given in Section 4.

2 Computational Method

The electronic band structure of Pt was calculated self-consistently by the augmented plane wave (APW) method [21]. The calculations were scalar relativistic and used the Hedin–Lundqvist exchange-correlation functional [22] of the local-density approximation (LDA). The computations were done for both the fcc and bcc structures of the metal; we used a mesh of 89 k points in the irreducible Brillouin zone for the fcc and 55 k points for the bcc structure. The total energy was calculated for six different lattice constants and the resulting volume dependence of the energy was fitted to a parabolic function [23].

The total energy of the system within the TB-SMA model [13–15] is given by

$$E = \sum_i \left(A \sum_{j \neq i} \exp\left(-p\left(\frac{r_{ij}}{r_0} - 1\right)\right) - \xi \sqrt{\sum_{j \neq i} \exp\left(-2q\left(\frac{r_{ij}}{r_0} - 1\right)\right)} \right) \quad (1)$$

Where the first sum over j represents a pair-potential repulsive term of the Born-Mayer type, and the second sum corresponds to the attractive band-energy term. In the above expression r_{ij} is the distance between atoms i and j, $r_0 = 2.78021$ Å stands for the calculated APW value of the nearest neighbor distance for Pt, and the sums over j include interactions up to fifth neighbors. There are four adjustable parameters, A, ξ, p and q in this scheme. These have been determined by simultaneous fitting to the APW total-energy curves as a function of volume for both the fcc and bcc structures. We note that, before performing the fitting procedure, we uniformly shifted the two curves, so that to match at the minimum of the fcc structure the experimental cohesive energy of Pt (E = 5.85 eV) [24].

As a test of the quality of our parameters, we have calculated the bulk modulus of Pt by using the method proposed in Ref. [23], as well as the elastic constants at room temperature. Details of the computational procedure for the elastic constants are reported elsewhere [17].

Using the above interatomic potential, we performed MD simulations in the microcanonical ensemble for a system of 2,880 particles arranged on an fcc lattice. The equations of motion were integrated by means of the Verlet algorithm and a time step of 5 fs. Periodic boundary conditions were applied in the space and two free surfaces parallel to the (111) planes were constructed. The latter were obtained by fixing the dimensions of the supercell size to a value twice as large as the thickness of the crystal in the direction normal to the surface. The phonon dispersion curves were calculated by Fourier transform of the velocity autocorrelation function [17].

The self-diffusion of adatoms and dimers on the (111) face of Pt was studied by putting them on each free surface of the slab. Simulations of 2 ns were performed in the temperature range of 300–2,000 K, in order to achieve reliable statistics for the determination of the diffusion coefficient. The latter was, furthermore, calculated using the Kubo integral of the velocity autocorrelation function [25]. The static activation energy for the diffusion process was obtained by a quasi-dynamic minimization method integrated in the MD code. Details of the above computational methods can be found in [6, 17, 25, 26].

3   Results and Discussion

In Fig. 1 we present the first-principles computed cohesive energy of Pt (with opposite sign) as a function of the volume in the fcc and bcc structures (solid lines) after the uniform energy shift, as described in Section 2. The dashed lines in the same graph show the results of the fit using the expression (Eq. (1)). From this procedure we have determined the potential parameters of Eq. (1), which are listed in Table 1.

The parameters of Table 1 have been used to calculate some physical quantities of Pt within the TB-SMA scheme. In Table 2 we report the calculated properties of the element in comparison with available experimental values. We find that the computed lattice parameter is within 0.3% of the experimental value [24]. In addition, the calculated values of the bulk modulus and the elastic constants C_{11} and C_{12} are in very good agreement with the measurements [27], comparable with the accuracy

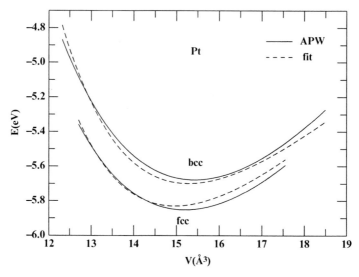

Fig. 1 Calculated cohesive energies of Pt (with opposite sign) as a function of volume. Solid lines correspond to the APW results; dashed lines refer to the results of fit (expression (1))

Table 1 Potential parameters of Eq. (1) for Pt

ξ (eV)	A (eV)	q	P
1.89706	0.09238	2.57947	16.00033

Table 2 Calculated properties of Pt within the present TB-SMA scheme in comparison with experimental data

Pt	Calculation	Experiment
Lattice constant (Å)	3.912	3.924 [24]
Bulk modulus (GPa)	303	288 [27]
C_{11} (GPa)	434	358 [27]
C_{12} (GPa)	238	254 [27]
C_{44} (GPa)	253	77 [27]
Coefficient of linear thermal expansion near room temperature (10^{-5} K^{-1})	1.02	0.89 [24]

of first-principles calculations, while the elastic constant C_{44} shows a rather large discrepancy. Furthermore, we have computed the temperature dependence of the lattice constant of Pt by using the MD simulations and we found that the coefficient of linear thermal expansion near room temperature is in very good agreement with the experimental value [24] (Table 2).

In Table 3 we report the computed phonon frequencies at the points X and L of Brillouin zone at 90 K, along with the corresponding experimental values [28]. We observe a fair accuracy of our model, especially for the transverse modes, while there is a slight overestimation of about 1.0 THz for the longitudinal modes. This is compatible with the inaccuracy found in our calculated value of C_{44}. In addition, we have computed the relative thermal relaxation of Pt (111) surface atoms in the

Table 3 Comparison between computed and experimental [28] phonon frequencies (in THz) at X and L boundaries of Brillouin zone for Pt at 90 K

	X (long)	X (trans)	L (long)	L (trans)
Computation	6.90	4.75	6.85	3.20
Experiment	5.80	3.84	5.85	2.90

Abbreviations: Long, longitudinal; trans, transverse modes.

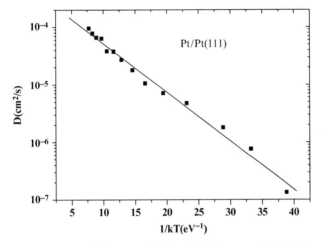

Fig. 2 Arrhenius diagram of the diffusion coefficient of Pt adatoms on the Pt(111) surface for the hopping diffusion mechanism in the temperature range 300–1,500 K. The solid line corresponds to least-squares fit

normal to the surface direction, with respect to bulk interlayer spacing at 300 K and we found a value of −1% in fair agreement with the experimental value of $1.1 \pm 0.5\%$ [29].

In the following we will discuss the 3-D adatom motion on the surface. In Fig. 2 we provide the Arrhenius diagram of the diffusion coefficient D of Pt adatoms on the Pt (111) surface for the hopping diffusion mechanism in the temperature range 300–1,500 K. The computed values follow a straight line with an energy barrier of $E_m = 194$ meV according to the expression:

$$D = D_0 \exp(-E_m/kT) \qquad (2)$$

where D_0 is the pre-exponential factor.

A careful analysis of the trajectories showed that at low temperatures the dominant diffusion mechanism is the simple hop to the neighbouring energy minima. On the other hand, at temperatures above 1,000 K, the adatoms hop with long and correlated jumps.

In Table 4 we give the computed migration energy E_m, the pre-exponential factor D_0 associated to the hopping diffusion mechanism, and the static energy barrier E_s.

Table 4 Static diffusion barriers E_s, migration energies E_m and pre-exponential factors D_0 of Pt adatom diffusion on the Pt (111) surface

Pt/Pt(111)	E_s(meV)	E_m(meV)	$D_0(10^{-3} \, cm^2/s)$
Present work	178	194 ± 6	0.36
Experiment (STM) [20]		260 ± 10	1
Experiment (FIM) [19]		260 ± 3	$2(\times 1.4^{\pm 1})$
Experiment (FIM) [30]		250 ± 20	
Ab-initio LDA, GGA [31]	420, 390		
Ab-initio, LDA, GGA [32]	330		
Ab-initio, LDA [33]	290		
EAM (semi-empirical) [7]	78		0.35
Effective medium theory (EMT) (semi-empirical) [34]	160		
TB-SMA (semi-empirical) [35]	176		

The experimental data of STM studies [20], FIM measurements [19, 30] and previous calculations or simulations [7, 31–35] are also listed in Table 4. Firstly, we note that there is an excellent agreement between our value for E_s and the static energy barrier from previous TB-SMA simulations [35], using parameters from fit to experimental data. Moreover, our static energy E_s agrees well with our computed migration energy E_m. The energy difference between the hcp and fcc sites of the adatom is only 5 meV. Secondly, our E_m is closer to the corresponding experimental values, compared to the previous semi-empirical computations [7, 34, 35] as well as to some first principles values [31, 32]. The computed value for D_0 agrees, furthermore, very well with the estimation of EAM calculation [7], and it is close to the experimental pre-exponential factors [19, 20].

We have also observed concerted exchange diffusion mechanisms at temperatures above 1,500 K, where there is a simultaneous motion of the adatom with two surface atoms belonging to nearest-neighbor rows. This has been also proposed in previous works using the EAM or TB-SMA methods in tandem with MD simulations [6, 35, 36]. These events require much higher energy than the hopping diffusion mechanism. At each temperature we have calculated the frequency of these events and then the corresponding diffusion coefficient, which is proportional to the frequency. In Fig. 3 we report the Arrhenius plot of the diffusion coefficient of Pt adatoms on the Pt (111) surface for the exchange diffusion mechanism at temperatures above 2,000 K. From a least-squares fit, we deduced a migration energy of 3.2 eV and a pre-exponential factor of 15 cm²/s. Our value for this diffusion barrier can be compared with the value of 2.1 eV computed by static calculations in Ref. [35].

Concerning Pt dimer diffusion on Pt(111) surface, we made a registration of the velocities of the center of mass of the dimer as a function of time for temperatures between 500 and 1,200 K and we have deduced the diffusion coefficient as explained in Section 2. In Fig. 4 we show the Arrhenius diagram of the diffusion coefficient of Pt_2 on Pt(111) surface. Working in the same way as for the single adatom diffusion, we obtain a migration energy for dimer diffusion $E_m = 320$ meV

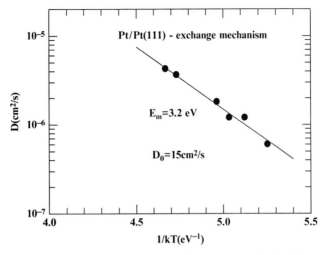

Fig. 3 Arrhenius plot of the diffusion coefficient of Pt adatoms on the Pt (111) surface for the exchange diffusion mechanism at temperatures above 2,000 K. The solid line corresponds to least-squares fit

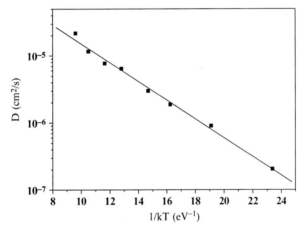

Fig. 4 Arrhenius diagram of the diffusion coefficient of a Pt dimer on Pt(111) for temperatures between 500 and 1,200 K. The solid line is least-squares fit

and a pre-exponential factor $D_0 = 3.8 \times 10^{-4}$ cm^2/s, which are listed in Table 5. In Table 5 we also provide the corresponding experimental data from FIM studies [19] as well as the static diffusion barrier from LDA calculations [37]. We note that the simulation results for Pt$_2$ diffusion are, within the experimental errors, in very good agreement with the measured values.

Finally, we have studied the dimer dissociation on Pt(111). In order to obtain the dimer dissociation energy we have performed two different procedures. The first one is static calculation. According to this method, we computed the difference between

Table 5 Migration energies E_m, and pre-exponential factors D_0 of Pt dimer diffusion, along with dissociation barriers E_{dis} of Pt_2 on the Pt(111) surface

$Pt_2/Pt(111)$	E_m (meV)	$D_0 (10^{-4} \, cm^2/s)$	E_{dis} (meV)
Present work	320 ± 20	3.8	685, 740
Experiment (FIM) [19]	370 ± 20	$1.9(\times 4.5^{\pm 1})$	
Ab-initio, LDA [37]	370		
Experiment (FIM) [38]			490 ± 10

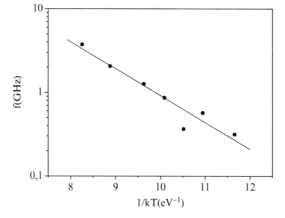

Fig. 5 Arrhenius diagram of the dissociation frequency of a Pt dimer on Pt(111) for temperatures between 1,000 and 1,400 K. The solid line is least-squares fit

the total energy of the slab with two well separated Pt adatoms on Pt(111) surface and the total energy of the slab with a Pt dimer on the surface. From this calculation we found $E_{dis} = 685$ meV. The second method is dynamical computation. According to this procedure, we executed several simulations for each temperature in the temperature range between 1,000 and 1,400 K. The aim was to obtain a mean time of dissociation as a function of temperature. The inverse of this time, the dissociation frequency, f, has been found to follow Arrhenius behavior against 1/kT. In Fig. 5 we provide the Arrhenius diagram of the dissociation frequency of Pt_2 on Pt(111). Applying the linear fitting procedure we obtain $f_0 = 1,500$ GHz and $E_{dis} = 740$ meV. The computed dissociation energies are reported in Table 5, along with the corresponding value of 490 meV from FIM experiments [38]. It should be mentioned that our calculated data overestimate the dissociation energy by 40%.

4 Conclusions

We have presented an interatomic potential of Pt in the framework of the second-moment approximation to the tight-binding model by adjusting to the volume dependence of the total energy, computed by first-principles APW calculations. We have applied this scheme to calculate the bulk modulus and elastic constants of the metal and we have obtained a good agreement with experiment. We also have performed molecular-dynamics simulations using the above semi-empirical potential and we have deduced the coefficient of thermal expansion at room temperature and the phonon spectrum at 90 K, where a satisfactory accuracy was obtained. We have computed, furthermore, the Pt(111) surface relaxations and we have obtained a fair agreement with experiment.

In addition, we have studied the self-diffusion of single adatoms and dimers on Pt (111) surface. The diffusion coefficient of Pt adatoms and dimers associated with the hopping diffusion mechanism presents Arrhenius behavior. The proposed model provides reliable values for the migration energies and pre-exponential factors of the hopping mechanism compared with STM and FIM studies. At high temperatures we have also observed a concerted exchange adatom diffusion mechanism where there is a participation of two surface atoms belonging to nearest-neighbor rows. The corresponding migration energy was found to be about 15 times higher than the energy of the hopping diffusion mechanism. We have finally obtained the dimer dissociation energy, which is somewhat overestimated compared with FIM measurements.

Despite the simplicity of the present scheme, it was found to be very useful by providing reliable results of diffusion of adatoms and dimers on Pt(111) surface.

Acknowledgement This study was supported by a PENED Grant No. 03–968 from the Greek General Secretariat for Research and Technology.

References

1. Tringides, M. C. (Ed.). Surface Diffusion: Atomistic and Collective Processes. Plenum, New York (1997).
2. Brune, H. Surf. Sci. Rep. **31**, 121 (1998) and references therein.
3. Tsong, T.T. Prog. Surf. Sci. **67**, 235 (2001).
4. Barth, J.V. Surf. Sci. Rep. **40**, 75 (2000).
5. Ala-Nissila, T., Ferrando, R., Ying, S.C. Adv. Phys. **51**, 949 (2002).
6. Papanicolaou, N.I., Papaconstantopoulos, D.A. Thin Sol. Films **428**, 40 (2003).
7. Liu, C. L., Cohen, J. M., Adams, J. B., Voter, A. F. Surf. Sci. **253**, 334 (1991).
8. Daw, M.S., Baskes, M.I. Phys. Rev. B **29**, 6443 (1984).
9. Daw, M.S., Foiles, S.M., Baskes, M.I. Mater. Sci. Rep. **9**, 251 (1993).
10. Finnis, M.W., Sinclair, J.E. Philos. Mag. A **50**, 45 (1984).
11. Ducastelle, F. J. Phys.(Paris) **31**, 1055 (1970).
12. Gupta, R.P. Phys. Rev. B **23**, 6265 (1981).
13. Tomanek, D.A., Aligia, A., Balseiro, C.A. Phys. Rev. B **32**, 5051 (1985).
14. Zhong, W., Li, Y. S., Tomanek, D. Phys. Rev. B **44**, 13053 (1991).
15. Rosato, V., Guillope, M., Legrand, B. Philos. Mag. A **59**, 321 (1989).
16. Cleri, F., Rosato, V. Phys. Rev. B **48**, 22 (1993).
17. Kallinteris, G.C., Papanicolaou, N.I., Evangelakis, G.A., Papaconstantopoulos, D.A. Phys. Rev. B **55**, 2150 (1997).
18. Papanicolaou, N.I., Kallinteris, G.C., Evangelakis, G.A., Papaconstantopoulos, D.A. Comp. Mat. Sci. **17**, 224 (2000).
19. Kyuno, K., Gölzhäuser, A., Ehrlich, G. Surf. Sci. **397**, 191 (1998).

20. Bott, M., Hohage, M., Morgenstern, M., Michely, Th., Comsa, G. Phys. Rev. Lett. **76**, 1304 (1996).
21. Mattheiss, L. F., Wood, J.H., Switendick, A. C. Methods Comput. Phys. **8**, 63 (1968).
22. Hedin, L., Lundqvist, B. I. J. Phys. C **4**, 2064 (1971).
23. Birch, F. J. Geophys. Res. **83**, 1257 (1978).
24. Kittel, C. Introduction to Solid State Physics. Wiley-Interscience, New York (1986).
25. Kallinteris, G.C., Evangelakis, G.A., Papanicolaou, N.I. Surf. Sci. **369**, 185 (1996).
26. Chamati, H., Papanicolaou, N.I.J. Phys.: Condens. Matter **16**, 8399 (2004).
27. Simmons, G., Wang, H. Single Crystal Elastic Constants and Calculated Aggregate Properties: A Handbook, second ed. MIT press, Cambridge, MA (1971).
28. Dutton, D.H., Brockhouse, B.N., Muller, A.P. Can. J. Phys. **50**, 2915 (1972).
29. Materer, N., Starke, U., Barbieri, A., Döll, R., Heinz, K, van Hove, M. A., Somorjai, G. A. Surf. Sci. **325**, 207 (1995).
30. Feibelman, P.J., Nelson, R.S., Kellogg, G.L. Phys. Rev. B **49**, 10548 (1994).
31. Mortensen, J.J., Hammer, B., Nielsen, O.H., Jacobsen, K.W., Nørskov, J.K. Springer Series in Solid- State Sciences, Okiji, A. (Ed.). Springer, Berlin, **121**, 173 (1996).
32. Boisvert, G., Lewis, L.J., Scheffler, M. Phys. Rev. B **57**, 1881 (1998).
33. Feibelman, P.J. Phys. Rev. Lett. **81**, 168 (1998).
34. Stoltze, P. J. Phys.: Condens. Matter **6**, 9495 (1994).
35. Bulou, H., Massobrio, C. Phys. Rev. B **72**, 205427 (2005).
36. Zhuang, J., Liu, L., Ning, X., Li, Y. Surf. Sci. **465**, 243 (2000).
37. Boisvert, G., Lewis, L.J. Phys. Rev. B **59**, 9846 (1999).
38. Kyuno, K., Ehrlich, G. Phys. Rev. Lett. **84**, 2658 (2000).

Gap-Townes Solitons and Delocalizing Transitions of Multidimensional Bose–Einstein Condensates in Optical Lattices

Mario Salerno, F. Kh. Abdullaev, and B.B. Baizakov

Abstract We show the existence of gap-Townes solitons for the multidimensional Gross–Pitaeviskii equation with attractive interactions and in two- and three-dimensional optical lattices. In absence of the periodic potential the solution reduces to the known Townes solitons of the multi-dimensional nonlinear Schrödinger equation, sharing with these the property of being unstable against small norm (number of atoms) variations. We show that in presence of the optical lattice the solution separates stable localized solutions (gap-solitons) from decaying ones, characterizing the delocalizing transition occurring in the multidimensional case. The link between these higher dimensional solutions and the ones of one dimensional nonlinear Schrödinger equation with higher order nonlinearities is also discussed.

Keywords Matter-waves · Gap-Townes soliton · Collapse · Delocalizing transition

1 Introduction

One interesting phenomenon occurring in ultracold atomic gases trapped in periodic potentials is the possibility to localize matter in states which can stay for a long time due to an interplay between nonlinearity, dispersion and periodicity. Such states (also called gap-solitons) have been observed in Bose–Einstein condensates (BEC) and in arrays of nonlinear optical waveguides [1–4]. For attractive atomic interactions in BEC and in absence of a periodic potential, stable localized solutions are possible only in a one-dimensional (1D) setting since in two (2D) and three (3D) dimensions the phenomenon of collapse appears [5]. More precisely, one observes that when the number of atoms exceeds a critical threshold, the solution collapses in a finite time (blow-up) while for number of atoms below the critical threshold

M. Salerno (✉)
Dipartimento di Fisica "E.R. Caianiello" and Consorzio Nazionale Interuniversitario per le Scienze Fisiche della Materia (CNISM), Universitá di Salerno, I-84081 Baronissi (SA), Italy
e-mail: salerno@sa.infn.it

F.Kh. Abdullaev and B.B. Baizakov
Physical-Technical Institute of the Uzbek Academy of Sciences, 100084, Tashkent, Uzbekistan

N. Russo et al. (eds.), *Self-Organization of Molecular Systems: From Molecules and Clusters to Nanotubes and Proteins*, NATO Science for Peace and Security Series A: Chemistry and Biology, © Springer Science+Business Media B.V. 2009

there is an irreversible decay of the state into background radiation. For the higher dimensional nonlinear Schödinger (NLS) equation, however, it is known that there exists an unstable localized solution, the so called Townes soliton [6], which separates decaying solutions from collapsing ones. Townes soliton, however, exists only for a single value of the number of atoms, being unstable against fluctuations around it (for slightly overcritical or undercritical number of atoms the solution collapses or decays, respectively). The situation is drastically changed in presence of an optical lattice (OL). To this regard, it has been shown that stable 2D and 3D solitons can exist in OLs both in BEC and nonlinear optics contexts [7–11]. Moreover, it is known that while the periodic potential can only marginally shift the critical value for collapse, it can substantially move the delocalizing transition curve, thereby increasing the soliton existence range in parameter space from a single point to a whole interval [12]. The typical situation with 2D and 3D BEC solitons in OLs is therefore the following: in the parameter space the stable localized solutions are confined from above by the collapse curve and from below by the delocalizing transition curve, thus, in contrast with the one dimensional case where there are no limits for the existence of localized states, strict limitations for soliton existence appear in multidimensional cases. From this point of view it is clear that for possible experimental observation of multidimensional BEC solitons the parameter design becomes very important. Since the collapse curve is only marginally affected by the periodic potential, to enlarge existence ranges of solitons it is of interest to give a full characterization of the delocalizing curve in parameter space.

The aim of this paper is just devoted to this, i.e. we characterize 2D and 3D delocalizing curves of gap-solitons in terms of an unstable solution of the multidimensional Gross–Pitaeviskii equation (GPE), which we call gap-Townes soliton. This solution can be viewed as a separatrix (it separates gap soliton states from extended (Bloch) states) and reduces to the known Townes soliton when the strength of the OL goes to zero. Similar solutions were found also for the 1D NLS equation with higher order nonlinearities in [13], where they were called gap-Townes solitons, and in [14] where they were termed Townes solitons. Conditions for the occurrence of the delocalizing transition phenomenon of one-dimensional localized modes of several nonlinear continuous periodic and discrete systems of the nonlinear Schrödinger type were also recently discussed in [15]. For the periodic multidimensional GPE the delocalizing curve has been characterized in [12] as the critical threshold for the existence of one bound state in an effective potential. The characterization given here, however, is more general since it is valid also for 1D NLS with higher order nonlinearities. To this regard we remark that in absence of confining potential the 2D and 3D GPE behaves similarly to the 1D NLS with quintic and septic nonlinearities, respectively. The interplay between dimensionality and nonlinearity has been used to investigate collapse in lower dimensional NLS on the basis of pure dimensional arguments. In particular, the critical condition for collapse has been characterized as $D(n-1)-4 = 0$, where n is the order of the nonlinearity in the equation and D is the dimensionality of the system [16]. In the following we take advantage of this interplay to construct approximate gap-Townes soliton solutions of the GPE with multidimensional separable OLs, in terms of products of exact gap-Townes solutions of the 1D NLS with higher nonlinearities. Remarkably, we find that, except for strengths of the optical lattices very small, our approach produces very accurate gap-Townes solutions of multidimensional GPE with OL, thus giving an evident computational advantages. The results obtained in this paper can be seen as a generalization of the existence of gap-Townes solitons in the quintic NLS discussed in [13, 14] to the case of the multidimensional Gross–Pitaeviskii equation.

We finally remark that the obtained results can also be applicable for photonic lattices with Kerr type of optical nonlinearity where the existence of a critical threshold for the lattice solitons has been observed [17].

The paper is organized as follows. In Section 2 we introduce the model equations and discuss the link between multidimensional GPE with a separable trapping potential and the corresponding 1D NLS equation with higher order nonlinearity. We use a self-consistent approach to approximate gap-Townes solitons of the GPE with products of exact gap-Townes soliton of the corresponding 1D NLS equation with higher order nonlinearity. In Section 3 we discuss the existence of localized solutions in the multidimensional GPE with OL by means of a variational approach (VA) and compare 2D and 3D results with those obtained from the VA applied to the quintic and septic NLS, respectively. In Section 4 we perform a numerical investigation of the existence (delocalizing) threshold for gap-Townes solitons of the 2D and 3D GPE. Finally, in the last section we briefly summarize our main results.

2 Model Equations and Existence of Gap-Townes Solitons

Let us consider the following Gross–Pitaevskii equation in d-dimensions ($d = 1, 2, 3,$) as a model for a BEC in an optical lattice [12]

$$i\psi_t + \nabla_d^2 \psi + \varepsilon \left[\sum_{i=1}^{d} \cos(2x_i) \right] \psi + \gamma |\psi|^2 \psi = 0, \tag{1}$$

where ∇_d^2 denotes the d-dimensional Laplacian, $\sum_{i=1}^{d} \cos(2x_i)$ denotes a square optical lattice with strength ε, γ is the coefficient of nonlinearity, and $x_i = x, y, z$ for $i = 1, 2, 3$, respectively. Here we will be mainly interested in cases $d = 2$ and $d = 3$. The existence of localized solutions of the multidimensional GPE with periodic potential and positive and negative nonlinearities (atomic scattering lengths), has been previously investigated both by variational analysis and by direct numerical simulations. In the following we concentrate on a topic which was not discussed in previous works, namely the existence of gap-Townes solitons in the multidimensional GPE and its link to the phenomenon of delocalizing transition. Due to the instability properties of these solutions it is difficult to find them without an analytical guide. To this regard we take advantage of the fact that the periodic potential is separable and in spite of the nonlinearity of the system we look for factorized stationary solutions of the form

$$\psi(x_1, ..., x_d) = \prod_{i=1}^{d} \phi_i(x_i) e^{-i\mu t}. \tag{2}$$

In 2D case ($d = 2$) the substitution of the factorized ansatz into Eq. (1) gives:

$$\frac{\phi_{1_{xx}}}{\phi_1} + \frac{\phi_{2_{yy}}}{\phi_2} + \varepsilon[\cos(2x) + \cos(2y)] + \gamma|\phi_1|^2|\phi_2|^2 = -\mu. \tag{3}$$

This equation can also be written as

$$\phi_{1_{xx}} + \varepsilon \cos(2x)\phi_1 + \frac{\gamma}{2}|\phi_2|^2|\phi_1|^2\phi_1 = -\mu_1\phi_1,$$

$$\phi_{2_{yy}} + \varepsilon \cos(2y)\phi_2 + \frac{\gamma}{2}|\phi_1|^2|\phi_2|^2\phi_2 = -\mu_2\phi_2, \tag{4}$$

with $\mu_1 = \mu_2 = \mu/2$. By assuming $\phi_1 = \phi_2 \equiv \phi$ and adopting a diagonal co-ordinate $x = y \equiv \xi$ we have that Eqs. (4) become equivalent to the following 1D eigenvalue problem

$$\phi_{\xi\xi} + \varepsilon \cos(2\xi)\phi + \frac{\gamma}{2}|\phi|^4\phi = -\frac{\mu}{2}\phi. \tag{5}$$

From this we see that there is a link between the 2D cubic NLS and the 1D quintic NLS which implies a rescaling of parameters as: $\gamma \to \frac{\gamma}{2}$, $\mu \to \frac{\mu}{2}$.

The above equations can be easily extended to the 3D GPE with periodic potential. In this case Eq. (5) will be replaced by the following 1D NLS equation with septic nonlinearity

$$\phi_{\xi\xi} + \varepsilon \cos(2\xi)\phi + \frac{\gamma}{3}|\phi|^6\phi = -\frac{\mu}{3}\phi, \tag{6}$$

from which we see that in this case parameters must be rescaled according to: $\gamma \to \frac{\gamma}{3}$, $\mu \to \frac{\mu}{3}$.

It is appropriate to mention that a factorized solution of the form Eq. (2) with the components solutions of the nonlinear eigenvalue problem Eq. (5) cannot be an exact solution of the 2D or 3D GPE, since, due to the nonlinearity, the problem is obviously not exactly separable. On the other hand, by imposing the coincidence of the solutions along the diagonal axis may be a constraint for a reasonable approximate solutions of the 2D and 3D problems, especially when the nonlinearity is small. In analogy with the 1D NLS with quintic nonlinearity investigated in Ref. [13] we expect that the delocalizing curve coincides with the existence curve of gap-Townes solitons for which it was shown that the critical number of atoms decreases with increasing the strength of the OL. This means that in a deep OL the effective nonlinearity required for the existence of a gap-Townes soliton is smaller and the problem may become effectively close to separable.

To check the correctness of this argument we construct factorized solutions (2) of the 2D GPE by means of a self-consistent method which allows to solve the 1D quintic NLS eigenvalue problem exactly (similar results can be obtained for the 3D case).

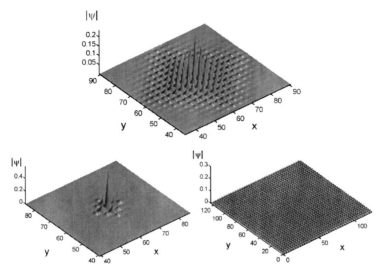

Fig. 1 Gap-Townes soliton (*top* panel), gap soliton (*lower left* panel) and Bloch state (*lower right* panel) of the GPE for parameter values $\varepsilon = 5$, $\gamma = 1$. The critical number of atoms in normalized units for the gap-Townes soliton is $N_c = 0.4261$. The gap soliton and the Bloch state in the lower panel are obtained for slightly overcritical ($N = 0.4347$) and undercritical ($N = 0.4176$) values, respectively

In the top panel of Fig. 1 we show a 2D gap-Townes soliton obtained from the product ansatz using exact (self-consistent) gap-Townes solitons of the corresponding quintic NLS equation in Eq. (5). The lower left and right panels show, respectively, the gap soliton and the extended Bloch state found at energy slightly below and slightly above (bottom of the lowest band) the one of the gap-Townes soliton. To check the reliability of the factorized ansatz we have computed the time evolution under the original 2D GPE equation using the factorized solution as initial condition. This is shown in Fig. 2 where the time evolution of the gap-Townes soliton in Fig. 1 (central panel) and the ones obtained for slightly overcritical and undercritical numbers of atoms are shown. We see that while the product solution constructed from the quintic NLS remains localized for a long time, a slight increase or decrease of the number of atoms produces shrinking or decay of the solution, respectively. This clearly shows the existence and role of gap-Townes solitons of the 2D GPE with optical lattice in characterizing the delocalizing threshold.

3   Variational Analysis and Existence of Localized States

The existence of localized states in multidimensional GPE with OL and in 1D NLS with higher order nonlinearities can be investigated by means of the variational approach. To this regard we first consider Eq. (1) and search for stationary solutions

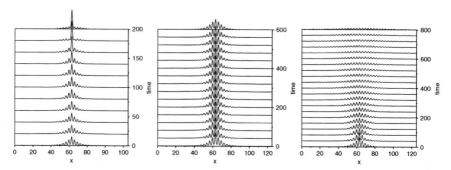

Fig. 2 Time evolution of the gap-Townes soliton shown in Fig. 1 with critical number of atoms (*central* panel) and for slightly overcritical (*left* panel) and undercritical (*right* panel) number of atoms

of the form $\psi = U e^{-i\mu t}$ and consider the Lagrangian density associated with the equation for the stationary field U

$$\mathcal{L} = \frac{1}{2}(\nabla_d U)^2 - \frac{\mu}{2}U^2 - \frac{\varepsilon}{2}\left[\sum_{i=1}^{d}\cos(2x_i)\right]U^2 - \frac{\gamma}{4}U^4. \tag{7}$$

By taking a Gaussian ansatz for U

$$U = A e^{-\frac{a}{2}\sum_{i=1}^{d}x_i^2}, \tag{8}$$

and performing spatial integration we obtain the following effective lagrangian $L_{eff} = \int \mathcal{L}dx$ for parameters A, a

$$L_{eff} = \frac{A^2}{2}\left(\frac{\pi}{a}\right)^{\frac{d}{2}}\left[\frac{d}{2}a - \mu - d\varepsilon e^{-1/a} - \frac{A^2\gamma}{2^{\frac{d}{2}+1}}\right], \quad d = 1, 2, 3. \tag{9}$$

Variational parameters A, a, indicating the amplitude and inverse width of the localized state, are linked to the number of atoms by the relation $N = A^2(\frac{\pi}{a})^{d/2}$. From the conditions of stationarity of the effective lagrangian $\partial L_{eff}/\partial a = \partial L_{eff}/\partial A = 0$, we get the following equations relating N with a and the chemical potential μ

$$\mu = \frac{d}{2}a - d\varepsilon e^{-1/a} - \frac{\gamma N}{2^{d/2}}\left(\frac{a}{\pi}\right)^{d/2}, \tag{10}$$

$$N = \frac{4\pi^{d/2}}{\gamma}\left(\frac{2}{a}\right)^{\frac{d}{2}-1}\left(1 - \frac{2\varepsilon}{a^2}e^{-1/a}\right), \quad d = 1, 2, 3.$$

In Fig. 3 we depict the (N, μ) curves obtained from the above transcendental equations for fixed values of ε and for the cases $d = 1, 2, 3$. We see that for the 1D case $dN/d\mu$ is always negative, this means, according to the Vakhitov–Kolokolov (V-K) criterion [18], that the solution exists and is stable for any value of ε without limitations on the effective nonlinearity $N\gamma$. In the 2D case, a threshold in $N\gamma$ appears

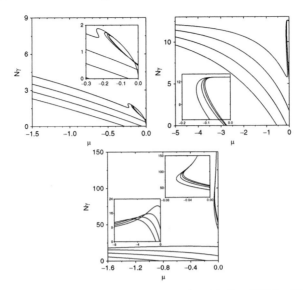

Fig. 3 *Left panel.* VA existence curves for the 1D case ($d = 1$ in Eqs. (15)) for different values of ε. The inset shows an enlargement for small values of μ. Curves refer to ε values increased in steps of 0.5 starting from $\varepsilon = 0.5$ (upper curve and most left curve in the inset) to $\varepsilon = 2.0$ (lower curve and most right curve in the inset). *Central panel.* Same as for the left panel but for the 2D case ($d = 2$ in Eqs. (15)). The inset shows an enlargement for small values of μ close to the threshold $N = 4\pi$. *Right panel.* Same as for the left panel but for the 3D case ($d = 3$ in Eqs. (15)). The upper inset shows an enlargement of the curves close to the $\mu = 0$ axis, while the lower inset displays the curves close to the $N\gamma = 0$ axis, on a larger scale

which is predicted by VA to be exactly 4π for $\varepsilon = 0$. Also notice that in this case $dN/d\mu < 0$ is still satisfied for most branch curves, meaning that the solution is usually stable. The situation is quite different in the 3D case, where there is no limiting threshold for existence but most of the curves display a positive slope meaning that the solution is unstable. In particular, from the upper inset of the right panel we see that for curves close to the $\mu = 0$ axis, $dN/d\mu$ change signs at $N\gamma \approx 70$ and becomes positive for higher values of $N\gamma$. The almost horizontal curves for lower values of $N\gamma$ are displayed in the lower inset of the figure, from which we see that $dN/d\mu$ changes from negative to positive after the curves have reached a maximum at values of μ which depend on ε.

Notice that for the 3D case the above equations predict for $\varepsilon = 0$ the number of atoms dependence on the chemical potential as

$$N = \frac{4\pi\sqrt{\pi}}{\gamma\sqrt{-\mu}}.$$

Similar dependence N vs. μ was previously obtained for stationary solutions of 3D NLS with cubic and quintic nonlinearity [19]. Also notice that according to V-K criterion this solution is unstable. From the condition $N \geq 0$, we obtain the limitation on the soliton width as

$$\frac{2\varepsilon}{a^2}e^{-1/a} \leq 1, \tag{11}$$

while for $\varepsilon \gg 1$ we obtain that the soliton exists if the width satisfies the condition

$$a < a_{c1} \approx \frac{1}{2(\log(\delta) + \log(\log(\delta)))}, \quad \delta = 2\sqrt{2\varepsilon}, \tag{12}$$

and $a > a_{c2} \approx \sqrt{2\varepsilon} - 1/2$. For $\varepsilon = 5$, for example, we obtain $a_{c1} = 0.2$, while the exact value is $a_e \approx 0.17$.

In view of the analogy of the 2D and 3D GPE with the 1D NLS with quintic and septic nonlinearity, respectively, it is of interest to compare the above VA equations with these cases. To this regard, we consider the 1D GPE with a high order nonlinearity of the type $\gamma|\psi|^{\alpha}\psi$ with $\alpha = 2, 4, 6$

$$i\psi_t + \psi_{xx} + \gamma|\psi|^{\alpha}\psi + \varepsilon\cos(2x)\psi = 0. \tag{13}$$

Using the same approach as before, one can readily show that the effective lagrangian in this case is

$$L_{eff} = \frac{A^2}{2}\left(\frac{\pi}{a}\right)^{1/2}\left[\frac{1}{2}a - \mu - \varepsilon e^{-1/a} - \frac{A^{\alpha}\gamma}{\left(\frac{\alpha+2}{2}\right)^{3/2}}\right], \tag{14}$$

from which the following VA equations are derived

$$\mu = \frac{a}{2} - \varepsilon e^{-1/a} - \frac{\gamma N^{\alpha/2}}{\left(\frac{\alpha+2}{2}\right)^{1/2}}\left(\frac{a}{\pi}\right)^{\alpha/4}, \tag{15}$$

$$N^{\alpha/2} = \frac{2a}{\alpha\gamma}\left(\frac{\alpha+2}{2}\right)^{3/2}\left(\frac{\pi}{a}\right)^{\alpha/4}\left(1 - \frac{2\varepsilon}{a^2}e^{-1/a}\right),$$

here $\alpha = 2, 4, 6$. We see that the case $\alpha = 2$ coincides with the case $d = 1$ considered above, and the case $\alpha = 4$ with the quintic VA equations derived in [13].

Notice that for $\varepsilon = 0$ Eq. (13) admits exact solutions also for $\alpha = 4, 6$. For the septic case, indeed, we have, using $\psi = ue^{imt}$, $m = -\mu > 0$, that

$$\psi = (\frac{4m}{\gamma})^{1/6}\text{sech}^{1/3}(3\sqrt{m}x), \tag{16}$$

is an exact solution with a norm

$$N = \frac{2^{1/3}\Gamma^2(\frac{1}{3})}{3\Gamma(\frac{2}{3})\gamma^{1/3}m^{1/6}}, \tag{17}$$

where $\Gamma(x)$ is the gamma function. The Hamiltonian for this solution is equal to

$$H = \int_{-\infty}^{\infty} (|\psi_x|^2 - \frac{\gamma}{4}|\psi|^8) dx = 0.44 \frac{m^{5/6}}{\gamma^{1/3}}.$$

From the above VA equations for the 2D case, one can derive the value of N_c for small values of $\varepsilon \neq 0$ as: $N_c = 4\pi(1 - 8\varepsilon \exp(-2))$. For $\varepsilon = 0.2$ we obtain $N_c = 9.845$ which is in reasonable agreement with the value 10.8 obtained from numerical simulations of the 2D case. We remark that for small values of ε the soliton is very extended in space and resembles a Bloch wave modulated with an envelope. In this case an effective mass approximation may be appropriate which allows to replace the GPE field equation with a nonlinear Schrödinger equation with effective mass m^* and nonlinearity β. In the 2D case we have

$$i u_t + \alpha(u_{xx} + u_{yy}) + \beta|u|^2 u = 0, \tag{18}$$

where $\alpha = m^*/m, m^{*(-1)} = (\partial^2 E/\partial k^2)_{k=0}$ and $\beta = (2\pi/L^2) \int d^2 r |\phi_{1,0}|^4$ (a similar equation can be written also for the 3D case). In this approximation the norm of the gap Townes soliton can be evaluated as $N = \frac{N_T}{\alpha\beta}$. For deep optical lattices ($\varepsilon > 5$) α can be approximated as $\alpha \approx \varepsilon^{1/4}$ and the norm $N \approx \frac{N_T}{\varepsilon^{1/4}\beta}$.

4 Numerical Study of Gap-Townes Solitons and Delocalizing Transitions

In this section we investigate the existence curve of gap-Townes solitons in the (N, ε) plane. As mentioned before, this curve coincides with the delocalizing transition curve which separates stable localized solutions from decaying ones. Its knowledge is therefore important for experimental investigations of multidimensional solitons. To this regard we remark that for the observation of multidimensional BEC solitons parameters should be chosen between the delocalizing and the collapsing curves. In the following we investigate the delocalizing curve by means of direct numerical integrations of the 2D and 3D GPE with periodic potential and by the corresponding 1D NLS systems with quintic and septic nonlinearities discussed above, respectively. The existence of gap-Townes solitons is then shown by direct numerical simulations of the multidimensional GPE using as initial conditions the above mentioned product states, which are found by solving the 1D GPE with higher nonlinearities by means of the self-consistent method described in [20]. In Fig. 4 we depict the existence curve of gap-Townes solitons of the 2D GPE for $\gamma = 1$ as obtained from numerical integrations of the 2D GPE. The corresponding curve obtained from the 1D quintic NLS approximation by means of a self-consistent approach is also shown. We see that for $\varepsilon > 1$ the 1D quintic NLS curve agrees very well with that of the 2D GPE, the deviations becoming evident only for strengths of the OL which are less than $\varepsilon \approx 1$ (one recoil energy). This fact can be easily understood from the observation that for a fixed value of ε there is one value of N for which the gap-Townes soliton exists and that by increasing ε the corresponding

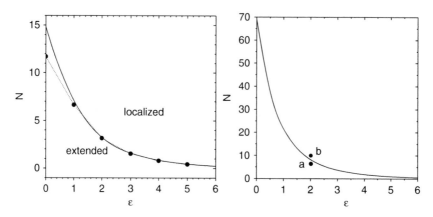

Fig. 4 *Left panel*: Existence curve of gap-Townes solitons of the 2D GPE for $\gamma = 1$. The continuous line represents the delocalizing curve as obtained from the 1D quintic NLS approximation. The dots joined by thin dotted line represent numerical results obtained from direct integrations of the 2D GPE. *Right panel*: Existence curve of gap-Townes solitons of the 3D GPE for $\gamma = 1$. The line represents the delocalizing curve as obtained from the 1D quintic NLS approximation, while the dots represent numerical results obtained from direct integrations of the 3D GPE

value of N decreases. This implies that for gap-Townes solitons in a strong OL the nonlinear interaction is effectively small, due to the potential barriers which prevent tunneling of matter into adjacent wells. On the contrary, in a shallow optical lattice the matter can easily tunnel through the barriers and the effective (attractive) non-linearity can be larger. Since the separability ansatz used to link the 2D GPE to the 1D quintic NLS works well when the nonlinearity is small, it is clear that a discrepancy can arise at small values of ε. The fact that the quintic NLS equation deviates from the 2D GPE only for $\varepsilon < 1$, however, makes the mapping between these two equations very convenient for practical calculations. A similar situation seems to be true also for the 3D case. In the right panel of Fig. 4 we depict the existence curve of gap-Townes solitons of the 3D GPE for $\gamma = 1$ as obtained from the 1D septic NLS approximation by means of a self-consistent approach. Due to the long computational times required in 3D simulations we have presented verification for only few points of the curve. A more complete analysis will require further investigation. In particular, the prediction of our analysis are checked by means of numerical integrations of the 3D GPE in correspondence of the two filled dots depicted in the right panel of Fig. 4 at $\varepsilon = 2$, just above and below the delocalizing curve obtained from the septic 1D NLS equation. In Fig. 5 we show the time evolution of the y-section of a 3D gap-Townes soliton obtained for $\varepsilon = 2$ in correspondence of these slightly undercritical (left panel) and overcritical (right panel) points. We see that, in analogy to what we observed in 2D case, the initial condition with an undercritical number of atoms leads to the complete delocalization of the state, while in the overcritical case a gap soliton state which remains stable over long time is formed. It is appropriate to mention that the exact profile of the 3D gap-Townes soliton, which separates these two behaviors is more difficult to obtain with the separability ansatz than for

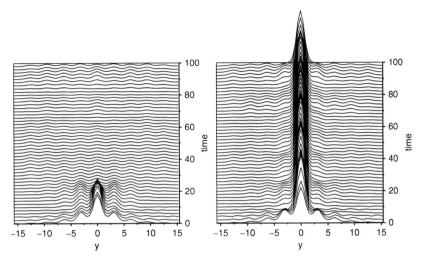

Fig. 5 Time evolution of the y-section of a 3D gap-Townes soliton obtained for $\varepsilon = 2$ with slightly undercritical (*left* panel) and overcritical (*right* panel) number of atoms, corresponding to points a and b in Fig 4, respectively

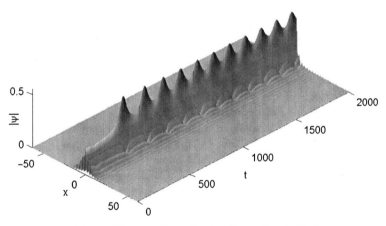

Fig. 6 Oscillations of the gap-Townes soliton when it collapses into the final gap state, according to Eq. (1) with initial state taken as the product solution constructed from two quintic NLS with a total norm increased by 1%. Shown is the $y = 0$ cross section of the 2D profile

the previous 2D case, since in the 3D case this ansatz appears to be less accurate. The signature of the gap-Townes soliton state, however, is very clear as one can see from the early stages of the time evolution depicted in Fig. 5 (notice from the left panel that the undercritical state remains stable up to a time $t = 20$ before starting to decay). This behavior strongly suggests the existence of gap-Townes solitons also in the 3D case.

We now address to another important property of the gap-Townes soliton which shows up in the transition to a gap soliton state for overcritical number of atoms. As it was already mentioned, even a slightly overcritical norm gives rise to a rapid shrinking of the gap-Townes soliton. Although at initial stage of evolution this behavior is similar to the collapse of ordinary Townes soliton, the final state is different. Specifically, gap-Townes soliton approaches the (final) gap state via long-lasting oscillations, as shown in Fig. 6. Each "reflection" from the broad state is accompanied by emission of linear waves, which can be viewed as a tunneling of matter from the localized mode into the extended one. This process also contributes to the damping of the oscillations. We remark that these oscillations, having a very regular behavior, could be used to detect the existence of gap-Townes solitons in a real experiment.

5 Conclusions

In this paper we have shown the existence of gap-Townes solitons in the multi-dimensional (2D and 3D) Gross–Pitaevskii equation with a periodic potential and discussed its link with the phenomenon of delocalizing transition. These solutions have the peculiarity of being unstable under small fluctuations of the number of atoms and separate localized (soliton like) states from extended (Bloch like) ones. The existence curve in the parameter space of this particular solution is very useful since it provides the lower threshold for the existence of localized states. The gap-Townes solitons discussed in this paper are a natural generalization of the Townes solitons of the nonlinear Schrödinger equation (without periodic potential). The existence of Townes solitons in a nonlinear glass sample modelled by the NLS equation was experimentally demonstrated in Ref. [21]. The fact that the transition from a gap-Townes soliton to a gap soliton is always accompanied by regular oscillations gives the possibility to indirectly observe the multidimensional gap-Townes solitons discussed in this paper in real experiments. In particular we expect these solitons to be observed both in multidimensional BECs in OL and in nonlinear optics systems, including 2D and 3D photonic crystals and arrays of nonlinear optical waveguides.

Acknowledgement MS acknowledges partial financial support from a MURST-PRIN-2003 and a MIUR-PRIN-2005 initiatives. FKhA and BBB wish to thank the Department of Physics "E. R. Caianiello" for hospitality and the University of Salerno for financial support.

References

1. S. Burger, K. Bongs, S. Dettmer, W. Ertmer, K. Sengstock, A. Sanpera, G. V. Shlyapnikov, and M. Lewenstein, Phys. Rev. Lett. **83**, 5198 (1999).
2. L. Khaykovich, F. Schreck, G. Ferrari, T. Bourdel, J. Cubizolles, L. D. Carr, Y. Castin and C. Salomon, Science **296**, 1290 (2002).
3. K. E. Strecker, G. B. Partridge, A. G. Truscott and R. G. Hulet, Nature **417**, 150 (2002).
4. B. Eiermann, P. Treutlein, Th. Anker, M. Albiez, M. Taglieber, K.-P. Marzlin, and M. K. Oberthaler, Phys. Rev. Lett. **91**, 060402 (2003).

5. C. Sulem and P. L. Sulem, *The Nonlinear Schrödinger Equation: Self-Focusing and Wave Collapse*, Springer, 1999.
6. R. Y. Chiao, E. Garmire, and C. H. Townes, Phys. Rev. Lett. **13**, 479 (1964).
7. B. B. Baizakov, V. V. Konotop, and M. Salerno, J. Phys. B: At. Mol. Opt. Phys., **35** , 5105 (2002).
8. N. K. Efremidis, S. Sears, D. N. Christodoulides, J. W. Fleischer, and M. Segev, Phys. Rev. E **66**, 046602 (2002).
9. E. Ostrovskaya and Yu.S. Kivshar, Phys. Rev. Lett. **90**, 160407 (2003).
10. B. B. Baizakov, B. A. Malomed, and M. Salerno, Eurphys. Lett. **63**, 642 (2003).
11. D. Mihalache, D. Mazilu, F. Lederer, B. A. Malomed, L.-C. Crasovan, Y. V. Kartashov, and L. Torner, Phys. Rev. A **72**, 021601(R) (2005).
12. B. B. Baizakov and M. Salerno, Phys.Rev. A **69**, 013602 (2004).
13. F. Kh. Abdullaev and M. Salerno, Phys. Rev. A **72**, 033617 (2005).
14. G.L. Alfimov, V.V. Konotop, P. Pacciani, Phys. Rev. A **75**, 023624 (2007).
15. H.A. Cruz, V.A. Brazhnyi, V.V. Konotop, and M. Salerno, Physica D (2008), doi:10.1016/j.physd.2008.09.008.
16. L. Berge, Phys. Rep. **303**, 259 (1998).
17. N. K. Efremidis, J. Hudock, D. N. Christodoulides, J. W. Fleischer, O. Cohen, and M. Segev, Phys. Rev. Lett. **91**, 213906 (2003).
18. N. G. Vakhitov, A. A. Kolokolov, Radiophys. Quant. Elect. **16**, 783 (1973).
19. N. Akhmediev, M. P. Das and A. V. Vagov, Int. J. Mod. Phys. B **13**, 625 (1999).
20. Mario Salerno, Laser Phys. **15**, No.4, 620 (2005).
21. K. D. Moll, A. L. Gaeta, G. Fibich, Phys. Rev. Lett. **90**, 203902 (2003).

Hydration Effects on Photophysical Properties of Collagen*

Agata Scordino, Rosaria Grasso, Marisa Gulino, Luca Lanzano', Francesco Musumeci, Giuseppe Privitera, Maurizio Tedesco, Antonio Triglia, and Larissa Brizhik

Abstract Collagen structure, in which water molecules mediate some networks of intra-chain and inter-chain hydrogen bonds, appears to be a promising model system to investigate in great detail the relationships between biological organization and the characteristics of Delayed Luminescence (DL), the phenomenon consisting of the prolonged ultra-weak emission of optical photons after excitation of the system by illumination. Samples of type I collagen molecules from tendons have been studied on varying their hydration state. Comparison of their dielectric properties with the DL response, along with the acquisition of excitation and emission spectra of tendon collagen, in its native and dried state have gathered additional information allowing to test hypothesis on the origin and/or the mechanisms of photoinduced DL emission. The peculiar structure of collagen, where a relevant role is played by the hydrogen bonded water network, suggested that collective excitations could be generated in this macromolecule. Upon hydration, changing of the phonon spectrum could significantly affect the type of the ground electron states which can be excited in the collagen, as evaluated by applying a variational method. Changing in the ground electron states could take into account for the different regimes of the DL decays, in turn.

Keywords Delayed luminescence · Collagen · Soliton state · Small polaron

A. Scordino (✉), R. Grasso, M. Gulino, L. Lanzano', F. Musumeci, and G. Privitera
Dipartimento di Metodologie Fisiche e Chimiche per l'Ingegneria, Catania University, viale A. Doria 6, I-95125 Catania, & Laboratori Nazionali del Sud—Istituto Nazionale di Fisica Nucleare, Via S. Sofia 44, 95123 Catania (Italy)
e-mail: ascordin@dmfci.unict.it

M. Tedesco and A. Triglia
Dipartimento di Metodologie Fisiche e Chimiche per l'Ingegneria, Catania University, viale A. Doria 6, I-95125 Catania, Italy

L. Brizhik
Bogolyubov Institute for Theoretical Physics, 14-b Metrologichna Str, 03680 Kyiv, Ukraine

*The experimental and theoretical data reported in this paper constitute part of the PhD thesis of Luca Lanzanò at the Catania University.

N. Russo et al. (eds.), *Self-Organization of Molecular Systems: From Molecules and Clusters to Nanotubes and Proteins*, NATO Science for Peace and Security Series A: Chemistry and Biology, © Springer Science+Business Media B.V. 2009

1 Introduction

Most of the biological and other materials emit, on being illuminated, optical photons characterized by an ultraweak intensity and relatively long duration. This low-level post-illumination emission is usually termed *Delayed Luminescence* (DL) and is something like 10^3–10^5 times weaker than fluorescence. Its first observations from biological systems date back to the 1950s, when Arnold and Strehler unexpectedly noticed a light emission from green plants lasting even for seconds after illumination [1]. Later on, the work of several researchers has proved that such post-illumination emission of light is typical not only of green plants, indeed it has been measured also in seeds, yeasts and mammalian cells and tissues [2].

In general DL is less 'popular' than fluorescence [3], probably because it is extremely polyphasic (the lifetime spectrum extends from about 10^{-7} s to more than 10 s, so that the signals have to be followed over many time decades) and because it is inherently a weak signal prone to noise contamination. Nevertheless a growing interest has been manifested toward DL phenomena especially after a long experimental work that has established that DL is a sensitive indicator of the biological state of the system. In some cases it has been possible to express this connection through an analytical relationship between some biological parameters and some parameters connected to the DL of the system under study. This particular aspect is of obvious interest for the applications.

DL emitted from biological samples is dependent on the kind of sample, on its state and on the characteristics of the illumination, still, some general characteristics have been observed which appear common to several systems [4]. The experimental dependence of DL intensity as a function of the time elapsed from the end of excitation is usually fitted by a hyperbolic decay or Bequerel empirical law $I = I_0/(1 + t/t_0)^m$. In Ref. [4] this is shown for DL from very different systems (unicellular algae, yeast cells, pepper seeds and tomato fruits) for values of the elapsed times ranging from 0.1 to 100 s. This property holds only approximately when the temporal range is increased, say, starting from about 10 μs. In this case, the trend can be fitted over the whole experimental range only taking into account the sum of more hyperbolas [5], their product or other empirical functions. We can therefore state that the decay of DL from biological system can be fitted by a hyperbolic trend in limited intervals of time, only.

It's worth to remember that the hyperbolic trend has been introduced to explain the non-exponential decay of luminescence in crystallophosphors, i.e. a kind of luminescence arising from an ordered spatial structure rather than from isolated molecule levels. The possibility to apply such a model, based on the long-range order present in solid state systems, also to biological systems, is generally undervalued due to the lack in the latter of such time-independent order [6]. On the other hand, it should be considered that biological systems contain ordered metastable structures with dimensions of the order of tens of nanometers and this could in principle influence the behavior of the electronic orbitals relative to the single molecules. In Ref. [6] this solid state approach has been validated by comparing some features of the DL from a simple unicellular organism with that from some solid state systems.

In order to better understand the properties of DL in the biological systems it is preferable to deal with relatively simple systems or pure substances. In this regard collagen is a protein which presents a well defined elementary structure but nonetheless it can be found hierarchically organized until a macroscopic level, as for instance in tendons, in order to fulfill its biological functions.

Moreover collagen is frequently selected as the model protein due to its special molecular structure [7] in order to study the role of water in living systems: protein hydration and protein induced water structuring determine chemical behavior. Cell biologists have proposed phenomenological models which suggest that water inside cells has properties differing from water in the bulk state [8].

Collagen in tendon is primarily a one-dimensional system, with long range order and capability of self-organization, so it has offered a great opportunity for fundamental biophysical research [9].

In order to explain the main properties of collagen Delayed Luminescence and its drastic changes when structural water is removed, we considered collagen as a solid-state system possessing long range order essentially in one direction. These macromolecular collagenous structures support existence of the collective electron states, in particular, solitons and small polarons, that can be excited in the photo-illumination process. We show that the kinetics and quantum yield of the DL depend significantly on the content of water in the sample and that these characteristics can be correlated with the properties of the ground electron states in macromolecular structure. Upon drying a sample, the phase transition takes place corresponding to different parts of the parametric diagram of the ground electron states, namely, to polaron and soliton parts. The properties of these two types of charge carriers differ significantly and this could take into account for the difference observed in the parameters of the DL collagen from native state to dry one.

2   Materials and Methods

2.1   Collagen from Tendon: Sample Preparation

The sample has been prepared according to the following procedure [10]. The tendons were provided by the slaughterhouse, immediately after the animals were butchered. Tendons were stripped of the external sheath and washed in bidistilled water. After washing, the tendons were cut into pieces, which were rinsed in bidistilled water and immersed in a 1 M solution of NaBr for 4 h. Then they were immersed in ether for 2 h, washed in four changes of bidistilled water, dipped in bidistilled water and stored at low temperature (about 11°C). This procedure allows to obtain a simplified structure composed only by collagen chains and bounded water. To perform the measurement the pieces were cut into slices perpendicular to the long axis of about 1 mm thick and about 10 mm diameter.

Slices of native tendon were then dehydrated gently at room temperature, keeping them flat by gently compressing them between rigid plastic plates covered by filter paper. A suitable holder system was designed to improve the natural convection of surrounding air, in order to reduce the inhomogeneity of the material during the drying process. The relative tendon water content h has been calculated using the following equation:

$$h = \left(W - W_{dry}\right)/W_{dry} \qquad (1)$$

where W is the weight of a sample at a certain humidity and W_{dry} is its weight after maximum dehydration. The samples were weighted prior and after DL measurement.

2.2 Collagen Gel Preparation

In order to compare DL from a collagen hydrogel with that of a solution of collagen the following protocol has been observed. The acid-soluble collagen from calf-skin (Collagen Solution, 3 mg/ml Ultra Pure Bovine, SIGMA C4243) has been mixed on ice with NaOH 0.1 M, in such a way to adjust PH value at about 7.5. After vortexing the collagen concentration was 2.7 mg/ml. Immediately after preparation, a volume $V = 25\,\mu l$ from the mixture has been pipetted into the sample holder (see Section 2.3) and incubated for 1 h at 37°C. In this way gels were rapidly formed. The 'sol' sample was obtained substituting the NaOH with an equal volume of bi-distilled water. In such a way, after mixing on ice and vortexing, the collagen concentration was again 2.7 mg/ml while PH didn't change appreciably. Also in this case, immediately after preparation a sampling volume $V = 25\,\mu l$ was pipetted into the sample holder and DL measurements were performed.

2.3 Delayed Luminescence Experimental Setup

DL measurements have been performed by using an improved version of the apparatus called ARETUSA setup [11, 12], whose general scheme is shown in Fig. 1.

The source of the experimental setup is represented by a high-intensity pulsed nitrogen laser (Laserphotonic LN203C) providing pulses at $\lambda = 337\,nm$, used to directly illuminate the sample or to pump a dye-laser which can emit at several wavelengths. The pulses of the laser have duration of a few nanoseconds and the maximum energy for every pulse is about $100\,\mu J$. During the measurements the laser pulse power can be checked by a standard energy-measuring head (Power Meter Ophir PE10-V2).

The detector of the setup is a photomultiplier tube (PMT) enhanced for single photon counting (Hamamatsu R-7206–1, multialkali, spectral response 300–850 nm) which can be cooled down to −30°C using a circulating cold liquid in

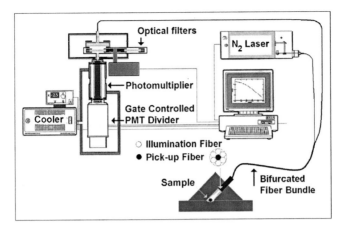

Fig. 1 Scheme of the detection setup

direct contact with its surface. In general during the laser pulse, used to excite DL in the sample, a large quantity of photons diffused by the sample would be able to reach the photocathode. This would normally damage in an irreversible way. In order to prevent the dimpling of the photomultiplier during the sample illumination, the setup is provided of an electronic system able to turn on and off the PMT in the microsecond range acting as an electronic shutter. Thus DL emission has been observed in a dynamic window ranging from about 10 μs and lasting until the signal was well above the background.

The low level of the emitted intensity in general does not allow to detect signals with a high spectral resolution. In our case, the spectral analysis is performed by a set of broad band (80 nm FWHM) Thermo-Oriel interference filters, put in a suitable wheel, between the sample and the photomultiplier.

The whole data acquisition process is performed and controlled by a personal computer through a multi-channel scaler (Ortec MCS PCI) plug-in card. A smoothing procedure is used in order to reduce random noise: experimental points are sampled in such a way that it results $\Delta t_i / t_i = $ constant, so that data will result equally spaced over a logarithmic time axis. The sensitivity factor of the apparatus, that is the product between the optical transmittance of the filtering system and the quantum efficiency of the photomultiplier, has been evaluated and used to correct the emission spectra.

To enhance the collection efficiency the excitation light is sent to the specimen through one branch of a double-branched fiber bundle whereas the photo-induced emission, coming out from the same irradiated specimen area, is collected by the inner part of the same bundle and carried up through a second branch to the detector.

In order to perform measurements on liquid samples, a novel kind of sample holder have been realized (Fig. 2). In this novel configuration the sample consists of a small liquid drop (volume $= 25\,\mu l$) with liquid sample suspended in. The drop is sustained only by contact with the border of a circular hole, performed on top of a hollow cylinder (Fig. 2). The fiber probe is held at a fixed distance slightly above

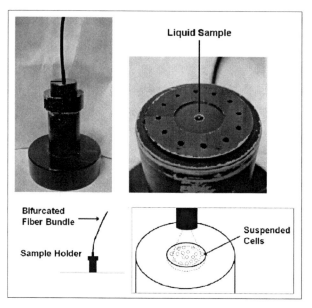

Fig. 2 Photographs (*up*) and schematic drawings (*down*) of the novel sample holder which has been used for measurements on liquid samples

the sample. Diameter of the hole is about 3.5 mm, larger than the spot generated by excitation light coming out from the fiber probe. This configuration allows to use small sample quantities and avoid unwanted background signals coming out from any solid material underlying liquid sample. In this case to evaluate the background of the setup, the signal obtained from a sample made of only bi-distilled water has been registered too.

3 Delayed Luminescence from Type I Collagen

Collagen is the major structural component of connective tissues. It is the most abundant protein in animals, comprising approximately one-third of the total protein by weight. At least 19 fibrillar and non-fibrillar, genetically different, types of collagen have been distinguished [13].

The monomeric building unit for the collagen fiber, referred to as a collagen molecule, is composed of three polypeptide chains, two identical α1 chains and one distinct α2 chain, assembled in a triple helix with a coiled coil conformation. The primary structure of this unit protein is made mostly of repeating Gly-X-Y triplets, where Gly is the aminoacid Glycine and X and Y are often the iminoacids Proline (Pro) and Hydroxiproline (Hyp). The role of the rod-like triple helix lies in its capacity to self-associate in a variety of forms as well as its ability to bind a wide range of ligands. The extensive hydrogen-bonded water network, together with the high content of sterically restricted imino acids, are the major contributors to the stabilization of triple helices, whereas electrostatic and hydrophobic interactions define intermolecular association and ligand binding. Moreover, mutations in the repeating Gly-X-Y sequences of triple helices have been shown to cause a variety of human diseases [14].

In contradistinction to other proteins, the collagen family possesses characteristic structural and chemical properties which permit its definitive identification. Collagen occurs in dense fibrous tissue of high tensile strength, as in tendons, or less tightly woven tissue, as in skin, or in more sparse distribution as in the loose connective tissue. The fibrous protein occurs in various hierarchies of size: fibers of micrometric diameter, fibrils with width of the order of tens to hundreds of nanometers, protofibrils (which are the unit columnar arrays that associate laterally to form fibrils) and finally the collagen macromolecules. In particular tendons are composed of crystalline arrays of type I collagen molecules as indicated by a number of X-ray diffraction studies [15, 16], and tendon collagen is considered a model system for the investigation of protein–water interaction in biological systems.

3.1 Spectral and Kinetic Properties of Collagen DL

Delayed Luminescence from bovine Achilles' tendon has been shown to depend strongly on the state of this system. Figure 3 reports the emission and excitation spectra of native and dry collagen sample. The emission spectrum has been obtained under excitation at the wavelength $\lambda_{exc} = 337$ nm and considering the total number of photons emitted in the observed time window. In the same graph the spectrum of fluorescence from the same sample is reported for comparison. The fluorescence spectrum, obtained under excitation at the same wavelength $\lambda_{exc} = 337$ nm and using a standard spectrophotometer, is peaked between 400 and 450 nm according to

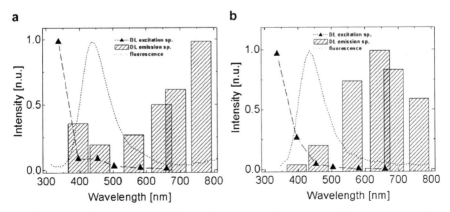

Fig. 3 DL excitation spectrum (▲), DL emission spectrum (histogram) and fluorescence emission spectrum (dotted line) for (**a**) native and (**b**) dry collagen

what is reported in Literature [17, 18]. The DL emission is clearly shifted towards longer wavelengths with respect to the fluorescence emission. The excitation spectrum for DL has been obtained fixing the emission wavelength at $\lambda_{em} = 763$ nm and is reported in Fig. 3a along with the other spectra. This spectrum decreases very rapidly towards 400 nm in a fashion similar to the excitation spectrum of collagen fluorescence [17].

The same kind of spectra have been acquired for samples of dry bovine Achilles' tendon collagen (FLUKA-27662, Collagen from bovine Achilles' tendon, lyophilized) and reported in Fig. 3b. It can be seen that the fluorescence emission spectrum and the DL excitation spectrum do not differ substantially from those shown in Fig. 3a. Instead the emission spectrum of DL from dry collagen appears to be shifted towards shorter wavelengths with respect to that of the native sample. The spectral analysis performed in native and dry collagen shows that there is a remarkable difference on the shape of the emission spectrum of DL whereas the fluorescence spectrum is unaltered in the two samples. Indeed, in native collagen, DL intensity increases almost monotonically towards near-infrared wavelengths while, in the dry sample, it has a peak between 600 and 700 nm. The different behaviour of fluorescence and DL has to be probably ascribed to the different origin of the emission, the first one being due to single fluorophores and thus being relatively insensitive to the collective structure of the system. On the other hand the emission spectra indicate that DL properties can be extremely sensitive to changes in the structure of the system, and this is even more evident by taking also into account the time dependence of the DL signal (see below).

The decay trends of five different DL spectral components have been reported in Fig. 4 for the native and dry samples respectively. The trends of the single components are not easily described by simple hyperbolic-like curves.

Significant differences can be observed in the relaxation kinetics of DL from native and dry collagen. Even though the time dependence is not simple to be analyzed, it is possible to note that the corresponding spectral components reported in Fig. 4 evolve versus time in a remarkably different manner. This is an indication that, in

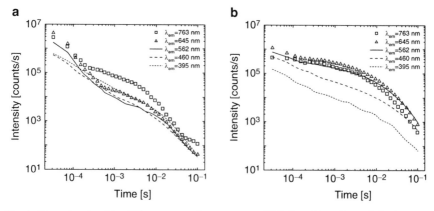

Fig. 4 Decay trends of the different spectral components of DL from (**a**) native and (**b**) dry collagen samples

the native and in the dry state, there is also a significant variation of the probability of decay of the excited states responsible for the DL emission, apart from variations in the spectra observed yet. From Fig. 4 it is also evident that different spectral components of DL, in both native and dry samples, do not follow the same relaxation kinetics. For instance, in the case of dry collagen, shorter wavelength components appear to follow a trend which is more similar to a power law (linear in a log-log scale) with respect to the one followed by the longer wavelength components.

3.2 Dependence on the Structure: Sol-gel Transitions

Collagen in tendon is hierarchically organized from molecules to fibers and water is involved in the interaction between collagen macromolecules so that the removal of water from tendon may alter noticeably its physical properties.

Measurements of DL on tendon collagen suggest that the structure of the system, strongly influenced by protein–water interaction, determines the spectral and kinetics properties of DL. This evidence agrees with the idea that DL originates from the excitation and subsequent decay of collective electronic states in the biological macromolecules (see Section 4.2) rather than from specific fluorophores.

According to this point of view, DL from collagen should be sensitive to experiments involving directly the state of aggregation of the collagen molecules. So DL has been investigated in collagen solutions and gels, which probably represent, also from the chemical point of view, the most simple and controllable example of collagenous systems. In particular the sol-gel transition has been considered.

In order to monitor the DL signal during the sol-gel transition, the sample of PH-adjusted collagen solution (see Section 2.2) has been left to polymerize at constant temperature in the measuring chamber. DL was measured at intervals throughout the gel formation.

The kinetics of collagen precipitation is generally described by temperature- or other parameter-dependent sigmoid-like curves, so two different incubation temperatures, T = 23°C and 30°C respectively, have been considered and in Fig. 5 the spectrally unresolved intensity of the DL (i.e. the signal detectable in the spectral sensitivity range of the photomultiplier, without any band-pass filter), integrated in the time interval $10 \, \mu s$–$0.1 \, s$, is reported as a function of the time elapsed from the sample preparation.

In both cases there is a lag phase, i.e. a period during which the relative increase of the signal per unit time is very low, followed by a sigmoid portion or growth period, in agreement with the general behavior reported in literature [19]. The two-step nature of collagen precipitation in solution has been discussed [19] in terms of a nucleation model. The first step, occurring in the lag period, consists mainly of the formation of nuclei whereas the second consists of the growth of these nuclei into fibrils. The time of half-growth $t_{0.5}$ of the sigmoid portion takes approximately the value $t_{0.5} = 265$ and $240 \, min$, for T = 23°C and 30°C respectively. These values appear quite long if compared with those reported in Ref. [19], however, as it has

Fig. 5 Total number of DL counts as a function of the polymerization time for samples kept at T = 23°C (dots) and T = 30°C (triangles)

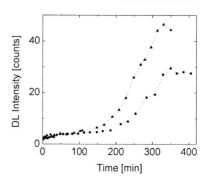

Fig. 6 Typical microscope image of a sample of collagen gel after drying. Bar length is 50 μm

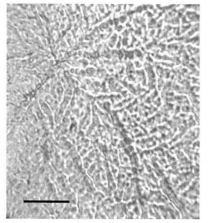

been pointed out yet, collagen polymerization kinetics is very sensitive to changes in various physical–chemical parameters, so that differences in the used protocol may justify the observed discrepancies.

Morphological characterization of the three-dimensional structure of the collagen network was out of the scope of the present work. However, as an additional control, at the end of the measurement the gel samples have been completely dried under a laminar flow hood in such a way that only a very thin film remained suspended in the sample holder. Such film could be easily observed with a standard optical transmission microscope, and ordered structures like those reported in Fig. 6 could be detected. The DL time decay from this dry sample has been measured as well, and reported in Fig. 7.

As shown in Fig. 5 when the system switches from the sol to the gel state, there is a dramatic increase of the DL signal. Such an increase is better expressed in Fig. 7 where the raw spectrally unresolved DL signals obtained for the 'sol' and 'gel' samples are reported. The data represent the average of measurements performed on four independent samples, after subtraction of the background signal. It appears that the signal from the collagen solution is very weak, slightly above the experimental background of the setup. On the other hand, it is clear that a remarkable

Fig. 7 Time dependence
of the spectrally unresolved
signal of DL coming from
a collagen solution (open
squares), a collagen-based
hydrogel (white dots) and the
same gel after drying (white
triangles). The DL decays
relative to samples of native
(black dots) and dry (black
triangles) tendon are reported
for comparison

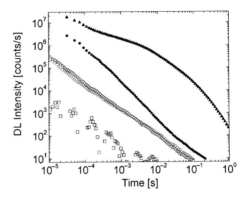

increase of the DL intensity accompanies the gel formation, i.e. the spontaneous creation of the more complex structure made up of molecules associated in the form of interconnected fibrils and fibers which characterizes the collagen network.

Due to the weakness of the signals, only the spectrally unresolved signal has been considered. The trend can be described in the first approximation as a power law decay of the form $I = A/t^m$ (corresponding to the tail of a hyperbolic decay $I = I_0/(1 + t/t_0)^m$ for $t \gg t_0$), with $m \approx 1.10$. The DL trend of the same sample after drying, also reported in Fig. 7, is not remarkably different, and fitting with the same function yields the value $m \approx 1.15$. Significant differences have not been detected in the decay curves as a function of the incubation time, neither as a function of different temperatures (fitting parameter m takes the value $m \approx 1.1$ for both gels formed at $T = 23°C$ and $30°C$).

These results show that even if the single protein monomers do not exhibit Delayed Luminescence, the self-assembled structure can instead give rise to a non-negligible signal. Indeed when monomers are brought together in a more complex structure a great number of covalent or non-covalent bonds come into play, giving rise to new kinds of electronic states which are suitable to describe the ground state of the system. Worth to note that while DL doesn't seem to change considerably after gel drying, a strong difference has been observed between DL from native and dry tendon (see Fig. 7). In the case of tendon, the structure is more complicated and its dehydration probably could not be assimilated to the gel drying. It's worth to note that, in the case of the gel, we are dealing with a system with a very high starting water content ($\sim$99%) but a final residual quantity which remains largely undetermined. Thus it's not possible to make a real comparison between dry gel and dry tendon, since that in only in the latter case the quantity of water present in the system can be monitored with enough precision, as we shall see in the following.

3.3   The Role of Water Content in the Complex Structure of Tendon

We pass now to examine the same collagen protein in the more complex, hierarchical structure of tendon, investigating in more detail the DL emitted by this system.

Several high resolution X-ray diffraction studies on native or molecular analogues of collagen have been performed [20–26] and it has been proposed that a water bridge network surrounds the collagen molecule. A molecular model consistent with these experimental evidences supports the existence of three categories of water bridges, as confirmed also by NMR experiments [27].

Accordingly to literature the most tightly bound water fraction consists of one highly immobilized water bridge per every three protein residues. Since the average mass per aminoacid in collagen is 91.2 Da, the water content corresponding to this tightly "bound water" fraction can be estimated as $h_b = 18\,\mathrm{Da}/(3 \times 91.2\,\mathrm{Da}) = 0.0658\,\mathrm{g_{water}/g_{protein}}$ [21, 28]. A second, less immobilized, fraction is represented by three additional "cleft water" molecules per tripeptide which reside in the groove-like depressions between the chains of the triple helix. In this case $h_c = 3 \times h_b = 0.197\,\mathrm{g_{water}/g_{protein}}$. Up to now we have considered four water molecules per tripeptide or a partial water content equal to $h_b + h_c = 0.263\,\mathrm{g_{water}/g_{protein}}$. These water molecules constitute linear chains which twist around forming a triple helix of water in the clefts of the collagen triple helix, according to the original hypothesis formulated by Berendsen [29].

The total water content of fully hydrated tendon in its native state has been measured [30] and is about $h_{tot} = 1.62\,\mathrm{g_{water}/g_{protein}} \cong 6 \times (h_b + h_c)$. So the remaining water constitutes a third fraction which has been shown to reside in the first "interfacial monolayer", corresponding to a water content $h_{im} = 5 \times 0.263 = 1.315\,\mathrm{g_{water}/g_{protein}}$. This layer corresponds to five additional chains of water per groove, so the total amount of water can be accounted for by considering about six chains of water per groove [27].

Indeed, removal of vicinal water, i.e. water that interacts with the protein, is expected to induce alterations in the structural properties of the protein and therefore in the spectral and/or kinetic properties of Delayed Luminescence (DL). In order to save time, only two spectral components ($\lambda_{em} = 460\,\mathrm{nm}$ and $\lambda_{em} = 645\,\mathrm{nm}$) of DL have been registered for every humidity value.

The decay trends of the DL intensity obtained for the two spectral components are reported in Figs. 8 and 9 respectively. Luminescence kinetics appears to be also sensitive to variations in the relative humidity of the tendon.

Only six curves, corresponding to six hydration level from native to dry state, have been reported for the sake of clarity. It appears that the drying process is accompanied by the change in the decays slope.

Moreover, the total number of DL counts increases drastically at a certain value of water content, and correspondingly a blue-shift is observed in the emission

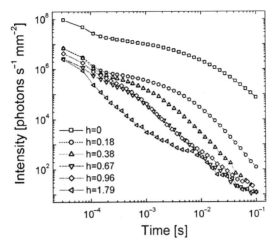

Fig. 8 Experimental time dependence of DL from bovine Achilles' tendon at different water content for the spectral component at $\lambda_{em} = 645\,nm$

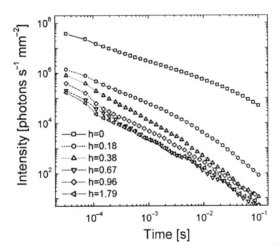

Fig. 9 Experimental time dependence of DL from bovine Achilles' tendon at different water content for the spectral component at $\lambda_{em} = 460\,nm$

spectra of dried samples with respect the native ones. More precisely (data not shown) it happens when the water content is close to the value corresponding to the tightly bound and cleft water forming the Berendsen linear chain [29].

Removal of this vicinal layer appears to affect the spectral properties of DL, and it is expected to affect strongly the dielectric properties of the collagen macro-molecules, too.

3.4 Delayed Luminescence and Dielectric Properties of Collagen

Several researchers have investigated the dielectric properties of biological systems over the years [31]. Biological materials are characterized by high dielectric permittivity at low frequency which falls off in distinct steps with increasing frequency. Three main dispersion regions are individuated: extra low frequencies (α-dispersion), radiofrequency (β) and microwave frequencies (γ), and in each region a different mechanism is responsible for the characteristic dispersion law. In order to understand the role of water network in collagen structure, measurements of Dielectric Spectroscopy (DS) have been performed on bovine Achilles' tendon as a function of the water content and variations in the dielectric permittivity can be compared with variations in the DL [10, 32].

The reported results can be summarized as follows. The dielectric permittivity of native tendon has a value of the order of 10^5 at 1 kHz and then decrease on increasing the frequency reaching a value of the order of 10^2 at 1 MHz, in agreement with previously reported results [33].

Dielectric permittivity at different frequencies depends on the relative water content h as shown in Fig. 10. For $h > 0.5$ dielectric permittivity is strongly dependent on the frequency but not on the water content. At higher frequencies the value of dielectric permittivity is comparable to that of water while for lower frequencies it becomes very large. For values of h between 0.1 and 0.5 there is a transition, more evident at low frequency, towards lower values of the dielectric permittivity.

This dependence of the dielectric permittivity on the water content can be compared with the analogous trend obtained from measurements of spectrally unresolved DL from bovine Achilles' tendon. Also in this case, in fact, a transition can be observed in the properties of DL moving from the native toward the fully dried state (see for instance Figs. 8 and 9), taking into account the strong variations in the total number of emitted photons and in the decay kinetics of the luminescence signal. In Fig. 11 the total number of emitted photons and the dielectric permittivity measured at 5 kHz are compared: it appears that on drying both quantities show an abrupt change when the more tightly bound water, corresponding to the Berendsen

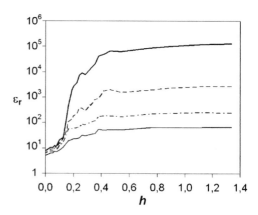

Fig. 10 Dielectric permittivity as a function of the collagen water content h for different values of the frequency: 500 Hz (thick solid line), 5 kHz (dashed line), 50 kHz (dot-dashed line) and 10 MHz (light solid line)

Fig. 11 Comparison of DL
and DS measurements as
a function of the tendon
water content: total number
of DL photons emitted in
arbitrary units (○) and inverse
of dielectric permittivity at
5 kHz (▲)

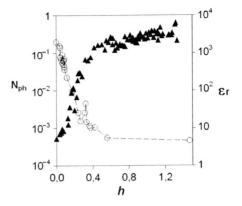

linear chain of water (see Section 3.3), is removed. This correspondence stresses once again the central role of hydration of the triple helix structure in determining the observable properties of collagen.

4 Modeling Delayed Luminescence from Collagen

The intriguing structures of biological systems have always attracted the interest of physicists, and since earlier times physical methods have been applied to investigate the nature of biological components [34, 35]. A living organism is made up of a large number of interwoven molecular networks, which primarily involve proteins, the macromolecules that enable and control virtually every chemical process that take place in the cell. Understanding how intra-molecular interactions determine the collective behavior of the protein, i.e. its native physical conformation and its correct functioning, is only one step in the more general problem of understanding how a living system, hierarchically organized from molecules to cells and tissues, actually works.

Several researchers have attempted to apply solid state theory to the problem of describing biological processes at the molecular level. The approach is partly inspired by the ordered structures observed in the biological world, as for instance the periodic structure of protein and DNA molecules, the translational symmetry observed in the chloroplast matrix, the quasi-crystalline arrangement of proteins in certain membranes, and so on.

Methods of solid state physics have been applied in particular in the investigation of the central issue in bioenergetics, that is the origin of the high efficiency of the storage and transfer of energy, electrons and protons within and between molecules.

The transfer of energy in protein structures has been discussed in the framework of the electronic states which describe the possible excited states of a molecular crystal [36, 37]. On the other hand, when dealing with a crystal the possibility of luminescence from collective electronic states should be also taken into account.

Since several experimental evidences indicate that Delayed Luminescence (DL) is dependent on the state of the biological system under consideration, it appears reasonable trying to connect this phenomenon with the participation of collective states of these structures.

4.1 Collective Excitations in Low-Dimensional Biological Systems

The original idea that a novel mechanism could be involved in the localization and transport of vibrational energy in proteins was put forward in 1973 by A. S. Davydov [38, 39].

According to his original idea, the vibrational energy of the $C == O$ stretching (also called Amide-I) oscillators that is localized on the alpha-helix structure acts via a phonon coupling effect to distort the helix structure. The helical distortion reacts to trap the Amide-I oscillation energy preventing its dispersion. This effect is called self-trapping of the excitation.

This idea was of great importance for biochemistry and later on a huge amount of theoretical and experimental work has developed around what has been called the Davydov's soliton [40–45].

The mathematical procedures involved in the analysis of Davydov's soliton are similar to those developed in the polaron theory.

The Davydov's soliton is propagated in the α-helix because of its particular structure, displaying the three chains of H-bonds connecting the peptide groups along the helix axis. This structure is a special case of quasi one-dimensional molecular crystal, as many other examples could be found among the biological macromolecules.

In order to recall briefly the main points of the theory, let assume that a molecular chain consists of N molecules, such as peptide groups in the polypeptide chain (unit cells), aligned along the z-axis at equilibrium distance R from one another and that internal (electronic or vibrational) excitations of isolated molecules are characterized by energy ε and electric dipole moment $\vec{d}$ at an angle θ to the z-axis.

The Hamiltonian for the collective excitations is thus written as the sum of three terms

$$H_{sol} = H_{ex} + H_{ph} + H_{int} \qquad (2)$$

containing the operator for internal excitation H_{ex}, the operator H_{ph} for the displacements of the molecules and the interaction operator H_{int}, expressed respectively as:

$$H_{ex} = \sum_n \left[(\varepsilon - D) B_n^+ B_n - J \left(B_{n+1}^+ B_n + B_{n-1}^+ B_n \right) \right] \qquad (3)$$

$$H_{ph} = \frac{1}{2} \sum_n \left[\frac{1}{M} p_n^2 + \kappa (u_n - u_{n-1})^2 \right] \qquad (4)$$

$$H_{int} = \chi \sum_n (u_{n+1} - u_{n-1}) B_n^+ B_n \qquad (5)$$

where the summation is carried out over the N molecules of the chain, B_n^+ and B_n denote respectively the creation and annihilation operators for the excitation on the nth site (the Amide-I vibration in the case of Davydov's soliton in α-helix), J is the resonance interaction energy, $-D$ is the energy of the chain deformation interaction, M is the mass of the molecule, κ is the elasticity coefficient for the chain, p_n is the momentum operator and χ is the exciton–phonon coupling parameter.

After some calculations it can be shown that the Hamiltonian Eq. (2) leads to the system of equations which can be reduced to a form of the non-linear Schrödinger equation, which is known to admit solutions in the form of solitary waves [37].

The main properties of DL in biological systems indicate that this phenomenon can be connected with coherent electron states in macromolecules. Biological cells contain several kinds of low-dimensional macromolecules as, for instance, alpha-helical proteins, actin and other cytoskeleton filaments, etc., whose structure is represented by arrays of parallel quasi-1D polypeptide chains formed by repeating peptide units. In general these chains are characterized by strong exchange and/or resonance interaction between neighboring molecules of the same chain and relatively weaker inter-chain interaction. Regarding the electronic structure, these macromolecules behave as semiconductor-like quasi-1D systems so that it appears reasonable to admit the existence of coherent collective (exciton or soliton) states. A model which connects DL to the formation of such states has been formulated and tested for simple biological systems able to perform photosynthesis [46–49].

4.2 Delayed Luminescence Arising from Collective States in the Collagen Macromolecules

The experimental evidences on Delayed Luminescence from biological systems are not confined to plants or organisms capable of photosynthesis. In view of these considerations, it appears interesting to test the validity of a general model, based on the formation of collective excited states in the biological macromolecules, able to describe at least some of the properties of DL on the basis of physical principles.

As it has been mentioned above, the primary structure of collagen is formed by the sequence $(Gly–X–Y)_n$, where the first component is the amino acid glycine (Gly) and X and Y are in most cases the iminoacids proline (Pro) and hydroxyproline (Hyp) respectively. The triple helix model for the secondary structure, containing three polypeptide chains, was proposed in order to accommodate these amino acid features and fit the fiber diffraction pattern. In this model, each polypeptide chain adopts a left-handed helix stabilized by a high iminoacid content. A set of three helical chains are staggered by one residue with respect to each other and are supercoiled about each other in a right-handed manner such that adjacent chains are linked by H-bonds between the groups $C == O$ and H–N of X and Gly residues respectively. The Gly–X–Y sequence requirement is generated by the close packing of the three chains with interchain hydrogen bonds, which can occur without distortion only if Gly is the every third residue [50]. Unlike α-helical and β-sheet protein structures, which are stabilized by the participation of every backbone carbonyl and amide group in N–H $\cdots$ C $=$ O hydrogen bonds, the triple-helix shows a serious deficiency in this regard. The only direct H-bonds of the type $\{(Gly)N–H \cdots C = O\{X\}$ leaves the carbonyl group of the Gly residues and the carbonyl of the residue in the Y position with no amide hydrogen bonding partners. In addition, the hydroxyl group of Hyp points out from the triple helix and cannot directly bind via hydrogen bond to any other group within the molecule. As a result, an extensive and ordered water network [51]) forms intra- and inter-chain hydrogen bonds with the available carbonyl and Hyp hydroxyl groups. Variations in the number of water molecules involved in water-mediated H-bonding are seen in the crystal structure, and the water-mediated H-bonds are not fully occupied. However, on the average, collagen binds in hydrophilic position one or more molecules of water [51].

In order to investigate the formation of collective excitations in the collagen macromolecule, one can model it as a simple quasi-1D molecular periodic system, taking into account the role of hydration. In the absence of water, the three peptides, Gly, X and Y, form the repeating unit of a strand in terms of the crystal structure. Every Gly forms a soft H-bond with the X of another chain. We can therefore model, in the first approximation, dry collagen as a single chain with a three-peptide unit cell with one optical mode of vibrations, which accounts for the direct hydrogen bond.

In the presence of water, a complex hydration network surrounds the collagen molecule. As a result, the unit cell of a hydrated collagen strand has more than one hydrogen bond, and, therefore, the phonon spectrum displays more than one optical branch (one intrinsic mode and others due to hydrogen bonding to water molecules). Moreover, the effect of biopolymer–water interaction on the collective modes of vibration of some proteins has been investigated [52–54]. Low-frequency collective acoustic modes have been determined experimentally and in some cases assigned to the collagen hydration monolayer [54]. Therefore, we model hydrated collagen as a simple chain with two optical modes at the low level of hydration and with optical and acoustic modes at sufficiently higher hydration level.

The variations of the phonon spectrum, induced by the different degree of hydration, are significant from the point of view of the type of the ground electron states which can be excited in such a chain, that represents our simplified collagen model. The ground state of a quasi-particle (exciton, electron or hole) interacting with optical and acoustic phonons in a one-dimensional chain, can be conveniently investigated using a variational method [55–58]. There are three types of ground states corresponding to the three principal approximations used in the Hamiltonian (Eq. (2)):

1. Weak coupling approximation, leading to an almost free quasiparticle state.
2. Strong coupling approximation, leading to the small polaron state which corresponds to the autolocalization of a quasiparticle within one lattice site and local chain deformation.
3. Large polaron approximation valid at the intermediate values of coupling constant and leading to the autolocalization of a quasiparticle within several lattice sites. Such states are described within the zero adiabatic approximation when kinetic energy of lattice vibrations is small, and correspond to the spontaneously

localized states with the broken translational symmetry. This case is especially interesting for one-dimensional systems in which the soliton-like states can exist and be responsible for charge and energy transport in the form of exceptionally stable non-spreading quasiparticles.

It is convenient to use two dimensionless parameters g and γ describing respectively the electron–phonon coupling strength and non-adiabaticity of the system. In the case of a quasi-particle interacting with optical phonon modes they are defined as:

$$g_{opt} = \frac{E_{bind}}{2J} = \frac{\chi^2_{opt}}{2J\hbar\Omega_0} \tag{6}$$

$$\gamma_{opt} = \frac{\hbar\Omega_0}{J} \tag{7}$$

where E_{bind} is the lowering of the quasi-particle energy due to electron-phonon interaction, χ_{opt} is the coupling constant for the optical mode, Ω_0 is the frequency of optical phonon mode, J is the resonance interaction energy. In the case of interaction with acoustic phonons they are defined as:

$$g_{ac} = \frac{\chi^2_{ac}}{2J\kappa} \tag{8}$$

$$\gamma_{ac} = \frac{\hbar V_{ac}}{2JR} \tag{9}$$

where χ_{ac} is the coupling constant for the acoustic mode, κ is the elasticity coefficient, R is the lattice constant and $V_{ac} = R\sqrt{\kappa/M}$ is the velocity of sound in the chain, with M being mass of a unit cell of the chain (note that following standard notation, χ_{opt} and χ^2_{ac}/κ have the same energy units). Then the ground electron state diagrams are plotted depending on the values of the two dimensionless parameters g and γ which correspond to the ratio between the characteristic energies of the phonon modes and electron band.

We estimated the values of the dimensionless parameters g and γ of these modes from the experimental and/or theoretical values of the physical constants appearing in Eqs. (6–9) which are reported in literature [37, 40, 53, 54, 59–61] and which are referred to collagen or other polypeptides.

For the optical modes we have that the dimensionless parameters are of the following order of magnitude (see Appendix for parameters evaluation): $g_{opt} \sim 0.5$ and $\gamma_{opt} \sim 50$, while for the acoustic mode the parameters have approximately the value $g_{ac} \sim 0.5 - 1.5$ and $\gamma_{ac} \sim 0.5$.

Following the numerical evaluation procedure described in Refs. [55–57], for each set of the dimensionless parameters g and γ, the most probable kind of state for a quasi-particle interacting with two phonon modes in a one-dimensional chain has been determined. In the ground state diagrams the three regions can be individuated: almost free quasi-particle state (I), large polaron or soliton state (II) and, small polaron state (III).

In the case of interaction with two optical modes, which models the dry collagen, the values estimated for the non-adiabaticity parameter γ_{opt} are too high to allow the formation of the spontaneous self-localized state. In Fig. 12a there is reported the ground state diagram obtained assuming that $g_{opt,2} = g_{opt,1}$ and $\gamma_{opt,2} = 0.2\,\gamma_{opt,1}$. According to the diagram, since $\gamma_{opt,1} \gg 1$ the spontaneous self-localization of the state appears highly improbable.

In the case of the quasi-particle interacting with one optical phonon mode and one acoustical phonon mode, which, in our simplified model, should correspond to hydrated collagen, the picture is quite different. Even if the optical mode is strongly non-adiabatic, the coupling with lower energy acoustic phonon modes opens the possibility to the formation of large polaron states inside a certain range of γ_{ac} and g_{ac} values. The ground state diagram obtained fixing the dimensionless parameters of the optical mode to the values $g_{opt} = 0.5$ and $\gamma_{opt} = 50$ is reported in Fig. 12b. According to our rough estimation of g_{ac} and γ_{ac} (shaded region in Fig. 12b), the

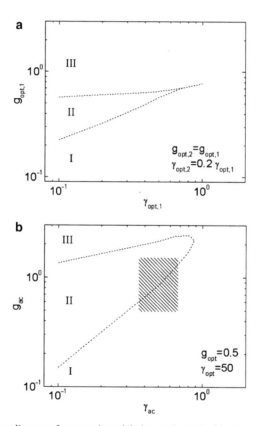

Fig. 12 Ground state diagrams for a quasi-particle interacting (**a**) with two optical phonon modes and (**b**) with one optical mode and one acoustical mode. (I) almost free quasi-particle, (II) large polaron (soliton), (III) small polaron. Fixed absolute or relative values of the dimensionless parameters g and γ are indicated in the figures following the notation used in the text

possibility of formation of a self-trapped large polaron (soliton) state in collagen can be taken into account as well as the possibility of a state transition during the de-hydration process.

5 Conclusions

The experimental studies reported in this work have shown that the response of biological systems to the stimulating light is sensibly related to the physiological state of the system and to the environment conditions and Collagen could represent a useful model system to test the hypothesis on the origin and/or the mechanisms of photoinduced DL emission. Particularly interesting has been the dependence of the DL from bovine Achilles' tendon at different level of dehydration on the hierarchical complexity of the system. In this respect there are relevant the results acquired during the polymerization process of collagen (the sol-gel transition): the spontaneous creation of the more complex structure is accompanied by increasing DL signals, so assessing the tight relationship between DL and ordered structures present at a mesoscopic level in biological sample.

This points out to the long range interactions observed in the organism, which, according to Fröhlich, lead to collective properties of the whole biological multi-component systems, thus creating the functional order. So the collectivization of electronic and/or molecular states in the macromolecular structures, in a way analogue to excitons in regular crystals, appears possible.

The observed correspondence between dielectric permittivity and DL photon emission behaviors as a function of the water content suggests that some features of collagen DL could be understood considering the possibility that collective excitations can be generated in this macromolecule. From the theoretical study reported in this paper it seems that the soliton excitation may be typical of native biological systems, and change in the type of collective excited state accompanies denaturation, as the loss of water during the drying process, which in turn affects the DL characteristics. This is of course just an intuition, to be explored and confirmed in future studies, but appears as a first possible picture. At least, the hope is that the present study could be useful as a small step in understanding the role of solitons among the energy excitations in biological systems and can contribute to assessing the possibility of using DL measurements in order to closely investigate the biophysics of a living organism as a whole complex system.

Appendix

First of all let's focus on the acoustic mode. In measuring vibrational spectra of type I collagen fibers, and the model polypeptide $(PPG)_{10}$ (prolyl-prolyl-glycine)$_{10}$, Middendorf et al. [54] observed, in the low frequency region, a maximum at $45\,cm^{-1}$.

In calculations of density of states for PGII (left-handed helix polyglycine II), the maximum due to longitudinal modes is seen at $13 \, \text{cm}^{-1}$ in the isolated molecule, but is displaced upward to $40 \, \text{cm}^{-1}$ when hydrogen-bonding to neighboring chains is taken into account. So they identified the $45 \, \text{cm}^{-1}$ peak in collagen and $(\text{PPG})_{10}$ spectra with this maximum for longitudinal acoustic modes in PGII-like chains hydrogen-bonded and supercoiled to form the collagen triple helix. On the other hand an acoustical band around $50 \, \text{cm}^{-1}$ was observed in hydrated samples, that corresponds to an acoustic band observed in pure H_2O ice. So the longitudinal acoustic phonon frequency in the polypeptide matches that of its associated interhelical ice shell, raising the possibility of coupling between solvent and polypeptide modes.

This value of the acoustic mode accords the observation of sound velocity in collagen. It was found [53] that the water of hydration "softens up" the long-wavelength excitations probed, in the sense that their velocity decreases from just under 4,000–2,500 m/s.

So considering the value of the average helical pitch of the collagen $a = 9.5 \, \text{Å}$, the characteristic frequency of phonon can be evaluated as $\nabla V_{ac}/a = 0.263 \times 10^{-21} \, \text{J} = 13.3 \, \text{cm}^{-1}$. Such frequency can be associated to a harmonic oscillator of mass of one peptide group, $M = 91 \, \text{amu}$, and elastic constant $\kappa = M V_{ac}^2/a^2 = 1.05 \, \text{N/m}$.

By assuming for the coupling constant for acoustical modes the value [37] $\chi_{ac} = 0.2 \div 0.3 \times 10^{-10} \, \text{N}$ and for the resonance interaction energy the value determined for parallel-chain pleated-sheet structure $J = 13.6 \, \text{cm}^{-1}$ [59], one gets:

$$\gamma_{ac} = \frac{1}{2J} \frac{\hbar V_{ac}}{a} = 0.5$$

and

$$g_{ac} = \frac{1}{2J} \frac{\chi_{ac}^2}{\kappa} = 0.71 \div 1.6$$

We will now consider the values for optical mode. According to Middendorf et al. [54] all collagen spectra are very complex in the high frequency region and the dry collagen spectra reveal significantly more intensity in the $600–800 \, \text{cm}^{-1}$ region. If one considers the intra-triple strand hydrogen bond, between glycine NH on one chain and proline CO on a neighboring supercoiled chain, the effect of triple-helical supercoiling is to shift the amide V mode of the Gly-Pro linkage, together with skeletal deformation and $C = O$ in-plane bending modes of the Pro-Pro linkages, downward to the region of $590 \, \text{cm}^{-1}$.

Theoretical normal mode analyses have predicted significant effects of minor angle variation and hydrogen bonding length on amide V modes. It raises the question of coupling between low frequency modes and high frequency peptide backbone modes. Low frequency dispersive modes can significantly alter local supercoiling and hydrogen bond length. Such anharmonic coupling may be an important mechanism of energy transport in biological systems [40], and amide V has been recently identified as a markedly anharmonic mode, and a candidate for such coupling, in acetanilide.

So considering such frequency for the optical phonon, $\nabla\Omega_{op} = 590\,\text{cm}^{-1}$ and the above said value for the resonance interaction energy determined for parallel-chain pleated-sheet structure [59], one gets:

$$\gamma_{op} = \frac{\hbar\Omega_{op}}{J} = 43$$

Instead, in order to evaluate the electron-phonon coupling constant, we assume the value $\chi_{op} = 6.2 \times 10^{-11}$ N for the coupling constant of the optical mode as determined for acetanilide (ACN) [60], and considering the elasticity of the hydrogen bond, $w = 13\,\text{N/m}$, one gets:

$$g_{op} = \frac{E_b}{J} = \frac{1}{J}\frac{\chi_{op}^2}{2w} = 0.55$$

for the non-adiabaticity constant.

References

1. W. A. Arnold (1986) Delayed light, glow curves and the effects of electric field. In "Light Emission by Plants and Bacteria" edited by Govindjee, J. Amesz, D. C. Fork, Academic, New York
2. A. Popp, K. H. Li, Q. Gu (editors) (1992) Recent Advances in Biophoton Research and its Applications, World Scientific, Singapore; F. Musumeci, L. S. Brizhik, M.-W. Ho (editors) (2002) Energy and Information Transfer in Biological Systems: How Physics Could Enrich Biological Understanding, World Scientific, Singapore; X. Shen and R. Van Wijk (editors) (2007), Biophotonics: Optical Science and Engineering for the 21st Century, Springer, New York (USA)
3. J. Lavorel, J. Breton, M. Lutz (1986) Methodological principles of measurement of light emitted by photosynthetic systems. In "Light Emission by Plants and Bacteria" edited by Govindjee, J. Amesz, D. C. Fork, Academic, Orlando, FL
4. A. Scordino, A. Triglia, F. Musumeci (2002) Delayed luminescence and cell's structure organization. In "Energy and Information Transfer in Biological Systems: How Physics Could Enrich Biological Understanding" edited by F. Musumeci, L. S. Brizhik, M.-W. Ho, World Scientific, Singapore
5. F. Musumeci, G. Privitera, A. Scordino, S. Tudisco, C. Lo Presti, L. A. Applegate, H. J. Niggli (2005) Discrimination between normal and cancer cells by using spectral analysis of delayed luminescence. Appl Phys Lett 86:153902–153905
6. A. Scordino, A. Triglia, F. Musumeci (2000) Analogous features of delayed luminescence from *Acetabularia acetabulum* and some solid state systems. J Photochem Photobiol B 56: 181–186
7. S. Leikin, V. A. Parsegian, W. Yang, G. E. Walrafen (1997) Raman spectral evidence for hydration forces between collagen triple helices. Proc Natl Acad Sci USA 94 (21): 11312–11317; E. Leikina, M. V. Mertts, N. Kuznetsova, S. Leikin (2002) Type I collagen is thermally unstable at body temperature. Proc Natl Acad Sci USA 99 (3): 1314–1318; P. L. Privalov (1982) Stability of proteins. Proteins which do not present a single cooperative system. Adv Protein Chem 35: 1–104
8. G. H. Pollack (2003) Cells, Gels and the Engines of Life. Ebner & Sons, Seattle
9. F. O. Schmitt (1959) Interaction properties of elongate protein macromolecules with particular reference to collagen (tropocollagen). Rev Mod Phys 31 (2): 349–358

10. M. Gulino, P. Bellia, F. Falciglia, F. Musumeci, A. Pappalardo, A. Scordino, A. Triglia (2005) Role of water content in dielectric properties and delayed luminescence of bovine Achilles' tendon, FEBS Lett 579: 6101–6104

11. Tudisco S, Musumeci F, Scordino A, Privitera G (2003) Advanced research equipment for fast ultraweak luminescence analysis, Rev Sci Instrum 74:4485–4489

12. S. Tudisco, A. Scordino, G. Privitera, I. Baran, F. Musumeci (2004) Advanced research equipment for fast ultraweak luminescence analysis: new developments, Nucl Instrum Meth A 518:463–464

13. D. J. Prockop, K. I. Kivirikko (1995) Collagens: molecular biology, diseases, and potentials for therapy. Annu Rev Biochem 64, 403–434

14. B. Brodsky, N. K. Shah (1995) The triple-helix motif in proteins. FASEB J 9:1537–1546

15. A. C. T North, P. M. Cowan and J. T. Randall (1954) Structural units in collagen fibrils. Nature 174:1142–1143

16. A. Miller and J. S. Wray (1971) Molecular packing in collagen. Nature 230: 437–439

17. I. Georgakoudi, B. C. Jacobson, M. G. Muller, E. E. Sheets, K. Badizadegan, D. L. Carr-Locke, C. P. Crum, C. W. Boone, R. R. Dasari, J. Van Dam, M. S. Feld (2002) NAD(P)H and collagen as in vivo quantitative fluorescent biomarkers of epithelial precancerous changes. Cancer Res 62: 682–687

18. T. Theodossiou, G.S. Rapti, V. Hovhannisyan, E. Georgiou, K. Politopoulos, D. Yova (2002) Thermally induced irreversible conformational changes in collagen probed by optical second harmonic generation and laser-induced fluorescence, Lasers Med Sci 17: 34–41

19. G. C. Wood, M. K. Keech (1960) The formation of fibrils from collagen solutions. 2. A mechanism of collagen-fibril formation. Biochem J 75:598–605

20. A. Rich, F. Crick. (1961) The molecular structure of collagen. J Mol Biol 3:483

21. G. N. Ramachandran (1967) In: Ramachandran GN, editor. Treatise on collagen. Academic, New York, pp. 103–183

22. J. Bella, M. Eaton, B. Brodsky, H. M. Berman (1994) Crystal and molecular structure of a collagen-like peptide at 1.9 A resolution. Science 266 (5182): 75–81

23. J. Bella, B. Brodsky, H. M. Berman (1995), Hydration structure of a collagen peptide. Structure 3: 893–906

24. R. Z. Kramer, L. Vitagliano, J. Bella, R. Berisio, L. Mazzarella, B. Brodsky, et al. (1998) X-ray crystallographic determination of a collagen-like peptide with the repeating sequence (Pro-Pro-Gly). J Mol Biol 280(4):623–638

25. R. Z. Kramer, J. Bella, P. Mayville, B. Brodsky, H. M. Berman (1999) Sequence dependent conformational variations of collagen triple-helical structure. Nat Struct Biol 6(5): 454–457

26. R. Z. Kramer, J. Bella, B. Brodsky, H. M. Berman (2001) The crystal and molecular structure of a collagen-like peptide with a biologically relevant sequence. J Mol Biol 311(eq1): 131–47

27. G. D. Fullerton, E. Nes, M. Amurao, A. Rahal, L. Krasnosselskaia, I. Cameron (2006) An NMR method to characterize multiple water compartments on mammalian collagen. Cell Biol Int 30: 66–73

28. G. D. Fullerton, M. R. Amurao (2006) Evidence that collagen and tendon have monolayer water coverage in the native state. Cell Biol Int 30: 56–65

29. H. J. C. Berendsen (1962) Nuclear magnetic resonance study of collagen hydration. J Chem Phys 36(12): 3297–3305

30. G. D. Fullerton, I. L. Cameron, V. A. Ord (1985) Orientation of tendons in the magnetic field and its effect on T2 relaxation times. Radiology 155(eq2): 433–435

31. H. P. Schwan (1957) Electrical properties of tissue and cell suspensions. Adv Biol Med Phys 5: 148–209; R. Pethig (1979), Dielectric and Electronic Properties of Biological Materials, Wiley, New York

32. M. W. Ho, F. Musumeci, A. Scordino, A. Triglia, G. Privitera (2002) Delayed luminescence from bovine Achilles' tendon and its dependence on collagen structure. J Photochem Photobiol B 66: 165–170

33. S. Gabriel, R. W. Lau, C. Gabriel (1996) The dielectric properties of biological tissues: II. Measurements in the frequency range 10 Hz to 20 GHz. Phys Med Biol 41: 2251–2269

34. J. R. Loofbourow (1940) Borderland problems in biology and physics. Rev Mod Phys 12: 267–358
35. H. Frauenfelder, P. G. Wolynes, R. H. Austin (1999) Biological physics. Rev Mod Phys 71: S419–S430
36. A. S. Davydov (1982), Biology and Quantum Mechanics, Pergamon, New York
37. S. Davydov (1985), Solitons in Molecular Systems, Reidel, Dordrecht
38. S. Davydov (1973) The theory of contraction of proteins under their excitation, J. Theor. Biol. 38:559–569
39. S. Davydov, N.I. Kislukha (1973) Solitary excitons in one-dimensional molecular chains. Phys. Stat. Sol. B 59: 465
40. A. Scott (1992) Davydov's soliton. Phys. Rep. 217: 1–67
41. L. Cruzeiro-Hansson (1994) Two reasons why the Davydov soliton may be thermally stable after all. Phys Rev Lett 73: 2927–2930
42. P. Ciblis, I. Cosic (1997) The possibility of soliton-exciton transfer in proteins. J Theor Biol 184: 331–338
43. Z. Ivic (1998) The role of solitons in charge and energy transfer in 1D molecular chains. Physica D 113: 218–227
44. L. S. Brizhik, L. Cruzeiro-Hansson, A. A. Eremko (1999) Electromagnetic radiation influence on nonlinear charge and energy transport in biosystems. J Biol Phys 24: 223–232
45. S. Caspi, E. Ben-Jacob (2000) Conformation changes and folding of proteins mediated by Davydov's soliton. Phys Lett A 272:124–129
46. L. Brizhik, F. Musumeci, A. Scordino, A. Triglia (2000) The soliton mechanism of the delayed luminescence of biological systems. Europhys Lett 52 (2): 238–244
47. L. Brizhik, A. Scordino, A. Triglia, F. Musumeci (2001) Delayed luminescence of biological systems arising from correlated many-soliton states. Phys Rev E 64: 031902
48. L. Brizhik, F. Musumeci, A. Scordino, M. Tedesco, A. Triglia (2003) Nonlinear dependence of the delayed luminescence yield on the intensity of irradiation in the framework of a correlated soliton model. Phys Rev E 67: 021902
49. L. Brizhik, F. Musumeci, A. Scordino, M. Tedesco, A. Triglia (2003) Delayed luminescence from biological systems within the Davydov soliton model. Ukr J Phys 48 (7): 699–703
50. B. Brodsky, N. K. Shah (1995) The triple-helix motif in proteins. FASEB J 9:1537–1546
51. B. Brodsky, J. A. M. Ramshaw (1997) The collagen triple-helix structure. Matrix Biol 15: 545–554
52. R. Harley, D. James, A. Miller, J. H. White (1977) Phonons and the elastic moduli of collagen and muscle. Nature 267: 285–287
53. M.-C. Bellissent-Funel, J. Teixeira, S. H. Chen, B. Dorner, H. D. Middendorf, H. L. Crespi (1989) Low-frequency collective modes in dry and hydrated proteins. Biophys J 56: 713–716
54. H. D. Middendorf, R. L. Hayward, S. F. Parker, J. Bradshaw, A. Miller (1995) Vibrational neutron spectroscopy of collagen and model peptides. Biophys J 69: 660–673
55. L. S. Brizhik, A. A. Eremko, A. La Magna (1995) The ground state of an electron or exciton in the Holstein model. Phys Lett A 200, 213–218
56. L. S. Brizhik, A. A. Eremko, A. La Magna, R. Pucci (1995) The ground state of an extra electron interacting with acoustic phonons in a molecular chain. Phys Lett A 205: 90–96
57. L. S. Brizhik, A. A. Eremko (1997) One-dimensional electron–phonon system at arbitrary coupling constant. Z Phys B 104: 771–775
58. L. S. Brizhik, K. Dichtel, A. A. Eremko (1999) Spontaneously localized electron states in a chain with electron–phonon coupling: Variational approach. J Supercond, 12: 239–241
59. N. Chirgadze Yu, N. A. Nevskaya (1976) Infrared spectra and resonance interaction of amide-I vibration of the parallel-chain pleated sheet. Biopolymers 15: 627–636
60. Careri G., Buontempo U., Galuzzi F., Scott A.C., Gratton E. and Shyamsunder E. (1984) Spectroscopic evidence for Davydov-like solitons in acetanilide. Phys Rev B 30: 4689–4702
61. C. Eilbeck, P. S. Lomdahl, A. C. Scott (1984) Soliton structure in crystalline acetanilide. Phys Rev B 30:4703–4712

Cluster Quantum Chemical Study of the Grignard Reagent Formation

A.V. Tulub and V.V. Porsev

Abstract The main stages of the Grignard reagent formation are described in a framework of quantum chemical cluster model. We have established two kinds of the adsorption of CH_3Hal on Mg_n clusters, one of which leads to radical formation and the second is responsible for radical free dissociate adsorption. The charge redistribution in cluster CH_3Mg_nHal result to the strong electrostatic interaction with ether and Grignard reagent formation without any activation barrier.

Keywords Grignard reagent formation · Magnesium clusters · Quantum chemical study

1 Introduction

The Grignard reaction has been known for more than a century and currently remains one of the most widely used reactions in synthetic organic chemistry. The structure of the Grignard reagent (GR) in different solutions is mostly known, some gaps, nevertheless, remain in the understanding, at the molecular level, of sequence of reactions, which final stage is the formation of GR. The activation of magnesium surface, which is normally covered by MgO, occurs through chemical or mechanical treatment followed by appearing the strongly disordered surface. In this connection the cluster model approach seems to be quite adequate for current numerical simulations, earlier it was applied in [1] for studying interaction between CH_3Cl and magnesium, Mg_n, clusters of different size. As a radical R, computations choose R = CH_3.

The monatomic model of Mg + CH_3Cl interaction predicts a too high activation energy barrier [2–4], with the increase in the size of Mg_n cluster for n ≈ 20 the reasonable value of the activation energy of about 10 kcal/mol [1] is close to that observed in the experiment [5]. The reaction of an alkyl halide with Mg_n represents an example of dissociate adsorption and leads to appearance of a new cluster, but not

A.V. Tulub (✉) and V.V. Porsev
Saint-Petersburg State University, Saint-Petersburg, Russia

N. Russo et al. (eds.), *Self-Organization of Molecular Systems: From Molecules and Clusters to Nanotubes and Proteins*, NATO Science for Peace and Security Series A: Chemistry and Biology, © Springer Science+Business Media B.V. 2009

to GR. Different experimental methods, including EPR, indicates that GR formation for alkyl halides involves the formation of radical intermediates, the later according to [6] are in the adsorbed form at the magnesium surface. The kinetic data are more consistent with the hypothesis of existence of free radicals in the solution [7], and literature therein. Presence of radicals is not a permanent condition for GR formation, they can be found in a very small amount as in the case of R $= C_6H_5$ (Ph).

Calculations were performed in the DFT approach: B3LYP/6-31 $+$ G$*$ with the GAMESS PC program package [8], in the unrestricted version. The changing of basic set from 6-31+G* to 6-311++G** did not give noticeable changes in geometries and energies. Activation barriers are in the problem of GR formation of great importance; their calculations represent a rather delicate problem sensitive to the atomic basis and to the method used to account the electron correlation. The B3LYP approximation has a tendency to underestimate barrier heights, from our own studies it can give by the fragmentation of the ionic systems wrong charge values [9] for the products. Different approximations in the field of DFT methods are systematically studied in the recent review article [10]. By comparing the theoretical predictions with experiment we prefer to use the difference of the activation barrier values.

Two main problems are remained unsolved and considered below, namely (1) the origin of radicals, (2) the role of ether in the reaction of GR formation. Numerical simulations have been performed in a framework of the Mg_n cluster model for $n = 1 \div 10$ for the RMgHal as a mostly known GR with R $=$ CH$_3$ (methyl), $\tilde{N}_6H_5$ (phenyl), as radicals and $(CH_3)_2O$ as ether for Hal $=$ F, Cl, Br. The Grignard reagent formation takes place only in the presence of ether; some number of ether molecules should be included in the quantum mechanical system. The interaction with two ether molecules is stronger compared with the one, it was also demonstrated by direct computations [11].

For the finding of reasonable sequence of reactions we compare first the binding energies of Mg_n cluster with CH_3Hal and with the two ether molecules. The geometric and energetic properties of Mg_n clusters are available from the literature [12, 13], the given geometric values were used below as starting for the finding of the optimized geometry for $Mg_n \cdot 2(CH_3)_2O$ as well as for CH_3Mg_nHal clusters. The clusters Mg_4 and Mg_{10} are chosen as the most stable among Mg_n for $n \leq 10$ [12]. Some chemical reactions with Mg_n with $n = 4$, were also described in [14].

The optimized geometry for $Mg_n \cdot 2(CH_3)_2O$ cluster is represented on Fig. 1. The binding energy for one ether molecule with magnesium Mg_n cluster is equal to 3.1 kcal/mol for $n = 4$ and 6.8 kcal/mol in the case of $n = 10$. The binding energies for Mg_n and CH_3Cl are much greater and equal correspondingly to 47.7 and 53.3 kcal/mol.

Keeping in mind this observation we consider main reactions in the following sequence:

1. $CH_3Hal + Mg_n \rightarrow CH_3Mg_nHal$
2. $CH_3Mg_nHal + 2(CH_3)_2O \rightarrow Mg_{n-1} + CH_3MgHal \cdot 2(CH_3)_2O$.

The second reaction would give by the suitable choose of the initial geometry of CH_3Mg_nHal cluster the desired product, the GR.

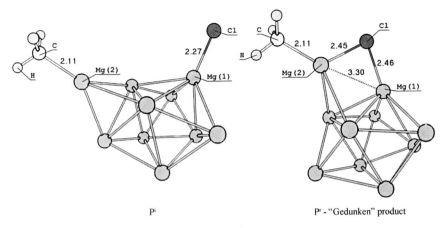

P^i P^r - "Gedunken" product

Fig. 1 Complexes $Mg_n \cdot 2(CH_3)_2O$

2 The Reaction $Mg_n + CH_3Hal$

The potential energy surface (PES) for $Mg_n + CH_3Hal$ reaction have in general some number of local stable states, one kind of the transition states (TS) for various $n \leq 21$ was discussed in [1] for Hal=Cl. We have found the reaction path with the TS described in [1] as well as the new TS [15]. The reality of obtained TS for $Mg_n + CH_3Hal$ reaction was verified by CI calculations at separate points of the reaction path in the space of about 2.7×10^5 configuration state functions (CSFs) obtained in the CASSF approach in the $6\text{-}31 + G^*$ basic set with ten active orbitals occupied by eight electrons. For the larger clusters DFT (B3LYP) approach was used with the full optimizations of the geometry. The correspondence between the products and reagents was determined by descending from a transition state along the intrinsic reaction coordinate. Two main reaction channels and corresponding TSs are described as ionic (TS^i) and the other as radical (TS^r):

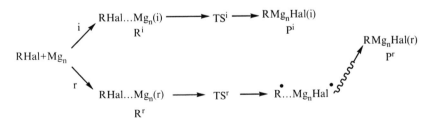

The main energetic quantities are: E_{reag} is energy of reagent formation; E_a is the activation energy, upper index indicate the channel; ΔE is the reaction energy,

$$E_{reag}^i = E(R^i) - E(Mg_n) - E(RHal)$$
$$E_a^i = E(TS^i) - E(R^i)$$
$$\Delta E^i = E(P^i) - E(R^i)$$
$$E_{reag}^r = E(R^r) - E(Mg_n) - E(RHal)$$
$$E_a^r = E(TS^r) - E(R^r)$$
$$\Delta E_1^r = E(R) + E(Mg_nHal) - E(R^r)$$
$$\Delta E_2^r = E(P^r) - E(R^r)$$

The both channels represent an example of the dissociation absorption of CH_3Cl on Mg_n cluster. It can be noted charge redistribution for fragments CH_3 and Hal and the negative total charge on Mg_n. Name "ionic" reflects the fact, that no radicals would be formed at this channel—only charge redistribution takes place.

The major feature of the transition state TS^i is that the symmetry axis of the methyl halide is parallel to some Mg–Mg bond. The second reaction path has quite another geometry of it's TS^r, C–Hal bond is perpendicular to the Mg–Mg bond formed by two neighbour to Hal magnesium atoms. The geometries of TS's are similar for all halogen and given below only for Hal = Cl, the both TS^i and TS^r are represented on Fig. 2. The geometric characteristics of the TS's are given in the Table 1.

The activation energy for CH_3Hal molecule in the radical channel E_a^r is approximately one half of the value E_a^i for the ionic reaction path, the difference is about 9–15 kcal/mol and remains fast the same for all Hal and clusters. This observation explains the appearance of radical intermediates as a permanent co-product in the reaction of GR formation. The activation energy is decreasing in a sequence F–Cl–Br for all clusters; the numerical values are given in the Table 2. We see also from the Table 2 the tendency of the activation energy decreasing, when the data for Mg_4 and Mg_{10} clusters are compared, the tendency corresponds to the conclusion derived in [1] for Hal = Cl as well as to experimental data. The large activation energy for F explains its little activity on the contrary to Br with its small activation energy.

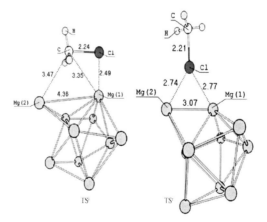

Fig. 2 Transition states for reaction of magnesium clusters with H_3Hal, n = 10, Hal = Cl

Table 1 Geometry of the transition states

Distances, Å	CH₃Hal, n = 4			CH₃Hal, n = 10			C₆H₅Cl, n = 4
	F	Cl	Br	F	Cl	Br	
RHal							
r(Hal–C)	1.40	1.81	1.96	–	–	–	1.76
TS^r							
r(Hal–C)	1.82	2.30	2.46	1.78	2.21	2.35	2.29
R(Hal–Mg₁)	2.16	2.82	2.86	2.12	2.77	2.87	2.71
r(Hal–Mg₂)	2.16	2.69	2.84	2.19	2.74	2.86	2.71
r(Mg₁–Mg₂)	3.24	3.19	3.19	3.07	3.07	3.07	3.19
TS^i							
r(Hal–C)	1.85	2.26	2.37	1.86	2.24	2.37	2.09
r(Hal–Mg₁)	1.99	2.69	2.86	1.94	2.49	2.65	2.67
r(C–Mg₁)	3.04	3.14	3.24	2.94	3.35	3.38	4.71
r(C–Mg₂)	2.83	2.81	2.74	2.84	3.47	3.32	2.84
r(Mg₁–Mg₂)	4.66	4.17	4.29	4.25	4.36	4.25	4.24

TS^r structure: C | Hal — with Mg(2) ———— Mg(1)

TS^r structure: C ⋮ Hal — with Mg(2) ———— Mg(1)

Table 2 Energetic values for different reaction path

Energy, kcal/mol	CH₃Hal, n = 4			CH₃Hal, n = 10			n = 4, C₆H₅Cl
	F	Cl	Br	F	Cl	Br	
Reactant (R)							
E_{reag}^i	−1.0	−0.1	−3.6	−2.1	−0.2	−5.7	≈0
E_{reag}^r	−1.0	−0.1	−3.5	−2.3	−0.5	−5.7	≈0
Transition state (TS)							
E_a^i	31.2	24.3*	19.3	25.9	18.9	14.6	22.0
E_a^r	15.7	13.7	7.5	12.3	9.7	3.7	19.8
Products (P)							
ΔE^i	−52.2	−51.4	−50.7	−49.4	−50.6	−50.2	−49.4
ΔE_1^r	−10.5	−0.5	1-1.5	−20.6	−10.2	−9.5	11.0
ΔE_2^r	−57.4	−47.7	−149.3	−63.9	−53.3	−53.8	−45.5

E_a^i(Hal = Cl. n = 4) = 23.7 kcal/mol [1]. All values corrected for ZPE.

The structure of the reaction product of ionic channel was obtained by descending from the transition state and differs from the structures suggested in [1]. The products for Hal = F, Cl, Br are very similar.

Transition to the final state can be associated in the case of the larger clusters with a number of locally stable states on the PES and it can be thought as a diffusion process with some characteristic time. That means that for a small time interval a halogen and the methyl group can be still kept together what is favorable for GR formation. Diffusion process takes place near the real metallic surface, the evaluation of characteristic time represents rather difficult problem.

The product state for Mg₁₀ cluster has an interesting structure, the dissociate absorption is accompanied by of some Mg–Mg bond breaking as represented of Fig. 3.

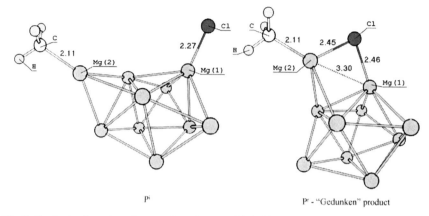

Fig. 3 Products of reaction of magnesium clusters with H_3Hal, $n = 10$, $Hal = Cl$

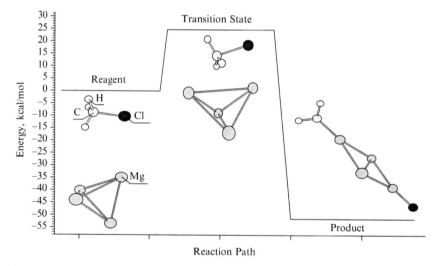

Fig. 4 Energetic scheme for ionic channel, $n = 4$, $Hal = Cl$

The same rearrangement seems to be valid for the larger clusters and resembles the surface destruction. The general view on ionic reaction path is represented on Fig. 4.

The general view of radical channel on the initial, intermediate and final states are given on Fig. 5 which contains at the final stage some "gedanken" product P^r which appearance can be explained as following. Consider the radical intermediates in which the radical CH_3 moves away from the cluster to a considerable distance, Fig. 5. The picture represents the possible events in a gas, not in the solution. In the later case the radical CH_3 would spend some time in the ether solution and then it can return back to his "home" cluster or to any other "foreign" cluster, the process corresponds to that, observed in the collision induced dynamical nuclear polarization (CIDNP). The radical CH_3 with a corresponding space orientation has

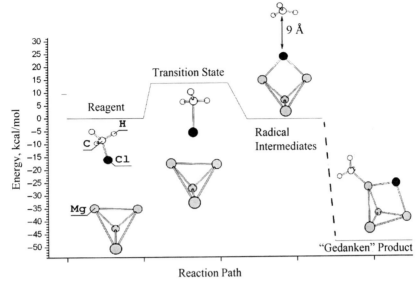

Fig. 5 Energetic scheme for radical channel, n = 4, Hal = Cl

Table 3 Effective charges for product state P^r in the radical channel

CH₃Hal	q(Hal)	q(CH₃)	q(Mg1)	q(Mg2)	q(Mgₙ)
Mg₄					
CH₃F	−0.968	−0.715	0.179	1.367	1.682
CH₃Cl	−0.868	−0.710	0.172	1.244	1.578
CH₃Br	−0.832	−0.713	0.166	1.204	1.545
Mg₁₀					
CH₃F	−0.978	−0.727	0.093	1.292	1.705
CH₃Cl	−0.861	−0.727	−0.017	1.150	1.589
CH₃Br	−0.819	−0.733	−0.039	1.099	1.552

a chance to be reabsorbed. Having in mind the interaction with ether we consider the case when the final product has the geometry represented in Fig. 3.

When one Mg atom is connecting with CH_3 as well as with the halogen, we have the CH_3MgHal group inside the total cluster. The charge distribution in this case is of a special interest and it is given in the Table 3 for the product state P^r.

3   Grignard Reagent Formation Without Radical Appearance

Consider the case Ph radical for which, as for the methyl radical, the two reaction paths where established, they are compared for TS for the ionic (left) and radical (right) channels on the Fig. 6.

The geometric properties of TSs are represented in the Table 1 and energetic in the Table 2. The replacement of CH_3 group by Ph radical leads to decreasing

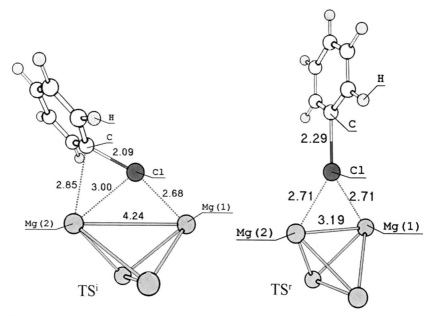

Fig. 6 Transition states for reaction of Mg_4 with phenyl chloride

activation energy of ionic channel to the value 22.0 kcal/mol. On the contrary the activation energy for the radical channel is increasing to 19.8 kcal/mol. The energy difference $\Delta E = (E_a^i - E_a^r)$ for the two channels is equal $\Delta E (CH_3Cl) = 10.6$ kcal/mol and for the phenyl radical $\Delta E (CH_3Cl) = 2.2$ kcal/mol in the case of $n = 4$. As result the ionic reaction path can be preferable for (Ph), the radicals would appear in a very small quantities if the same tendency remains for the larger clusters.

It can be awaited in real conditions that the both reaction path can be realized, much is depended upon the details in structure of radical, the radical channel explains as before the appearance of co-products [7], and references therein.

4 Energetic Characteristics of Radical Channel

The radical reaction

$$CH_3Hal + Mg_n \rightarrow CH_3 + Mg_nHal$$

is characterized by the energy release ΔE^{rad}

$$\Delta E^{rad} = E(CH_3) + E(Mg_nHal) - E(CH_3Hal) - E(Mg_n),$$

for which calculation it is necessary to find first the energy $E(Mg_nHal)$. The optimized geometric parameters for the later cluster are given on Fig. 7 for Hal = Cl.

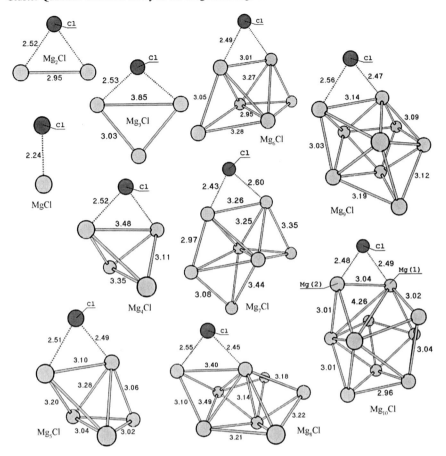

Fig. 7 Geometry of clusters Mg_n Hal, Hal = Cl

The interaction with one Mg atom is energetically unfavourable, the detail picture for different PES obtained in a framework of CI calculations contains examples of avoided crossings for the singlet as well as for the triplet electronic states, as it is demonstrated on Fig. 8 [4]. The energy difference between these states remains small at the TS in the case of magnesium clusters and remembering the mentioned underestimation of barrier heights in the B3LYP approximation for the singlet PES we can not exclude the possibility of crossing. It was a suggestion that at least two magnesium atoms should be responsible for the appearance of radicals [16].

The energy dependence ΔE^{rad} upon the size of Mg_n cluster is represented for Hal = F, Cl, Br at Fig. 9. The dependence ΔE^{rad} is similar for of the phenyl radicals to that for the methyl radical. The energy shift for $R = C_6H_5$ is to be noted, it has the important consequence for the understanding of radical particles absence.

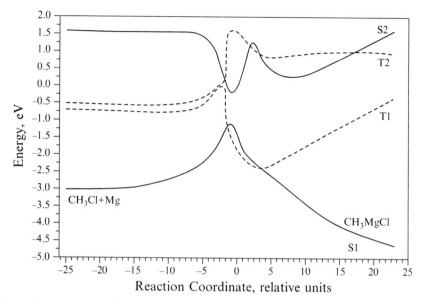

Fig. 8 Crossing of ground singlet (S1) and ground triplet (T1) states on PES along with intrinsic reaction coordinate of $CH_3Cl + Mg \rightarrow CH_3MgCl$. CI approach in basis set 6–31G*, in space about $1.4 \cdot \times 10^4$ CSFs for singlet states and $2.1 \cdot \times 10^4$ CSFs for triplet states

5 Interaction with Ether

We have already demonstrated that the interaction Mg_n with ether is relative weak. The question arise about the properties of a new cluster (CH_3Mg_nHal) which are responsible for it quite noticeable interaction with ether. Consider the charge value on the Mg(2) atom of P^r at Fig. 3. Mg(2) is strongly bound to the both CH_3 and Hal and has large positive charge, what creates some kind of active site on the cluster. Weak interaction of Mg_n with ether would be replaced in the case of Mg_nCH_3Hal by strong one due to the negative effective charge on ether oxygen.

Our next step is the interactions of methylmagnesium chloride with the ether [17]. Some increasing of the interaction energy is known when one dimethyl ether molecule is replaced by two ether molecules. The reaction

$$CH_3Mg_nCl + 2(CH_3)_2O \rightarrow CH_3MgCl \cdot 2(CH_3)_2O + Mg_{n-1} \rightarrow GR \ldots Mg_{n-1},$$
$$GR = CH_3MgCl \cdot 2(CH_3)_2O - \text{Grignard reagent.}$$

is very sensitive to an initial relative position of ether molecule and CH_3Mg_nCl cluster. It was found some domain in the initial geometry parameters responsible for the ether–CH_3Mg_nCl interaction process developing without any activation barrier. The optimized geometry structures are represented on Fig. 10 and give an idea for the step-by-step elimination of GR from Mg_{n-1} cluster.

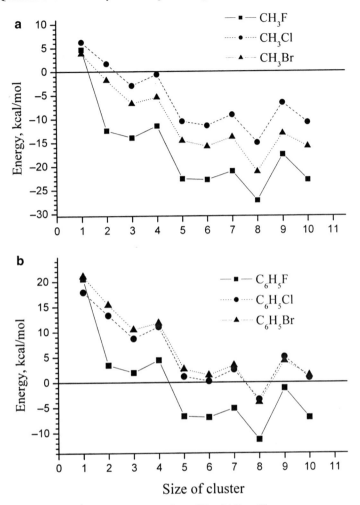

Fig. 9 Dependence ΔE^{rad} on cluster size. (a) R = CH$_3$; (b) R = Ph

In the "transition state" the group CH$_3$MgCl·2(CH$_3$)$_2$O becomes partly isolated from Mg$_{n-1}$ cluster what is in a correspondence with the proposal of Bronsted-Evans-Polanyi on the largerly resemlence of TS of the reaction to the product state, see Fig. 10. The increase of the cluster size is non favorable for GR formation, as it follows from Fig. 11. This statement reflects the known fact that the increase of magnesium cluster leads to the increase of the binding energy per one atom [12, 13] and as consequence to the decrease of the solvation energy of GR. We note once more that the charge redistribution, Table 3 in reactant CH$_3$Mg$_n$Cl with the suitable positions of CH$_3$ and Cl is responsible to the strong interaction with ether and GR formation. The numerical simulations are performed for gas phase, GR in a solution is to be thought in general as surrounded not only by two, but by some number of ether molecules.

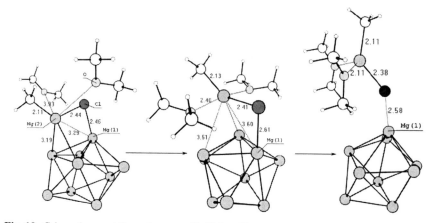

Fig. 10 Grignard reagent formation, n = 10, Hal = Cl

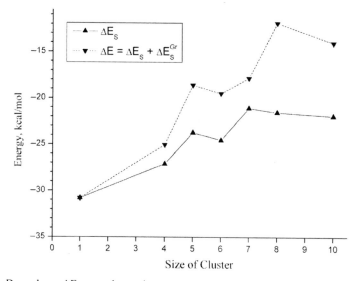

Fig. 11 Dependence ΔE_{solv} on cluster size

6 Outlook

We have realized from the given scenario the following points.

1. Two kinds of the dissociate adsorption of CH_3Hal on Mg_n clusters.
2. Appearance of active sites on RMg_nHal clusters, leading to strong electrostatic interaction with ether.
3. GR can arise due the interaction RMg_nHal with ether without activation barrier.
4. Radical-free GR formation for some organic halides, aryl for example.

The activity of different halogens as well as radicals is in general in a correspondence with experimental data. Only the simplest GR was considered above, it is of interest to investigate also more complicated GR structures as well as more complicated surfaces, bimetallic instead of monometallic.

Our work is dedicated to the memory of Alexander S. Davydov, whose works in the theory of wave propagation in molecular crystals, in liquid water and in biopolymers are so important for the understanding of self-organization processes. A. S. Davydov was a great teacher, what is clear not only from his pioneer investigations, but also due to a number of books, among them is "Quantum Mechanics", it is still the best known in the world. His fascinating lectures as well as his bright personality remain in the memory of peoples, which knew him personally; among them is one of the author (A.V.) of this work.

References

1. Jasien P.G., Abbondondola J.A. Journal of Molecular Structure (Theochem), **671**, 111–118 (2004).
2. Davis S.R. JACS, **113**, 4145–4150 (1991).
3. Tulub A.A. Russian Journal of General Chemistry, **72**, 948–953 (2002).
4. Tulub A.V., Porsev V.V., Tulub A.A. Doklady Akademmi Nauk, **398**, 502–505 (2004). English translation: Doklady Physical Chemistry, **398**, Part 2, P. 241–244 (2004).
5. Beals B.J., Bello Z.I., Cuddihy K.P., Healy E.M., Koon-Church S.E., Owens J.M., Teerlinck C.E., Bowyer W.J. The Journal of Physical Chemistry A, **106**, 498–503 (2002).
6. Walborsky H.M., Zimmermann C. JACS, **114**, 4996–5000 (1992).
7. Garst J.F., Soriaga M.P. Coordination Chemistry Reviews, **248**, 623–652 (2004).
8. Schmidt M.W., Baldridge K.K., Boatz J.A., Elbert S.T., Gordon M.S., Jensen J.H., Koseki S., Matsunaga N., Nguyen K.A., Su S.J., Windus T.L., Dupuis M., Montgomery J.A. Journal Computational Chemistry, **14**, 1347–1363 (1993).
9. Tulub A.V., Simon K.V. Journal of Structural Chemistry, (Russian), **48**, S86–S100 (2007).
10. Zhao Y., Truhlar D.G. Accounts of Chemical Research, **41**, 157–167 (2008).
11. Tammiku-Taul J., Burk P., Tuulmets A. The Journal of Physical Chemistry A., **108**, 133–139 (2004).
12. Lyalin A., Solov'ev. I.A., Solov'ev. A.V., Greiner W. Physical Review A, **67**, 063203 (2003).
13. Jellinek J., Acioli P.H., Garciá-Rodeja J., Zheng W., Thomas O.C., Bowen Jr. K.H. Physical Review B, **74**, 153401 (2006).
14. Tjurina L.A., Smirnov V.V., Barkovskii G.B., Nikolaev E.N., Esipov S.E., Beletskaya I.P. Organometallics, **20**, 2449–2450 (2001).
15. Porsev V.V., Tulub A.V. Doklady Akademii Nauk, **419**(1), 71–76 (2008); English translation: Doklady Physical Chemistry, **419**, Part I, 53–58 (2008).
16. Jasien P.G., Dykstra C.E. JACS, **107**, 1891–1895 (1985).
17. Porsev V.V., Tulub A.V. Doklady Akademii Nauk, **409**, 634–638 (2006); English translation: Doklady Physical Chemistry, **409**, Part 2, 237–241 (2006).

Index

Printed in the United States
147391LV00002B/40/P